LONGEVITY

LONGEVITY

THE BIOLOGY AND DEMOGRAPHY OF LIFE SPAN

James R. Carey

PRINCETON UNIVERSITY PRESS PRINCETON AND OXFORD

Published by Princeton University Press, 41 William Street,
Princeton, New Jersey 08540
In the United Kingdom: Princeton University Press, 3 Market Place,
Woodstock, Oxfordshire OX20 1SY

Library of Congress Cataloging-in-Publication Data

Carey, James R.
Longevity : the biology and demography of life span / by James R. Carey.
p. cm.
Includes bibliographical references (p.).
ISBN 0-691-08848-9 (alk. paper)—ISBN 0-691-08849-7 (pbk. : alk. paper)
1. Longevity. I. Title.
QP85 .C283 2002 2003
612.6'7—dc21 2002025294

British Library Cataloging-in-Publication Data is available

This book has been composed in Times Roman

Printed on acid-free paper. ∞
www.pupress.princeton.edu
Printed in the United States of America

10 9 8 7 6 5 4 3 2 1

DEDICATED TO MY CHILDREN

Bryce, Ian, and Meradith

FOR THEIR STRENGTH OF CHARACTER

". . . we find less talk of life as an exercise in endurance, and of death in a hopeless cause; and we hear more of life as a seeking and a journeying."

—R. W. SOUTHERN

Contents

Figures

Tables

Preface and Acknowledgments

THE SEEDS for the research upon which much of this book is based were sown at a National Institute on Aging (NIA)-sponsored workshop titled "Upper Limits to Human Life Span" held at the University of California, Berkeley, in March 1987. One of the most important issues that emerged from this workshop was the paucity of information in the literature on actuarial aging (i.e., life table data) on nonhuman species: despite the thousands of life tables in the ecology and gerontology literature, the vast majority were based on a few score or a few hundred individuals. These small initial numbers thus provided virtually no information on mortality rates at older ages in any species. Immediately after this workshop James Vaupel (who had also attended it) and I agreed to explore the possibility of collaborating on a project designed to construct a large-scale life table of the Mediterranean fruit fly, a species that is reared in vast numbers at Moscamed, a medfly factory near Tapachula, Mexico. The director at that time was one of my former Ph.D. students, Pablo Liedo. A year after the idea for this large-scale life-table study was conceived we submitted a proposal to NIA as part of a larger program project (headed by James Vaupel). This program was subsequently funded, and the preliminary results were reported in a "News and Review" article in *Science* magazine (Barinaga 1992). A formal paper was published in 1992 titled "Slowing of mortality at older ages" (Carey, Liedo, Orozco, and Vaupel 1992; see also companion paper by Curtsinger et al. 1992). This set the stage for a subsequent funding cycle and a series of new research questions.

The second funding cycle of the program project began in 1994 and focused initially on concerns that were raised in "Letters to the Editor" in *Science* (Kowald and Kirkwood 1993; Nusbaum et al. 1993; Robine and Ritchie 1993; Olshansky et al. 1993; Gavrilov and Gavrilova 1991), most notably whether deceleration was due to density effects, whether the phenomenon was general and thus observed in other species, and if medflies subjected to a wide range of different environmental manipulations would also exhibit mortality deceleration at older ages. Later in this 5-year funding cycle (mid-1990s) we began to shift emphasis from life table studies involving tens or hundreds of thousands of flies in group cages to experiments involving the mortality and daily reproduction of individual flies. At that time we also began efforts at constructing a large database on the record life spans of vertebrates. The project was funded for third cycle starting in 1999 and therefore, the medfly life table project is in the 13th year of continued funding.

There are several reasons why I am hopeful that the proposed book will contribute new knowledge to the biology of aging and the biodemography of life span. First, the book reflects the overall continuity of the project from inception to present. Indeed, it is the first nonedited book on the biology of aging and longevity that focuses on a single-model system in which the experiments were designed by the same group of core scientists (i.e., myself, Vaupel, Liedo, and other key colleagues), in which the data were gathered in the same laboratory by essentially the same group of technicians (with very little turnover), and in which the results were analyzed, written up, and published quickly in uniformly high-quality journals.

Second, the overall data set on which the collective studies described in the book are based is truly unprecedented. The complete database currently consists of age- and sex-specific mortality data for over 5.6 million individuals (i.e., data quantity) in which the same types of cages, larval and adult food, physical conditions, and collection procedures were used from the beginning of the studies in 1989 to the present (i.e., data quality).

Third, the science of aging is such a rapidly moving field that knowledge of many aspects in the medfly system described in this book—including experimental designs, interpretation of results, and summaries and syntheses—may be helpful for studies using other model systems (e.g., nematode, *Drosophila*, yeast, rodents). In particular, the results presented in the book will sharpen the experimental focus and frame the thinking of researchers who use more costly vertebrate systems.

Fourth, the book will serve as an introduction to experimental demography and biodemography for a wide range of biologists not specializing in aging science (e.g., ecologists, population biologists, demographers, actuaries). There appears to be no other source of advanced concepts in life table and mortality analysis for biologists, and I am unaware of any ecology text that goes much beyond elementary life table analysis.

Fifth, the book contains several new syntheses including (i) biodemographic principles that capture the general concepts such as slowing of mortality, male-female mortality differentials, and the indeterminacy of life span; (ii) analytical models that make explicit the relationship between life history traits such as reproduction and longevity or mortality; and (iii) longevity theory that outlines a general model in social organisms describing how longevity extension coevolves with sociality. My focus at the end of the book is primarily on human longevity.

Sixth, the book contains studies that were motivated as much by the conventional *hypothesis-driven* questions (tests of specific hypotheses such as upper life-span limits) as by *discovery-driven* science, which was first characterized in the context of genome research (Aebersold et al. 2000; Idelker et al. 2001) but which I believe applies equally well to some of the research described in this book. Indeed, one of the most exciting outcomes of using huge initial numbers in life table studies was the discovery of nuances in mortality trajectories at both young and old ages that then required hypothesis testing. It is discovery science because the research involves unknown biological and demography territory.

Much of the writing for this book was done over the past ten years in the form of refereed, coauthored publications. This series of research papers started with the 1.2 million medfly life table study (Carey et al. 1992) and ended with a paper containing a conceptual overview of life span (chapter 10) that I initially presented at the workshop "Life Span: Evolutionary, Ecological and Demographic Perspectives" held in Santorini, Greece, in May 2001. The final version of the article that became chapter 10 was not completed until early 2002 (Carey 2003). The book organization and content reflect the progression of ideas and the evolution of concepts that emerged from my thinking about the determinants of mortality, survival, and life span over these years. One of the most influential articles that shaped my thinking about the integration of longevity and mortality concepts was by the late gerontologist George Sacher (1978), "Longevity and aging in vertebrate evolution." In this paper Sacher introduced the concept of the "biology of the finitude"—a triad consisting of the biology of life span, aging, and death. In my mind this "finitude" idea provides conceptual continuity by framing life span in a functional (life history) context, aging in a mechanistic context, and death in a finality context. Thus the concept introduced in the Sacher article helped me to integrate different sections of the book—the results of studies on mortality dynamics presented in chapters 3–7 are woven into many of the general biodemographic principles presented in chapter 8, which, in turn, are used as a foundation to develop theories and concepts on longevity and life span presented in chapters 9–10. In other words, Sacher's ideas provide a bridge between the allied but different conceptual domains of mortality, longevity, and life span (I consider the "biology of death" in this book in the final chapter only briefly; however, this neglected area is ripe for development of a stand-alone research program in aging).

I thank members of Duke University-based, NIA-funded program project (originally titled "Oldest-old Mortality" and directed by James Vaupel) including Kaare Christensen, James Curtsinger, Lawrence Harshman, Aziz Khazaeli, Valter Longo, Kenneth Manton, Cynthia Owens, Linda Partridge, Deborah Roach, Marc Tatar, and Anatoli Yashin for their input over the past decade and especially for their friendship. I am particularly grateful to Pablo Liedo for the important role he played as the senior scientist and collaborator in Tapachula; to both Jane-Ling Wang and Hans Müller for their statistical help and remarkable insights; to Debra Judge for her superb contributions as coauthor of several key papers on the principles of biodemography (chapter 8) and the general theory of longevity (chapter 9); and to Nikos Papadopoulos for the huge job of collating and reformatting the scores of figures included in the book. I express my deep appreciation to Robert Arking and Marc Mangel for taking the time to read and critique the entire book manuscript and to the many colleagues with whom I coauthored papers used in this book, including William Capra, J.-M. Chiou, Byron Katsoyannos, Nikos Kouloussis, Xieli Liu, Brad Love, Dina Orozco, Nikos Papadopoulos, Scott Pletcher, Daniel Promislow, D. Wu, Zhang Yi, and Ying Zhang. I thank my colleagues at the UC Berkeley Center for the Economics and Demography of Aging for their insights and inspiration throughout the years, including Ronald Lee, John Wilmoth, Shripad Tuljapurkar, and especially Kenneth Wachter for his interest and support from the very beginning of the program project. I also deeply appreciate the strong support of both Rose Li and Richard Suzman and the research funding from the National Institute on Aging. I thank Linda Truilo for her meticulous and thorough editing. Lastly, it is a pleasure to acknowledge with gratitude the encouragement and patience of Samuel Elworthy at Princeton University Press. I thank them all, truly.

Although I was lead author and writer on most of the research papers used in this book, James Vaupel (JV) and Hans Müller (HM) were the lead authors and writers of four papers from which substantial portions were used, including section 3.4 on demographic selection (JV), several subsections on modeling in chapter 5 on density effects (JV), section 6.1 on early mortality surge (HM), and section 7.2 on reproductive clock (HM).

James Vaupel has provided input at all levels and has made a difference in how I approach demography in particular and science in general. This book could not have been written had he not taken interest in the program project, recognized the potential of the medfly as a model system, and provided intellectual stimulation and camaraderie. I owe my greatest thanks to him.

Permissions

Section 2.2 reprinted from: Carey, J. R. and Liedo, P. 1999. Measuring mortality and reproduction in large cohorts of the Mediterranean fruit fly, in: H. Sternberg and P. S. Timiras (ed.), Studies of Aging, pp. 111–124, Berlin: Springer-Verlag, Copyright © 1999 with permission from Springer-Verlag.

Section 2.3 reprinted from: Carey, J. R. 1999. Population study of mortality and longevity with Gompertzian analysis, in: B. P. Yu (ed.), Methods in Aging Research, pp. 3–24, Boca Raton: CRC Press, Copyright © 1999 with permission from CRC Press, and Carey, J. R. 2001. Insect biodemography. Annual Review of Entomology 46:79–110, Copyright © 2001 with permission from Annual Reviews, Inc.

Section 3.2 reprinted from: Carey, J. R., Liedo, P., Orozco, D., and Vaupel, J. W. 1992. Slowing of mortality rates at older ages in large medfly cohorts. Science 258:457–461, Copyright © 1992 with permission from the American Association for the Advancement of Science.

Section 3.4 reprinted from: Vaupel, J. W. and Carey, J. R. 1993. Compositional interpretations of medfly mortality. Science 260:1666–1667, Copyright © 1993 with permission from the American Association for the Advancement of Science.

Section 3.5 from: Carey, J. R. and Liedo, P. 1995. Sex-specific life table aging rates in large medfly cohorts. Experimental Gerontology 30:315–325, Copyright © 1995 with permission from Elsevier Science, and Carey, J. R., Liedo, P., Orozco, D., Tatar, M., and Vaupel, J. W. 1995. A male-female longevity paradox in medfly cohorts. Journal of Animal Ecology 64:107–116, Copyright © 1995 with permission from Blackwell Publishing.

Section 4.1 reprinted from: Carey, J. R., Liedo, P., Müller, H.-G., Wang, J.-L., and Vaupel, J. W. 1998. A simple graphical technique for displaying individual fertility data and cohort survival: case study of 1000 Mediterranean fruit fly females. Functional Ecology 12:359–363, Copyright © 1998 with permission from Blackwell Publishing.

Section 4.2 reprinted from: Carey, J. R., Liedo, P., Müller, H.-G., Wang, J.-L., and Chiou, J.-M. 1998. Relationship of age patterns of fecundity to mortality, longevity, and lifetime reproduction in a large cohort of Mediterranean fruit fly females. Journal of Gerontology: Biological Sciences 53A:B245–B251, Copyright © 1998 The Gerontological Society of America. Reproduced by permission of the publisher.

Section 4.3 reprinted from: Carey, J. R., P. Liedo, L. Harshman, Y. Zhang, H.-G. Müller, L. Partridge, and J.-L. Wang. 2002. A mortality cost of virginity at older ages in female Mediterranean fruit flies. Experimental

Gerontology 37:507–512, Copyright © 2002 with permission from Elsevier Science.

Section 4.4. reprinted from: Papadopoulos, N. T., J. R. Carey, B. I. Katsoyannos, N. A. Kouloussis, H.-G. Müller, and X. Liu. 2002. Supine behaviour predicts time-to-death in male Mediterranean fruit flies. Proceedings of the Royal Society of London: Biological Sciences (in press), Copyright © 2002.

Chapter 5 reprinted from: Carey, J. R., Liedo, P., and Vaupel, J. W. 1995. Mortality dynamics of density in the Mediterranean fruit fly. Experimental Gerontology 30:605–629, Copyright © 1995 with permission from Elsevier Science.

Section 6.1 reprinted from: Müller, H.-G., Wang, J.-L., Capra, W. B., Liedo, P., and Carey, J. R. 1997. Early mortality surge in protein-deprived females causes reversal of sex differential of life expectancy in Mediterranean fruit flies. Proceedings of the National Academy of Sciences 94:2762–2765, Copyright © 1997.

Section 6.2 reprinted from: Carey, J. R., Liedo, P., Müller, H.-G., Wang, J.-L., Love, B., Harshman, L., and Partridge, L. 2001. Female sensitivity to diet and irradiation treatments underlies sex-mortality differentials in the Mediterranean fruit fly. Journal of Gerontology: Biological Sciences 56A:B1–B5, Copyright © 2001 The Gerontological Society of America. Reproduced by permission of the publisher.

Section 6.3 reprinted from: Carey, J. R., Liedo, P., Müller, H.-G., Wang, J.-L., and Chiou, J.-M. 1999. Mortality oscillations induced by periodic starvation alter sex-mortality differentials in Mediterranean fruit flies. Journal of Gerontology: Biological Sciences 54A:B424–B431, Copyright © 1999 The Gerontological Society of America. Reproduced by permission of the publisher.

Section 7.1 reprinted from: Carey, J. R., Liedo, P., Müller, H.-G., Wang, J.-L., and Vaupel, J. W. 1998. Dual modes of aging in Mediterranean fruit fly females. Science 281:996–998, Copyright © 1998 with permission from the American Association for the Advancement of Science.

Section 7.2 reprinted from: Müller, H.-G., Carey, J. R., Wu, D., and Vaupel, J. W. 2001. Reproductive potential determines longevity of female Mediterranean fruit flies. Proceedings of the Royal Society, London B 268:445–450, Copyright © 2001.

Section 7.3 reprinted from: Carey, J. R., P. Liedo, L. Harshman, X. Liu, H.-G. Müller, L. Partridge, and J.-L. Wang. 2002. Food pulses increase longevity and induce cyclical egg production in Mediterranean fruit flies. Functional Ecology (in press), Copyright © 2002 with permission from Blackwell Publishing.

Chapter 8 reprinted from: Carey, J. R. and Judge, D. S. 2001. Principles of biodemography with special reference to human longevity. Population: An

English Section 13:9–40, Copyright © 2001 with permission from Institut National d'Etudes demographiques.

Chapter 9 reprinted with the permission of the Population Council, from James R. Carey and Debra S. Judge, "Life Span Extension in Humans is Self-Reinforcing: A General Theory of Longevity," Population and Development Review, vol. 27, no. 3 (September 2001), 411–436.

Chapter 10 reprinted with the permission of the Population Council, from James R. Carey, "Life Span: A Conceptual Overview," Population and Development Review (in press).

Figure 10.5c re-drawn with permission: Miller J. D. 1997. Reproduction in sea turtles. In The Biology of Sea Turtles, ed. P. L. Lutz, J. A. Musick, pp. 51–81. Boca Raton: CRC Press, Copyright © 1997.

Figure 10.5b reprinted from: Riedman, M. Pinnipeds: Seals, Sea Lions and Walruses, with permission from University of California Press, Copyright © 1989, The Regents of the University of California.

1

Introduction

1.1. The Problem

Some of the most profound questions in biology are those concerned with the nature and origin of both life span and aging—equal in stature to those involving the genesis of life, of sex, and of human consciousness. One of the most important reasons for studying aging is because it is basic to life and its endpoints—morbidity and death. As Gavrilov and Gavrilova (1991) note, it will never be possible fully to understand the nature and origin of life without understanding the nature and origin of both its constraints and its limits. Another reason for exploring the large mystery of aging is that slowing the aging process in humans could yield powers to retard senescence, to preserve youthfulness, and to prolong life greatly (Kass 1983). This would have vast and far-reaching affects on all of our important social institutions and fundamental beliefs and practices inasmuch as it is not possible to change one segment of society without affecting the entire network of relations.

Despite the intense interest in human aging, virtually nothing is known about why some individuals live to middle age and others live to extreme ages. Indeed in humans, the life-style recommendations that follow from biomedical and social studies of the elderly are unremarkable—do not smoke, use alcohol in moderation, exercise, avoid fatty diets, shed excessive weight, and minimize risk of accidents (Christensen and Vaupel 1996). Although this strategy of identifying individual factors associated with extended longevity in humans may eventually provide valuable insights for improving quality of life and reducing mortality risks, most gerontologists believe that the evolutionary and biological determinants of longevity can only be understood through the use of comparative demography and of model experimental systems such as yeast, nematodes, fruit flies, and laboratory rodents. This book is about studies using both of these approaches: (1) a large-scale Mediterranean fruit fly experimental system used to construct life tables, which, in turn, are brought to bear on questions concerning the nature of aging and longevity; and (2) comparative demography of life span using large-scale databases containing information on both vertebrates and invertebrates.

My broad goal is to collate, integrate, and synthesize the results of over a decade of research on both actuarial aging in the medfly and the comparative demography of life span and to interpret these findings in the context of human aging. Specific goals for the book are (1) to present the major conceptual, empirical, and analytical results from medfly studies on longevity and mortality using graphical arguments and actuarial techniques; (2) to integrate concepts related to the science of aging at the level of the whole organism from demography, gerontology, and insect biology; (3) to identify general biodemographic principles including those concerned with senescence, mortality, and longevity as well as conceptual aspects of life span and maximal ages; and (4) to situate the biodemographic findings in the context of human aging and to use these fundamental principles both as a foundation for the emerging field of biodemography and as a framework for considering the future of human life span.

1.2. The Epistemological Framework

1.2.1. Mortality and Aging as Fundamental Processes

The results of studies on the biology of death, mortality, longevity, and life span using animal models such as the medfly are as relevant to humans as are those on basic aspects of inheritance in *Drosophila* flies (Ashburner 1989; Jazwinski 1996) and on development in nematode worms (Hengartner 1995; Thomas 1994). In these cases emphasis is placed on studying the basic process rather than on studying the specific outcome (Carey 1997). For example, eye color in *Drosophila* has little to do with eye color in humans; but geneticists and evolutionary biologists have made major advances in understanding genetic aspects of populations such as drift, dominance, sex linkage, mutation rates, and selection by studying the changes in the frequency and inheritance patterns of these traits in experimental fly populations. Similarly, studies of fly mortality provide important insights into the nature of many fundamental actuarial processes important to demography: whether differential rates of aging underlie the gender differences in longevity; whether Gompertz mortality rates are manifestations of universal senescence "laws"; whether animals possess definitive life-span limits; and whether physiological changes at the individual level influence both local (short age periods) and lifetime patterns of cohort mortality.

I believe that answers to these basic actuarial questions are important to biodemography for several reasons: (1) they provide a frame of reference for interpreting actuarial data for both human and nonhuman species; (2) they serve as a stimulus for new approaches to studying aspects of human mortality such as the gender gap or the existence of life-span limits; (3) they provide a biological context for predicting possible changes in mortality trajectories in situations where human data are sparse or less reliable such as for mortality trajectories at the most advanced ages; and (4) mortality studies on nonhuman species can provide "proof of principle" for alternative hypotheses concerning the underlying causes of changes in the age trajectory of mortality, such as demographic heterogeneity versus physiological changes at the individual level.

1.2.2. Model Systems and Actuarial Patterns

One of the main stumbling blocks to the serious use of model systems in studying actuarial aging has been the mistaken belief that, because causes of death in humans are unrelated to causes of death in invertebrates (e.g., nematodes, fruit flies), little can be learned from detailed knowledge of age-specific mortality in these model species. This perspective is based on the "theory of the underlying cause" in public health and medicine—if the starting point of a train of events leading to death is known (e.g., cancer), death can be averted by preventing the initiating cause from operating (Moriyama 1956). For aging research the problem with this perspective is that death is seen as a single force—the skeleton with the scythe. A more apt characterization that applies to deaths in all species is given by Kannisto (1991), who notes that deaths are better viewed as the outcome of a crowd of "little devils"; individual potential or probabalistic causes of death, sometimes hunting in packs and reinforcing each other's efforts, at other times independent. Inasmuch as underlying causes of death are frequently context-specific and are difficult to distinguish from immediate causes, and given that their post-mortem identification in humans is often arbitrary (and in invertebrates virtually impossible), we find that studying the causes of death often provides little insight into the nature of aging. If aging is considered as a varying pattern of vulnerability to genetic and environmental insults, then the most important use of model species in aging research is to interpret their age patterns of mortality as proxy indicators of frailty.

1.3. Importance of Scale

1.3.1. Historical Background

One of the most important conclusions of the National Institute of Aging's workshop on "Upper Limits to Human Life Spans," held at UC Berkeley in 1987, was that data on mortality at advanced ages on nonhuman species was lacking. For example, a review of the literature on life tables on several hundred species of arthropod revealed that the vast majority of studies were based on less than fifty individuals. While these small numbers provide reasonable estimates of life expectancy at birth for cohorts, it is not possible to estimate mortality rates from data derived from small cohorts because so few individuals remain alive at the older ages. Even the widely cited classic life-table studies suffer from this problem, including those by Pearl and Parker (1924) on *Drosophila*, Leslie and Ransom (1940) on voles, Leslie and Park (1949) on flour beetles, Evans and Smith (1952) on the human louse, Pearl and Miner (1935) on several "lower" organisms, Deevey (1947) on a wide range of invertebrates and vertebrates in the field, and Birch (1948) on insects. In general, the biological, ecological, and gerontological literature contains perhaps several thousand life tables on a wide variety of species but collectively these life tables contribute very little to knowledge of age-specific mortality rates. In particular, they contribute virtually nothing to knowledge of age-specific mortality at the most advanced ages.

1.3.2. Large-scale Medfly Life Tables

A universal assumption made by most biologists and gerontologists is that mortality rates increase with age at the same exponential rate over all mature age classes (Gompertz 1825). Because no one seriously challenged this assumption, constructing mortality schedules required only that researchers monitor mortality in a relatively small cohort at younger ages, fit a straight line to the logarithm of these rates, and extrapolate to the older ages. That the logarithm of these rates did not increase linearly with age was simply not open to question.

Perhaps the main reason that no one previously challenged the gerontological canon that mortality rates increase exponentially at older ages in most species was a practical one—large numbers of individuals of any species are both expensive and difficult to rear. Enormous amounts of time, money, and effort are required to construct the mortality schedule for a cohort of even a few thousand laboratory rodents. For example, it is estimated that the maintenance costs for a single mouse is $1/day. Thus monitoring a cohort of 1,000 mice throughout their life times would cost nearly $1 million. But even these studies would provide little information on mortality rates at the oldest ages since only 100 mice would be alive when 90% of the original cohort was dead, and there would only be 10 individuals alive when 99% of the cohort was dead. Moreover, the numbers are halved when questions about mortality sex differentials are addressed. Insects are less expensive to rear than rodents but are still relatively costly. This is because a considerable amount of space is needed for rearing and a full-time staff must be hired that is dedicated exclusively to rearing.

Gaining access to essentially unlimited numbers of medflies at the medfly rearing facility in Mexico removed the main logistical obstacle to gathering mortality data on a large scale. Even if an insect rearing program would have been developed exclusively for the studies discussed in this book, the scale could not possibly have matched the industrial scale of the Moscamed medfly rearing program. Consequently the mortality studies would have consisted of perhaps a 1,000 insects reared from each of 1,000 different batches of diet rather than over 100,000 flies reared from 8 to 10 different batches, as was the case with access to the factory flies. A large amount of variation in mortality rates between cages and trials is eliminated by having essentially unlimited numbers of same-aged flies at any time.

1.3.3. Experimental Principles

Studies were initiated according to three operational principles. *First*, focus on only a small number of basic questions. For example, the initial project focused on two straightforward questions: "What is the trajectory of mortality at the most advanced ages?" and "What are the sex-mortality differentials?" *Second*, conduct studies on a large scale. The scale of the database from each study provided a rich source database for subsequent analyses. This proved to be extremely important because of the nature of mortality measurement—large initial cohorts become small at older ages due to attrition. Thus extraordinarily large initial cohorts provide enough survivors to measure mortality at the most advanced ages. *Third*, keep both the data and the data-gathering simple. We required technicians to record only two pieces of information on each fly—sex and age of death. This simplicity minimized the likelihood of error. However, the simplicity of the original question helped to reinforce the main goal, foster a clear sense of purpose at all research levels, and promote simplicity and a sense of purpose in design and execution of the project.

1.3.4. Overview of the Medfly Mortality Database

Inasmuch as a substantial part of this book is based on the results of large-scale life table studies of the medfly, it will be useful to review briefly the database that serves as its empirical foundation. A summary of the main experiments, species used, and number of individuals is presented in table 1.1. Three aspects of this table merit comment. The first and most obvious characteristic of the database is experimental scale—the majority of studies used anywhere from 100,000 to 1.2 million individuals. This scale provided considerable statistical power with respect to differences within and between treatments, between sexes, and over many age classes including the most advanced ages. A second characteristic of this database is the range of conditions under which adult mortality and longevity were measured, including

TABLE 1.1
Large-scale Mortality Studies on Mediterranean Fruit Fly and Other Tephritid Fruit Flies or Parasitoid

Species/Study Description	*n*	*Reference(s)*
Mediterranean fruit fly (*Ceratitis capitata*)		
1. Large-scale life table study (groups)	1,200,000	Carey, Liedo, Orozco, Tatar, and Vaupel 1995; Carey, Liedo, Orozco, Tatar, and Vaupel 1992
2. Large-scale life table study (solitary)	50,000	Carey et al. 1995; Carey et al. 1992
3. Effects of initial density	215,000	Carey et al. 1995; Carey et al. 1992
4. Effects of sterilizing irradiation	425,000	Carey et al. 1995; Carey et al. 1992
5. Old vs. new medfly strain	200,000	Liedo and Carey 2000
6. Sugar vs. full (protein) diets	416,000	Müller et al. 1997b
7. Larval density effects on adults	232,000	Liedo and Carey (unpublished)
8. Periodic starvation	400,000	Carey, Liedo, Müller, Wang, and Chiou 1999
9. Irradiation-by-diet effects	536,000	Carey, Liedo, Müller, Wang, Love, Harshman, and Partridge 2000
10. Catastrophic starvation	200,000	Liedo and Carey (unpublished)
11. Sexes reared separately (virgins)	120,000	Carey, Liedo, Harshman, Zhang, Müller, Partridge, and Wang 2002b
12. Reproductive history of individuals	1,000	Carey, Liedo, Müller, Wang, and Chiou 1998a; Carey, Liedo, Müller, Wang, and Vaupel 1998b
13. Dual modes of aging	500	Carey, Liedo, Müller, Wang, and Vaupel 1998c
14. Periodic alteration of diet (sugar-full)	800	Carey, Liedo, Harshman, Liu, Müller, Partridge and Wang 2002a
Subtotal (medfly)	3,996,300	
Life table studies of other fruit fly species and a parasitoid		
15. *Anastrepha ludens* (Mexican fruit fly)	1,100,000	Vaupel et al. 1998

TABLE 1.1 (*continued*)

Species/Study Description	*n*	*Reference(s)*
16. *A. obliqua* (West Indian fruit fly)	300,000	Vaupel et al. 1998
17. *A. serpentina* (Sapote fruit fly)	340,000	Vaupel et al. 1998
18. *D. longicaudatis* (parasitoid wasp)	28,000	Vaupel et al. 1998
Subtotal (other species)	1,768,000	
GRAND TOTAL	5,764,300	

Note: See cited reference(s) for technical details and study objectives. Number of individuals denoted by *n*.

cage densities, cage types (group vs. solitary), medfly strains, larval rearing density, diet types, diet periodicity, starvation conditions, mate availability, irradiation effects, reproductive timing, and male behavior. These manipulations provided important insights into the plasticity of mortality, the determinants of longevity, and sex-mortality differentials. A third characteristic of the database is that it includes the results of large-scale life table studies, not only on the medfly but also on 3 different tephritid fruit flies and a parasitoid wasp. This is important in the context of comparative biodemography and enabled us to answer the general question "How robust are the findings from the medfly studies?"

1.4. Overarching Themes

Three broad themes emerged from the medfly research that cut across many aspects of aging research and that we use as conceptual anchors for the book. One overarching theme derived from the research project is *absence of species-specific life span limits*. As will be shown in subsequent chapters, evidence for the veracity of this concept is found in the results of virtually every large-scale life table study on the medfly and related species. The slowing of mortality rates at advanced ages in all studies suggested that it is not possible to specify a specific life-span limit to the medfly and, by implication, to that of any species.

The second overarching theme of the book is that *the mortality response of males and females is context-specific*. Although this concept seems both intuitive and obvious due to behavioral and physiological differences between the sexes, there are surprisingly few experimental studies that include not simply measures of the sex-specific mortality response but also in-depth analyses of sex-specific differences. Reasons for the paucity of information and analyses on sex differences are, in large part, related to the particulars of the model systems used in aging research. For example, sex differentials are not measured in the yeast model because reproduction in yeast is primarily asexual, nor in the nematode model because *C. elegans* is hermaphroditic, nor in the *Drosophila* model because fruit fly researchers are primarily concerned with genetic rather than demographic differences, nor in the rodent models because the primary research area in rodents is dietary restriction and males are used mainly in this research because many mouse and rat researchers believe that the more complicated response of females to dietary restriction would be difficult to interpret.

The third major theme that became apparent from the results of the medfly research is the "*biodemographic linkages between longevity and reproduction.*" Reproduction is the cardinal function of all living organisms and, therefore, it is not surprising that reproduction and the length of life are related. However, this relationship is often not straightforward since what constitutes a "reproductive unit " is not necessarily the single offspring. For example, in mammals there is a physiological cost of gestation but there is also a cost in the depletion of primordial follicles (i.e., atresia).

1.5. Organization of the Book

The book unfolds in a hierarchical sequence consisting of three clusters of chapters. I introduce the book in the first cluster: chapter 1 ("Introduction") describes the conceptual, empirical, and analytical details, and chapter 2 ("Operational Framework") provides a brief survey of experimental and analytical concepts and methods. These chapters contain a quick overview of the main themes in aging and biodemography research, background information on the medfly, a description of the facility where the NIA-funded research was conducted and the basic experimental protocols, and a review of demographic techniques and concepts. The second cluster of chapters (3–7) contains experimental and/or theoretical results of the medfly studies. In chapters 3 ("Mortality Deceleration") and 4 ("Reproduction and Behavior") I present the results of descriptive studies on mortality, longevity, reproduction, and what my colleagues and I have termed "supine behavior" in the medfly. These baseline studies include the results from the original large-scale life table study on 1.2 million medflies and reproductive patterns obtained from studies of the reproductive history of 1,000 individual females. The results from research described in these chapters provided the foundational work for posing hypotheses about underlying mechanisms that framed the research on mortality and longevity that is described in chapters 5 ("Mortality Dynamics of Density"), 6 ("Dietary Effects"), and 7 ("Linkages between Reproduction and Longevity"). The results of research contained in these chapters all involved different types of environmental manipulations. The last cluster of chapters is concerned with syntheses, idenfication of general biodemographic principles, and theory. In chapter 8 ("General Biodemographic Principles") I outline a total of seventeen fundamental principles derived from either the results of medfly research presented in previous chapters or from the general literature. I then build on the concept of principles but focus on those pertaining to "humans as primates" as well as those pertaining to humans in contemporary societies. Chapter 9 (*A General Theory of Longevity*) consists of a theoretical paper that my colleague Debra Judge and I wrote arguing that life span extension in humans and other social species is self-reinforcing. In chapter 10 ("Epilogue: A Conceptual Overview of Life Span") I provide a synopsis of the main abstract and conceptual elements concerned with research on aging and longevity including hypotheses on the future of human longevity.

2

Operational Framework

2.1. Background

2.1.1. The Mediterranean Fruit Fly

The Mediterranean fruit fly (*Ceratitis capitata*), commonly known as the medfly, belongs to the dipteran family Tephritidae referred to as "true" fruit flies—a group of about 4,000 species distributed throughout most of the world (Christenson and Foote 1960). Members of this group lay eggs *in* intact fruit, using their sharp ovipositor, rather than *on* decaying fruit as do their distant relatives the gnat-sized vinegar flies in the family Drosophilidae (also referred to as pomace flies). Although the most common drosophilids used for research in genetics and development (e.g., *Drosophila melanogaster, D. subobscura*) are also known as "fruit flies," the two dipteran families are separate and distinct. Other tephritid species include the Mexican fruit fly (*Anastrepha ludens*), the Caribbean fruit fly (*A. suspensa)*, the Queensland fruit fly (*Dacus tryoni)*, and the apple maggot fly (*Rhagoletis pomonella)*.

The life cycle of the medfly is typical of most tephritids (Hagen et al. 1981). Adult mated female medflies deposit their eggs within host fruit, the eggs hatch into larvae within 1 to 2 days and begin feeding on the pulp of the fruit. Within 1–2 weeks, the larvae complete their development and exit the fruit to pupate in the ground. After about 2 weeks, the adults emerge from the pupae and begin to forage for food and mates. Adults become sexually mature in about 1 week, they mate, and the females begin to lay eggs anywhere from 7 days to 2 months later, depending on host, food, and temperature. Lifetime egg production ranges from 800 to 1200 eggs/female with an expectation of adult life of 3–6 weeks (Carey 1989). The medfly is thought to have evolved in Africa (Bateman 1972), but is currently distributed in the Mediterranean regions of Europe, the Middle East, western Australia, Hawaii, and Central and South America.

2.1.2. Medfly as a Model Species

Although historically the medfly has not been used as a model species for aging research as extensively as *Drosophila*, the experimental protocols and procedures for its study at the whole-organism level are well developed due to its economic importance. Methods for rearing this species have been improved as a result of advances in the "sterile insect technique" (Hendrichs et al. 1995; Schwarz et al. 1989; Vargas 1989). Whereas *Drosophila* are ideal genetics research models due to their genetic variability, abundance of genetic markers, large salivary-gland chromosomes, and ease of culturing, medflies are well suited as demographic research models because of their large size, robustness, and ease of sex determination. Medflies also make good demographic models because, unlike *Drosophila*, which lay their eggs directly in moist adult food, medfly adults will feed on a dry mixture of sugar and yeast but lay their eggs in a separate medium such as artificial (agar) hosts (Freeman and Carey 1990) or organdy mesh (Vargas 1989). This is important because reproduction can be manipulated independent of diet. That is, egg laying can be stimulated or suppressed by introducing or removing the egging medium while keeping adult food present. This is not as easily done with *Drosophila*.

2.1.3. The Moscamed Mass-rearing Facility

The medfly research reported in this book was conducted at the Moscamed medfly mass-rearing facility in Metapa, a small village located about 20 km from the city of Tapachula in the state of Chiapas, Mexico. The facility was constructed in 1979 as a joint enterprise funded by the U.S. Department of Agriculture and the Mexico Ministry of Agriculture to rear large numbers of medfly adults for sterilization and subsequent release as a tactical component of a program to prevent the spread of the pest further into Mexico. The factory employs several hundred workers and is the size of a large warehouse with over 60,000 sq. ft. of floor space. An average of 500 million medfly pupae are produced there weekly. The large-scale production capability was the overriding reason for using the medfly for life table studies inasmuch as the factory provided the program with essentially unlimited numbers of pupae at any time.

2.2. Empirical Methods

2.2.1. Cages

Four different types of cages were used to gather data on mortality alone or on mortality and reproduction together (Carey and Liedo 1999). The cages included (1) a "cassette-type" for monitoring large number of flies in a single cage; (2) 1-oz condiment cups for monitoring medflies in solitary confinement; (3) Falcon 24-cell units (3.5-ml tissue cells) for monitoring flies in solitary confinement; and (4) small plastic bottles for housing pairs of flies and for monitoring reproduction in individual females. Trade-offs exist in the use of each type of cage. For example, experiments using larger initial numbers are possible when flies are grouped in cassette-type cages, but such an arrangement preempts data gathering on individual reproduction. In contrast, space and labor requirements are large for experiments involving the use of individual pairs of flies, but because of the absence of density effects the conditions within each cage are more favorable for fly survival. In addition, data can be gathered on individual reproduction.

The cassette-type cages were made of a 15-cm × 45-cm × 70-cm aluminum frame covered with mesh. A 2.8-cm in diameter plastic pipe, plugged at the distal end and fitted with an open L-joint at the other end, was placed at middle height of the cage. Stiff filter paper was inserted into a slot cut into the top of the pipe. The pipe was filled with water, and the flies obtained water from the saturated filter paper. Water was added through the L-joint, so it was not necessary to remove the whole pipe. Blocks of solid food were placed outside the cage, on the screened cage top, and flies could feed through the screen. Cassette cages were designed to be long and narrow rather than square in order to maximize the per-fly surface area for resting and preening. Cages were placed on specially adapted racks with light uniformly distributed.

Single-pair cages were essentially modified 500-ml plastic containers, the tops of which were fitted with organdy mesh. Females inserted their ovipositors through the mesh at the front for egg laying. Freshly laid eggs fell onto the water-saturated black filter paper lining the petri dish at the front of the cage. Water was provided by a vial with a dental wick, which was introduced through a 1.5-cm in diameter hole made on the back of the container. Newly eclosed adults rather than pupae were introduced into the cage. Dead males were replaced with a virgin male of the same age taken from an all-male cohort set up in an aquarium-type cage at the same time as the main experiment. The cages fitted with the egg collecting devices were placed on lighted shelves. Fly mortality and egg laying for each fly was monitored daily.

2.2.2. *Adult Rearing*

Two types of adult foods were used in the studies: plain solid sugar, or sucrose, and a 3:1 solid mixture of yeast (enzymatic yeast hydrolysate) and sugar, which was referred to as "full diet." Food type has significant effects on survival and reproduction. For example, fecundity is low and female early mortality is high when medflies are maintained on a sugar-only diet relative to a full diet (Hagen 1952; Hagen and Finney 1950; Jacome et al. 1995; Müller et al. 1997). The typical procedure was as follows. Pupae were collected at random from a batch that started emerging within a 1–2 day period. Adult size was standardized using pupal sorting machines (Schwarz et al. 1989). Adults were carefully collected with a mouth aspirator and placed in the cage through the hole for the water vial, then the vial with water and the dental wick were replaced. Dead males were replaced with a same-aged male from the same batch. For studies on reproduction involving single-pair cages, the dish with the eggs was removed daily and counted under a dissecting microscope. A hygrothermograph recorded daily max-min temperature and relative humidity and fluorescent lights with commercial timers used to control photoperiod at a 12:12 light:dark cycle.

2.2.3. General Experimental Procedure

To determine the initial number, N_0, of live adults desired per cage, a rule-of-thumb starting point was to estimate the minimum surface area needed by individual flies for resting and preening (e.g., 1-cm^2, 2-cm^2, etc.). The estimate for the medfly was 1.3 to 2-cm^2 per fly whereas (for perspective) the estimate for the larger Mexican fruit fly, was 3-cm^2 per fly. The initial number of flies per cage was obtained by dividing this total by the per-capita space requirement. The desired number of newly emerged adults were placed in the cage in one of two ways. The first was simply to place the exact number of adults in the cage using either a mouth aspirator or an adapted car-vacuum cleaner. This method was used for setting up single-sex cages. The second way was to estimate the emergence/eclosion rate of pupae of age x, E_x. The estimated number of pupae placed in a cage to obtain the desired number of newly emerged adults was determined by the formula $N_P = N_0/E$ where N_P denotes the number of pupae, N_0 the initial number, and E the pupal eclosion rate. The amount of pupae required to achieve the desired initial density was placed in a petri dish inside the cage and removed 24 hours later to ensure that the ages of all individuals within a cage were similar. All dead individuals were removed each day at a fixed time (0800–1200 hr) by brushing them into an opening in the bottom of the cage. Dead flies were sorted by sex and counted.

2.3. Analytical Methods

2.3.1. The Life Table

The life table is one of the most important tools in biogerontological research because it serves as a framework for organizing age-specific mortality data, provides detailed, transparent descriptions of the actuarial properties of a cohort, and generates simple summary statistics such as life expectancy that are useful for comparisons. There are two general forms of the life table, both of which are relevant to biogerontology. The first is the *cohort life table*, which provides a longitudinal perspective in that it includes the mortality experience of a particular cohort from the moment of birth through consecutive ages until none remains in the original cohort. The second basic form is the *current life table*, which is cross sectional. This table assumes a hypothetical cohort that is subject throughout its lifetime to the age-specific mortality rates that prevail for the actual population over a specified period and are used to construct a synthetic cohort. Cohort (longitudinal) life tables are frequently used in laboratory studies of relatively short-lived species (insects, rodents), whereas current (cross-sectional) life tables are most often used in studies of long-lived species such as humans.

2.3.2. *Life Table Construction and Interpretation*

The five main functions of the life table are defined and their formulae are presented in table 2.1 and include the following:

l_x—the proportion of all newborn surviving to age x
p_x—the proportion of individuals alive at age x that survive to $x + 1$
q_x—the proportion of individuals alive at age x that die prior to $x + 1$
d_x—the proportion of all newborn that die in the interval x to $x + 1$
e_x—the expected lifetime remaining to the average individual age x

A partial life table on the medfly (Carey, Liedo, Orozco, and Vaupel 1992) is presented in table 2.2 and reveals that: (1) less than 4% of the original cohort survives to 40 days (i.e., $l_{40} = 0.03784$; (2) age-specific mortality is extremely high around 40 days ($q_{40} = 0.12125$) but decreases at older ages (e.g., $q_{120} = 0.05128$), (3) 1-of-8 or 12.5% of all flies die in the 3-day interval 20–22 days (i.e., $d_{20} + d_{21} + d_{22} = 0.125$); and (4) expectation of life at age 90 ($e_{90} = 27.0$ days) is greater than expectation of life at eclosion ($e_0 = 20.8$ days). This reflects the leveling off and decrease in mortality at advanced ages that will be discussed in the next chapter.

2.3.3. *Additional Life Table Parameters and Relationships*

CENTRAL DEATH RATE

The parameter central death rate, also known as the age-specific death rate, denoted m_x, is defined as the number of deaths occurring in a specified period in a specific age category divided by the population at risk. The central death rate is not a probability but rather an observed rate—the number of individuals that die relative to the number at risk. It is essentially a

TABLE 2.1
Main Life Table Functions

Function	*Notation*	*Equation*	*Description*
Cohort survival	l_x	$\frac{N_x}{N_0}$	fraction of initial cohort surviving to age x where N_x denotes number alive at age x
Age-specific mortality	q_x	$1 - \frac{l_{x+1}}{l_x}$	fraction alive at age x that die prior to $x+1$
Age-specific survival	p_x	$\frac{l_{x+1}}{l_x}$	fraction alive at age x that survive to $x+1$
Expectation of life at age x	e_x	$\frac{1}{2} + \frac{l_{x+1} + l_{x+2} + \ldots + l_\omega}{l_x}$	average days remaining to an individual age x
Frequency distribution of deaths	d_x	$l_x - l_{x+1}$	fraction dying in interval x to $x + 1$

Source: Carey 1998.

TABLE 2.2
Life Table Parameters for Mediterranean Fruit Fly

Age	*Number Living*	*Fraction Surviving*	*Age-Specific Survival*	*Age-Specific Mortality*	*Frequency of Deaths*	*Expectation of Life*
x	N_x	l_x	p_x	q_x	d_x	e_x
(1)	*(2)*	*(3)*	*(4)*	*(5)*	*(6)*	*(7)*
0	1,203,646	1.00000	1.00000	0.00000	0.0000	20.8
1	1,203,646	1.00000	0.99856	0.00144	0.0014	19.8
2	1,201,913	0.99856	0.99599	0.00401	0.0040	18.9
3	1,197,098	0.99456	0.99492	0.00508	0.0050	17.9
4	1,191,020	0.98951	0.99362	0.00638	0.0063	17.0
5	1,183,419	0.98320	0.99247	0.00753	0.0074	16.1
6	1,174,502	0.97579	0.99023	0.00977	0.0095	15.3
7	1,163,026	0.96625	0.98768	0.01232	0.0119	14.4
8	1,148,693	0.95434	0.98358	0.01642	0.0157	13.6
9	1,129,836	0.93868	0.97816	0.02184	0.0205	12.8
10	1,105,164	0.91818	0.97018	0.02982	0.0274	12.1
⋮	⋮	⋮	⋮	⋮	⋮	⋮
20	575,420	0.47806	0.90772	0.09228	0.0441	8.3
21	522,319	0.43395	0.90320	0.09680	0.0420	8.0
22	471,756	0.39194	0.89976	0.10024	0.0393	7.9
⋮	⋮	⋮	⋮	⋮	⋮	⋮
40	45,544	0.03784	0.87875	0.12125	0.0046	7.0
41	40,022	0.03325	0.87207	0.12793	0.0043	6.9
42	34,902	0.02900	0.86986	0.13014	0.0038	6.8
⋮	⋮	⋮	⋮	⋮	⋮	⋮
90	91	0.00008	0.94505	0.05495	0.0000	27.0
91	86	0.00007	0.98837	0.01163	0.0000	27.5
92	85	0.00007	0.92941	0.07059	0.0000	26.8
⋮	⋮	⋮	⋮	⋮	⋮	⋮
120	39	0.00003	0.94872	0.05128	0.0000	18.7
121	37	0.00003	0.97297	0.02703	0.0000	18.7
122	36	0.00003	0.91667	0.08333	0.0000	18.2
⋮	⋮	⋮	⋮	⋮	⋮	⋮
164	4	0.00000	0.50000	0.50000	0.0000	4.0
165	2	0.00000	1.00000	0.00000	0.0000	6.5
166	2	0.00000	1.00000	0.00000	0.0000	5.5
⋮	⋮	⋮	⋮	⋮	⋮	⋮
172	0	0.00000	0.00000	1.00000	0.0000	0.0

Source: Carey 1998.

weighted average of the force of mortality between ages x and $x + 1$. The relationship between m_x and q_x is

$$m_x = \frac{q_x}{1 - \frac{1}{2} q_x} \quad \text{and} \quad q_x = \frac{m_x}{1 - \frac{1}{2} m_x}. \tag{2.1}$$

For example, the central death rate at age 10 for the medfly (see table 2.3) is $m_{10} = 0.03027$ whereas the probability of dying in the age interval 10 to 11 is $q_{10} = 0.02982$. The parameter central death rate is used in computing another important actuarial parameter discussed in the next section—the life table aging rate.

LIFE TABLE AGING RATE

Horiuchi and Coale (1990) introduced to the demographic literature the parameter life table aging-rate (LAR), denoted k_x and defined as the rate of change in age-specific mortality with age. The measure is based on *relative* rather than *absolute* rate of change in mortality with age. The formula is given as

$$k_x = \ln(m_{x+1}) - \ln(m_x), \tag{2.2}$$

where m_x denotes the central death rate discussed in the previous section. Life table aging rate is an age-specific analog of the Gompertz parameter, b, since it is a measure of the slope of mortality with respect to age. But unlike the Gompertz parameter, which assumes constancy of the mortality slope typically over a large age interval, LAR examines the change over short intervals. Example computations of k_x from the medfly data in table 2.2 for ages 5 and 40 are $k_5 = 0.26098$ and $k_{40} = 0.05723$, which indicate that mortality is changing at over 26% per day at day 5 but less than 6% per day at day 40. Additional perspectives for LAR applied to the medfly and to the bean beetle, *Callosobruchus maculatus*, are presented in Carey and Liedo (1995b) and Tatar and Carey (1994a and b).

MORTALITY RATIOS

Manton and Stallard (1984) describe a mortality crossover as an attribute of the relative rate of change and level of age-specific mortality rates in two populations: one group is "advantaged" (i.e., experiences lower relative mortality) and the other "disadvantaged" (i.e., experiences higher relative mortality). The disadvantaged population must manifest age-specific mortality rates markedly higher than the advantaged population through middle age, at which time the rates change. Crossovers in mortality occur due to differences in rates of aging at the individual level, and to demographic selection

TABLE 2.3

Selected Mortality Parameters and Formulae, Change Indicators, and Scaling Models

Parameter/Model	*Description*	*Notation*	*Formula/Model*
General Mortality Parameters			
Force of mortality	Instantaneous mortality rate	μ_x	$-\frac{dl_x}{l_x dx}$
	Estimation formula (#1)	μ_x	$-\ln p_x$
	Estimation formula (#2)	μ_x	$-\frac{1}{2}(\ln p_{x-1} + \ln p_x)$
	Estimation formula (#3)	μ_x	$\frac{1}{2n}\ln_e\left(\frac{l(x-n)}{l(x+n)}\right)$
Central death rate	Number dying at age x relative to number at risk	m_x	$\frac{q_x}{1-\frac{1}{2}q_x}$
Life table aging-rate	Rate of change in age-specific mortality with age	k_x	$\ln(m_{x+1}) - \ln(m_x)$
Mortality smoothing	Smoothing for discrete form	$\hat{q}_x$	$1-\left[\prod_{y=x-n}^{x+n} p_x\right]^{-(n+1)}$
	Smoothing for continuous form	$\hat{\mu}_x$	$\frac{1}{n+1}\sum_{y=x-n}^{x+n}\mu_y$

Average daily mortality	Daily mortality given expectation of life, e_0	$\overline{\mu}$	$\frac{1}{e_0}$
Entropy	Days gained per averted death	H	$\frac{\sum_{x=0}^{\omega} e_x d_x}{e_0}$
Mortality Change Indicators (Cohort A vs Cohort B)			
#1a	Mortality increase/decrease (relative)	—	$\frac{\mu_x^A}{\mu_x^B}$
#1b	Mortality increase/decrease (absolute)	—	$\mu_x^A - \mu_x^B$
#2a	Survival increase/decrease (relative)	—	$\frac{l_x^A}{l_x^B}$
#2b	Survival increase/decrease (absolute)	—	$l_x^A - l_x^B$
#3a	Life-days gained/lost at age x (relative)	—	$\frac{e_x^A}{e_x^B}$
#3b	Life-days gained/lost at age x (absolute)	—	$e_x^A - e_x^B$
Mortality Scaling			
Proportional	Age-independent scaling (δ = scaling factor)	$\hat{\mu}_x$	$(1 + \delta)\mu_x$
Age-specific	Age-dependent scaling (δ_x = scaling factor at age x)	$\hat{\mu}_x$	$(1 + \delta_x)\mu_x$

Source: Carey 2001.

where individuals with high mortality are selected out early for one population and therefore the more robust individuals survive to the older ages. The general formula for the mortality ratio of two cohorts, A and B is

$$R_x = \frac{\mu_x^A}{\mu_x^B}, \tag{2.3}$$

where R_x denotes the ratio at age x. Mortality at age x is higher in cohort A than in cohort B if $R_x > 1$, lower in cohort A than in cohort B if $R_x < 1$, and the same in cohort A and cohort B at age x if $R_x = 1$.

AVERAGE LIFETIME MORTALITY

The inverse of life expectancy at birth, e_0, is the average mortality experienced by the cohort, denoted $\overline{\mu}$. Or more generally, the inverse of life expectancy at age x, e_x, is the average mortality experienced by the cohort beyond age x denoted $\overline{\mu}_x$. That is,

$$\overline{\mu} = \frac{1}{e_0} \tag{2.4}$$

and

$$\overline{\mu}_x = \frac{1}{e_x}. \tag{2.5}$$

Example values of $\overline{\mu}_x$ from the medfly data presented in table 2.2 for ages 0, 50, and 100 days reveal the increase, peak, and decrease in overall pattern of age-specific mortality: $\overline{\mu}_0 = 0.048$, $\overline{\mu}_{50} = 0.150$, and $\overline{\mu}_{100} = 0.036$. In other words, average daily mortality experienced by the cohort beyond ages 0, 50 and 100 days was approximately 5%, 15%, and 4%, respectively.

2.3.4. *The Mortality Function and Its Importance*

The life table provides five different expressions or functions that describe the mortality and survival experience of a cohort. Because each of the functions can be independently derived from the original cohort data and all but expectation of life can be used to derive the other functions, it is often inferred that no single function has precedent. Although this is true algebraically, it is not accurate biologically, demographically, or actuarially. The age-specific mortality schedule—the series of probabilities that an individual alive at age x dies prior to age $x + 1$—serves as the actuarial foundation for all other functions (Carey 1998). The basic role of mortality is evident by considering the following. First, death is an "event" indicating a change of state from living to dead, a failure of the system. In contrast, survival is a "nonevent" inasmuch as it is a continuation of the current state. This orientation toward events rather than nonevents is fundamental to the analysis of risk.

Second, an individual can die due to a number of causes such as an accident, a predator, or a disease. Therefore mortality rates can be disaggregated by cause of death and thus shed light on the biology, ecology, and epidemiology of deaths, the frequency distribution of causes, and the likelihood of dying of a particular cause by age and sex. This concept of "cause" obviously does not apply to survival.

Third, the value of mortality rate at a specified age is independent of demographic events at other ages. In contrast, cohort survival rate (l_x) to older ages is conditional upon survival to each of the previous ages, life expectancy at age x (e_x) is a summary measure of the consequences of death rates over all ages greater than x, and the fraction of all deaths (d_x) that occur at young ages will determine how many individuals remain to die at older ages. This independence of mortality rate relative to events at other ages is important because age-specific rates can be directly compared among ages or between populations that, in turn, may shed light on differences in relative age-specific frailty or robustness.

Fourth, a number of different mathematical models of mortality have been developed (e.g., Gompertz, Weibull, Logistic) that provide simple and concise means for expressing the actuarial properties of cohorts with a few parameters. Therefore the mortality and longevity experience of different populations can be more readily compared.

2.3.5. *The Force of Mortality*

The force of mortality at age x, denoted μ_x, is the instantaneous mortality rate representing the limiting value of the age-specific mortality rate when the age interval to which the rate refers becomes infinitesimally short (Pressat 1985). It is given as

$$\mu_x = \frac{dl_x}{l_x dx} \tag{2.6}$$

and

$$l_x = l_0 \exp\left\{-\int_0^x \mu_y dy\right\}, \tag{2.7}$$

where l_x is the life table survival rate to age x and l_0 is the radix.

Also known as the instantaneous death rate and hazard rate, μ_x is preferred over age-specific mortality (q_x), gerontology, and demography because it is not bounded by unity, it is independent of the size of the census (age) intervals, and it forms the argument of numerous parametric mortality functions. Three formulae that are commonly used for computing μ_x include

$$\mu_x = -ln\, p_x, \tag{2.8}$$

$$\mu_x = -\frac{1}{2}(ln\, p_{x-1} + ln\, p_x), \tag{2.9}$$

and

$$\mu_x = -\frac{1}{2n}\ln_e\left(\frac{l_{x-n}}{l_{x+n}}\right), \tag{2.10}$$

where n denotes the band width. For example, μ_x is computed using 3 age classes if $n = 1$, 5 age classes if $n = 2$, and so forth. The expressions for μ_x are based on mortality rates in 2 or more adjacent age classes. The relationship between μ_x and the other measures of mortality and death is this: μ_x denotes an instantaneous mortality rate that applies to each moment of the specified interval. In contrast q_x denotes the probability of death over a discrete age interval and m_x denotes the death rate, which is a density function and not a probability, per se.

2.3.6. Smoothing Age-specific Mortality Rates

Because of the binomial noise present in age-specific mortality schedules due to small numbers or to environmental variation, it is often useful to smooth mortality rates for plotting. A formula for computing the running geometric mean of an age-specific mortality schedule, denoted $\hat{q}_x$, is given as

$$\hat{q}_x = 1 - \left[\prod_{y=x-n}^{x+n} p_x\right]^{-(n+1)}, \tag{2.11}$$

where $p_x = 1 - q_x$ and n denotes the "width" of the running geometric average. The analytical counterpart for the running mean of force of mortality, denoted $\hat{\mu}_x$, is

$$\hat{\mu}_x = \frac{1}{n+1}\sum_{y=x-n}^{x+n}\mu_y. \tag{2.12}$$

More sophisticated methods for smoothing hazard rates using locally weighted least-squares techniques are described in Müller et al. (1997a) and Wang et al. (1998).

2.3.7. The Gompertz Model

The Gompertz model (Chiang 1984; Gompertz 1825) suggests that the number of living organisms decreases in geometrical (exponential) progression as age increases in arithmetical (linear) progression. The model implies that the probability of dying from age increases exponentially after age 15 (Finch 1990; Pressat 1985); it is given as

$$\mu_x = ae^{bx}, \tag{2.13}$$

where μ_x denotes mortality at age x, a denotes the initial rate of mortality, and b is the exponential mortality rate coefficient. The Gompertz model is used widely in demography and gerontology for fitting mortality schedules of nonhuman species (Finch 1990) and for comparing rates of actuarial aging between and among different species (Comfort 1979; Lamb 1977). It is used in demography for smoothing the mortality functions of recorded life tables (Pressat 1985), for developing model life tables (Coale and Demeny 1983), and for projecting numbers of persons living to older ages (Bell et al. 1992). The Gompertz equation is generally recognized by most demographers and gerontologists as an empirical model and not a "law" (Finch 1990).

2.3.8. Below-threshold Mortality

The estimation of age-specific mortality rates brings with it a variety of statistical challenges that can bias conclusions about many contemporary issues in demography, ecology, and evolutionary biology (Promislow et al. 1999). At the early-age and late-age boundaries of mortality trajectories, when either mortality or survival are close to zero, mortality behaves as a threshold character and becomes difficult to measure accurately (Gaines and Denny 1993). Consider a cohort with initial sample size of $N_0 = 50$ individuals, with a true mortality rate in the youngest age class of $\mu_x = 0.001$. What is the probability of observing one or more deaths in this cohort at a given age x? There is a less than 5% chance of observing any mortality in the first age-class. In fact, to determine accurately that the true mortality rate is significantly greater than zero at the 95% confidence level, a sample size of $N = 10{,}000$ is needed. Thus the point estimate of this so-called "below-threshold mortality" will be highly inaccurate (Promislow et al. 1999), as shown schematically in figure 2.1. This concept of below-threshold mortality is essentially a demographic sampling error and can bias estimates of rates of actuarial aging, age at onset of senescence, costs of reproduction, and demographic tests of evolutionary models of aging (Promislow et al. 1999).

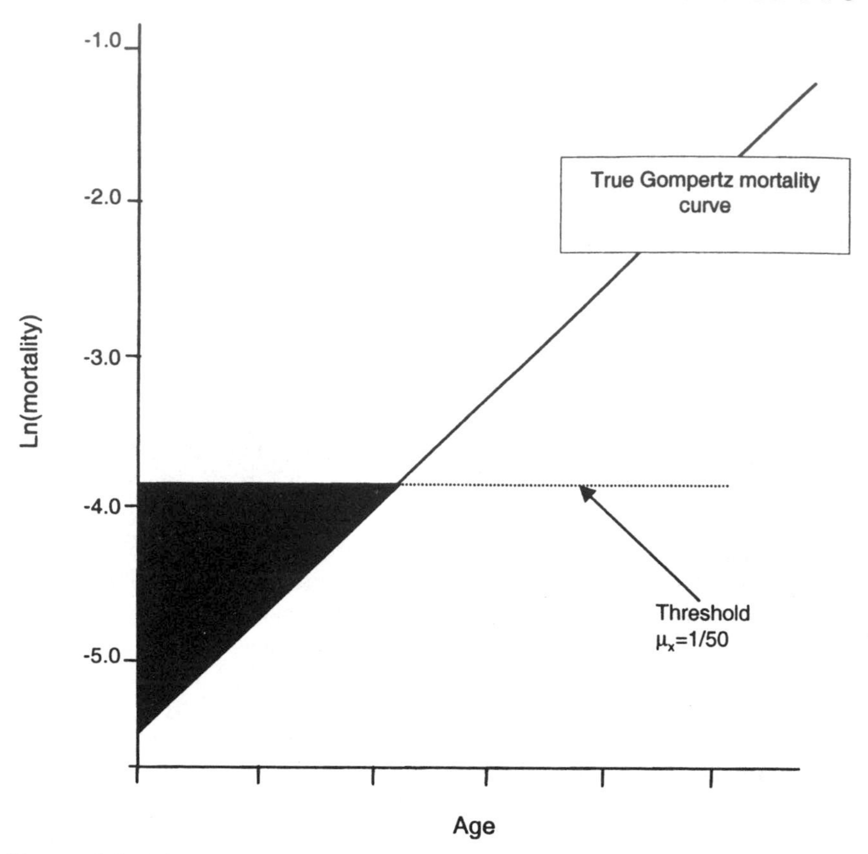

FIGURE 2.1. Schematic diagram of the "threshold-mortality" concept, redrawn from Promislow et al. 1999. The solid line represents a Gompertz mortality curve for a cohort of N = 50 individuals. The dotted line represents the threshold at which mortality and $\mu = 1/50$, below which mortality cannot be observed, and the shaded region shows the part of the Gompertz curve that we will accordingly not be able to estimate due to insufficient numbers of deaths.

2.3.9. Peak-aligned Averaging

An interesting and important characteristic of mortality in some species such as the medfly is that it both increases and decreases with age thus creating local as well as lifetime peaks. Because the timing of these peaks may differ slightly between cohorts due to both chance and subtle differences in environmental conditions, Müller et al. (1997a) introduced a simply statistical technique aimed at avoiding biases associated with averaging peaks across cohorts since averaged peaks tend to "flatten out" due to the variability in the timing of peaks for individual cohorts.

The technique consists of three steps: (1) estimate the location (age) of the hazard rate peak, denoted $\hat{\theta}_j$, where j is the cohort index; (2) obtain the estimated average peak location for all cohorts, $\hat{\theta}$, by averaging all individual cohort peak locations:

$$\hat{\theta} = \frac{1}{N}\sum_{j=1}^{N} \hat{\theta}_j, \tag{2.14}$$

and (3) transform the age coordinate x for each of the N cohorts using the transformed age:

$$x_j' = \frac{x\hat{\theta}}{\hat{\theta}_j}. \tag{2.15}$$

This time-scale transformation maps all individual peak locations $\hat{\theta}_j$ to $\hat{\theta}$. An example application of this technique is illustrated in Müller et al. (1997a) where local maxima (i.e., "shoulders") in the age-specific mortality schedules around day 10 for 33 female medfly cohorts were peak-aligned.

2.4. Summary

There are at least three important epistemiological outcomes of the medfly studies:

ACTUARIAL SCALE

Whereas biologists and gerontologists considered a "large" life table experiment as one that used an initial number of a few hundred individuals, the medfly oldest-old project redefined 'large-scale' for fruit fly studies as meaning the use of initial numbers in the tens of thousands, hundreds of thousands, or even millions. One of the main reasons for initial numbers on this scale never being previously considered was that the empirical and conceptual importance of age-specific mortality was not fully acknowledged. There was no reason to use large initial numbers if researchers were only concerned about differences in either life expectancy or cohort survival, since these values are less dependent on sample size (Carey 1993).

EXPERIMENTAL METHODOLOGY

The increase in experimental scale required new cage designs from small box-like cages where the emphasis was on "living space" (i.e., volume) to what were termed "cassette-like" cages where one of the main criteria was surface area for resting. In addition, it was necessary for the cages used for the large-scale mortality studies to provide ready access to technicians for removing dead flies without harming or killing live flies.

ANALYTICAL CONCEPTS

The shift of emphasis from determining the mean to measuring age-specific mortality at all ages encouraged the use and development of new analytical concepts and techniques that had seldom been applied to nonhuman actuarial data. The most important development in this context was the focus on the overall age pattern of mortality, which then fostered the development of techniques for peak alignment, mortality smoothing, awareness of subdetectable mortality, sex-mortality differentials, and mortality cross-overs.

3

Mortality Deceleration

IT USED to be generally assumed for most species that mortality rates increase monotonically at advanced ages. The results of the study described in this chapter revealed that mortality rates leveled off and decreased at older ages in a population of 1.2 million medflies maintained in cages of 7,200 and in a group of approximately 48,000 adults maintained in solitary confinement. Consequently life expectancy in older individuals increased rather than decreased with age. These results cast doubt on several central concepts in gerontology and the biology of aging: (i) that senescence can be characterized by an increase in age-specific mortality, (ii) that the basic pattern of mortality in nearly all species follows the same unitary pattern at older ages, and (iii) that species have absolute life-span limits.

3.1. Background

Although age-specific mortality rates are used by gerontologists, demographers, and biologists in a number of interrelated ways, including quantifying senescence in populations (Finch 1990), comparing species (Gavrilov and Gavrilova 1991), and inferring species-specific life-span limits (Fries 1980, 1983), surprisingly the pattern of age-specific mortality is well known only for *Homo sapiens*, and even data for humans are sparse after age 85. For 48 species scattered across various phyla, Finch (1990) estimated the level of mortality at the age of sexual maturity and the increase in the mortality rate with age, but he warned that the estimates "should be considered first approximations within a twofold range." The estimates depend on the untested assumption that mortality increases at the same rate from sexual maturity to advanced old age; as Finch (1990) noted, "there is no a priori reason why mortality rates should conform to functions of this type." The number of observations of age at death for any nonhuman species is small. In a typical study of mortality, the life spans of some 20 to 50 individuals are observed in laboratory or field setting; only rarely has mortality in several thousand individuals been monitored. When only a few hundred individuals are observed, the pattern of age-specific mortality at older ages, when perhaps 90 percent of the population is dead, is beyond the scope of definitive study. It is unknown whether the pattern of mortality at advanced ages is typically one of high and increasing mortality, as among humans; moderate and constant mortality; or some other pattern.

Cohorts that totaled one million flies were used in the study under discussion to ensure that mortality rates at older ages would be based on large numbers. This information was used to determine whether mortality rates increased at advanced ages, whether the Gompertz mortality model fit the mortality data at older ages, and whether patterns of age-specific mortality implied an upper life-span limit. Three separate trials were conducted. In experiments 1 and 2, death rates were monitored in over 20,000 medflies maintained in solitary confinement. In experiment 3, over 1.2 million medflies were maintained in groups of approximately 7,200.

3.2. Slowing of Mortality at Older Ages

The results of the large-scale life table studies by Carey, Liedo, Orozco, and Vaupel (1992) revealed that life expectancy at eclosion was highest for medflies maintained in solitary confinement and lowest for medflies maintained in groups (table 3.1). Life expectancy decreased only slightly in the interval between the age when 10% of the original cohort remained alive to the age when 1% of the original cohort remained alive. It then increased at the oldest ages. In all three experiments the life expectancy for flies at the age when 90% of the original cohort was dead was similar to the life expectancy for individuals still alive at the age when 99.9% of the original cohort was dead. Life expectancy for medflies in each experiment either remained the same or increased with age at advanced ages (> 45 days old).

Constant or increasing life expectancies with age can occur only if the underlying age-specific mortality rates are also constant or are decreasing at older ages. This pattern of mortality rate decrease was observed for flies maintained in solitary confinement (experiments 1 & 2) and for flies maintained in groups (experiment 3) (see figures 3.1 and 3.2). Rate of change in mortality slowed in each of the 167 cages of flies in experiment 3 at older ages. There was little overlap in the distributions of rates of change in mortality with age between medfly cohorts 10 days old and cohorts 30 and 45 days old (see figure 3.3). In virtually all cohorts the rate of change in mortality at older ages slowed down, leveled off, or decreased.

As flies aged, mortality rates decreased from a positive rate to a negative one in each experiment. For example, mortality rates decreased in the interval from 20 to 35 days in experiment 1, in the interval from 40 to 55 days in experiment 2, and in the interval 60 to 100 days in experiment 3. Mortality rates in all cages increased at age 10 days but mortality rates decreased in over half of all cohorts after 30 and 45 days. Mortality rates were not monotonic; rather they increased and decreased with age. Slowing of the change

TABLE 3.1
Number Alive, Age (Days), and Remaining Life Expectancy (e_x) of Medflies in Each of 3 Experiments

*Proportion Remaining Alive**	*Experiment 1 (Cups)*			*Experiment 2 (Cells)*			*Experiment 3 (Cages)*		
	No.	*Age*	e_x	*No.*	*Age*	e_x	*No.*	*Age*	e_x
1.0	21,204	0	30.6	27,181	0	28.2	1,203,646	0	20.9
0.1	2,1204	64	19.5	2,718	45	13.7	120,365	33	7.3
0.01	212	103	15.8	272	79	12.9	12,036	50	6.7
0.001	21	135	18.2	27	106	13.7	1,204	64	9.7
0.0001	2	170	34.5	3	117	86.8	120	86	24.8
0.00001							12	146	11.3
0.000001							1	165	6.5

Note: Survival rates at which these three parameters are reported differ by four to six orders of magnitude starting at 100% survival (proportion = 1.0) for flies in all three experiments to survival at one-millionth (proportion = 0.000001) of the original cohort in experiment 3. The greatest ages attained by flies in each of the experiments are as follows: experiment 1 = 216 days, experiment 2 = 241 days, and experiment 3 = 171 days (Carey, Liedo, Orozco, and Vaupel 1992).

*Day on which the number living was ≤ the specified proportion. The last two flies died on the same day. Thus, the age corresponding to this proportion remaining alive represents the age at which the initial cohort of 1.2 million were reduced to the last two flies.

of mortality with age resulted in daily mortality rates that were uniformly low for the oldest flies in all three experiments. For example, average daily mortality for the last 1,000 medflies was 4% in experiment 2 and 6% in experiments 1 and 3.

Mortality rates were similar at the most advanced ages for flies maintained under different physical and biological conditions. Flies maintained in solitary confinement were subject to conditions that minimize mortality risk. Activity was restricted by the small cage size, there was no mating and little egg laying. There was also minimal mechanical wear and no stress due to crowding. In contrast, flies held in groups of 7,200 were subject to conditions that increase mortality risk—large cage-size for flying, mating, some egg laying, mechanical wear, and considerable stress due to crowding (Aigaki and Ohba 1984a; Aigaki and Ohba 1984b; Carey, Krainacker, and Vargas 1986; Luckinbill et al. 1988; Partridge 1986, 1987; Partridge and Andrews 1985; Partridge and Fowler 1992; Partridge and Harvey 1985, 1988; Ragland and Sohal 1973; J. M. Smith 1958; Sohal and Buchan 1981).

There are numerous reasons why mortality rates may slow, remain constant, or decline with age among older individuals in various species, including the possibility that repair mechanisms at older ages can compensate for damage at younger ages. Slowing of the rate of change in mortality with age may also be an artifact of compositional change in the cohort, resulting from

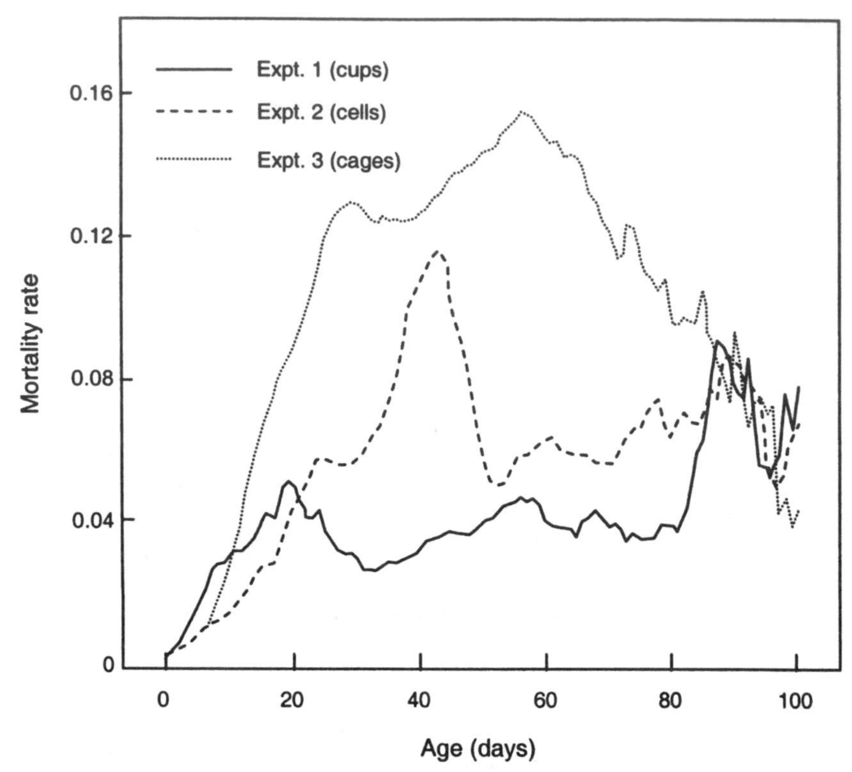

FIGURE 3.1. Age-specific mortality rates for experiment 1 in which 21,204 medflies were maintained in individual cups (top); experiment 2 in which 27,181 medflies were held individually in tissue cells (middle), and experiment 3 in which 1,203,646 medflies were held in cages of approximately 7,200 each (bottom) (Carey, Liedo, Orozco, and Vaupel 1992).

heterogeneity in mortality patterns within the population of genotypes and phenotypes (Vaupel et al. 1979). As the population ages it becomes more and more selected because individuals with higher death rates will die out in greater numbers than those with lower death rates, thereby transforming the population into one consisting mostly of individuals with low death rates. Therefore the possibility exists that death rates fell at older ages, not because of a decrease in the risk of dying at the individual level, but due to heterogeneity at the cohort level.

The reliability theory of aging also sheds light on possible reasons for the late-life mortality deceleration and decline. Gerontologists Garvilov and Gavrilova (1991, 2001) developed mathematical arguments showing that actuarial aging is a direct consequence of systems redundancy. Thus cohorts such as the medfly may exhibit Gompertz-like mortality patterns at younger

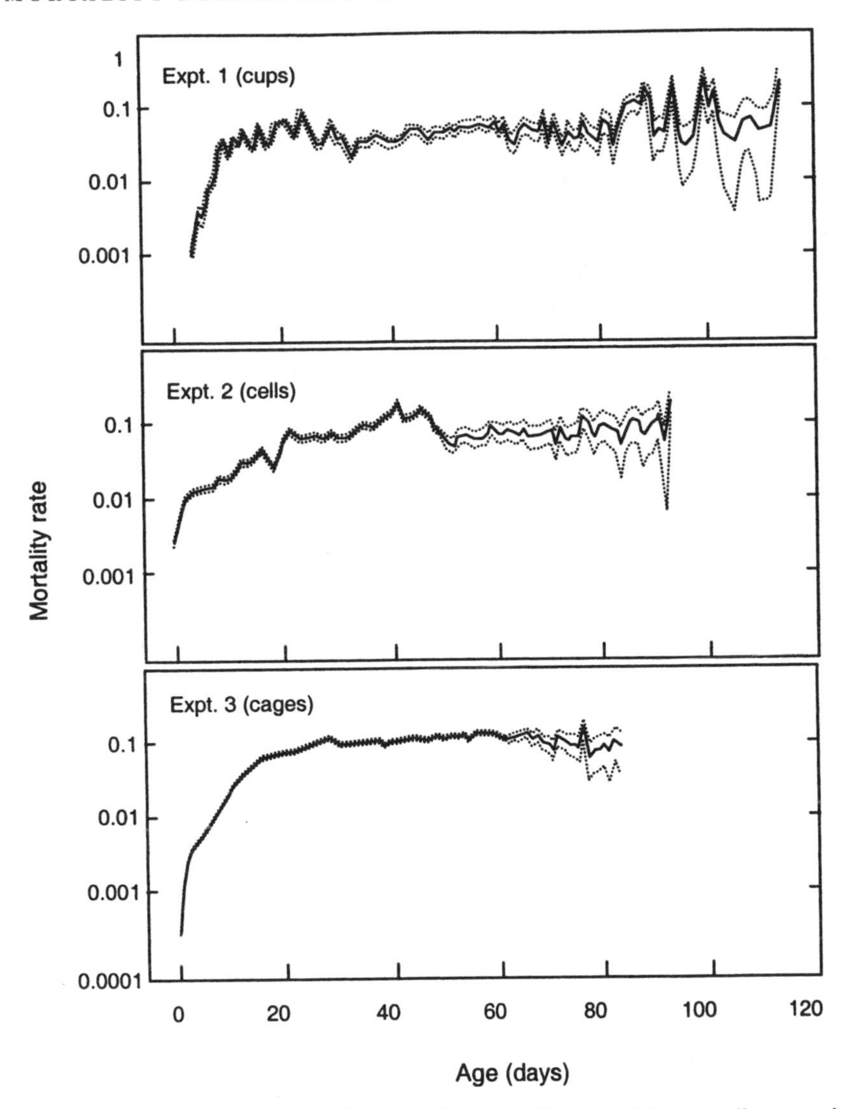

FIGURE 3.2. Smoothed age-specific mortality rates for 3 medfly mortality experiments plotted on a linear scale. In all three experiments the upper 95% confidence limit when mortality rates were declining was lower than the lower 95% confidence limit at the mortality high point. Thus the observed declines were significantly different than a pattern of level mortality rates. The total number of individuals remaining alive at age 100 days for experiments 1 (cups), 2 (cells), and 3 (cages) was 307, 31, and 62, respectively. Thin bounding lines are the 95% confidence intervals (Carey et al. 1992).

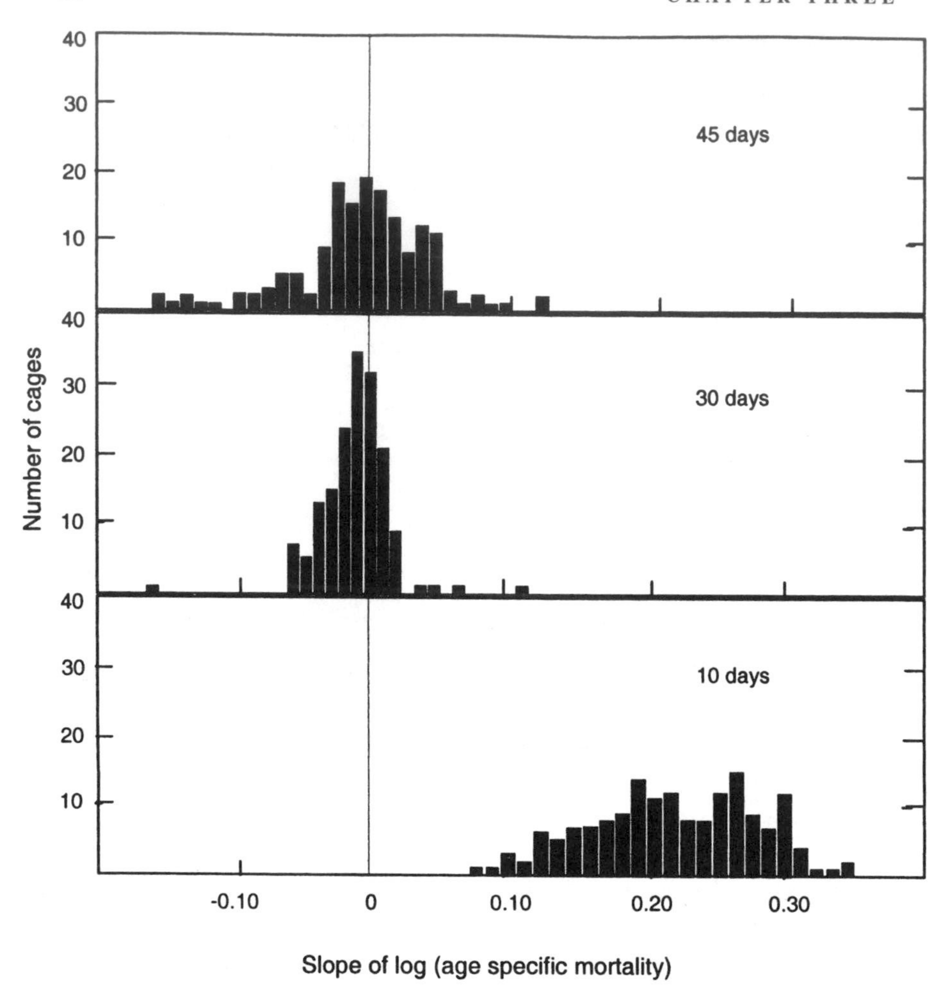

FIGURE 3.3. Distribution of the estimated slopes of the logarithms of age-specific mortality at 10, 30, and 45 days in each of 167 medfly cages with an average initial density of 7,200 adults (experiment 3). Slope estimates were made with a least-squares linear regression of the smoothed mortality curve for each cage centered on the specified age ± 5 days (Carey et al. 1992).

ages and deceleration at older ages due to *redundancy exhaustion*. The mortality decline in late life can be explained by system redundancy in nonidentical components with different failure rates or what these authors refer to as *age-induced population heterogeneity*.

Leveling off of mortality at older ages was reported in studies of other insects, including *Drosophila* (Pearl et al. 1923), houseflies (Rockstein and

Lieberman 1959), medflies (Krainacker et al. 1987), and bruchiid beetles (Tatar et al. 1993; Tatar and Carey 1994a, 1994b, and 1995). Also, several studies have suggested a slowing of the rate of increase of mortality with age for humans at ages greater than 85 years (Economos and Miquel 1979; Gavrilov and Gavrilova 1991; Kannisto 1988, 1991; Kannisto et al. 1994; Thatcher 1992).

3.3. Implications of Mortality Deceleration

The medfly mortality pattern casts doubt on three central concepts in gerontology and the biology of aging and mortality. One concept is that senescence can be operationally defined and measured by the increase in mortality rates with age. Because mortality rates fluctuate up and down around a rough average over most of their life spans, medflies' lives, according to this definition, are characterized by alternating periods of positive and negative senescence. It is questionable whether it is helpful to define the word senescence in this way. Another concept that is not consistent with the data is that the basic pattern of mortality at adult ages in nearly all species follows the same unitary pattern described by the Gompertz model (i.e., exponential increase). The finding that medfly age-specific mortality is not described by this model at old ages provides direct empirical evidence that Gompertz's law does not hold in all populations (Fries 1980, 1983, 1987). Finally, the data are inconsistent with the concept that species can be characterized by their species-specific life spans as measured by (i) the oldest age attained, even in relatively small populations of one hundred or fewer individuals, or (ii) a pattern of age-specific mortality tending toward unity at the maximal age. Different maximum life spans were observed in these experiments and none of the trajectories of age-specific mortality tended toward unity as would have been expected if a species-specific life span limit existed. Furthermore, if small samples were taken, very different maximum life spans would be observed. It is possible to estimate life expectancy but medflies appear not to have a characteristic life span.

The results of the studies by Carey, Liedo, Orozco, and Vaupel (1992) have two methodological implications. One is that it may not be possible to determine the mortality pattern of a species from data on one hundred or even fewer individuals (figure 3.4). Only with 20,000 or 30,000 (experiments 1 and 2) and or with more than a million individuals (experiment 3) was it possible to determine the pattern of medfly mortality through advanced ages. The second implication is that survival curves are poorly suited for summarizing mortality patterns. Survival curves are useful in studying survival—what proportion of the initial cohort is alive at a certain age. It is, however, very difficult to discern the pattern of mortality rates by looking

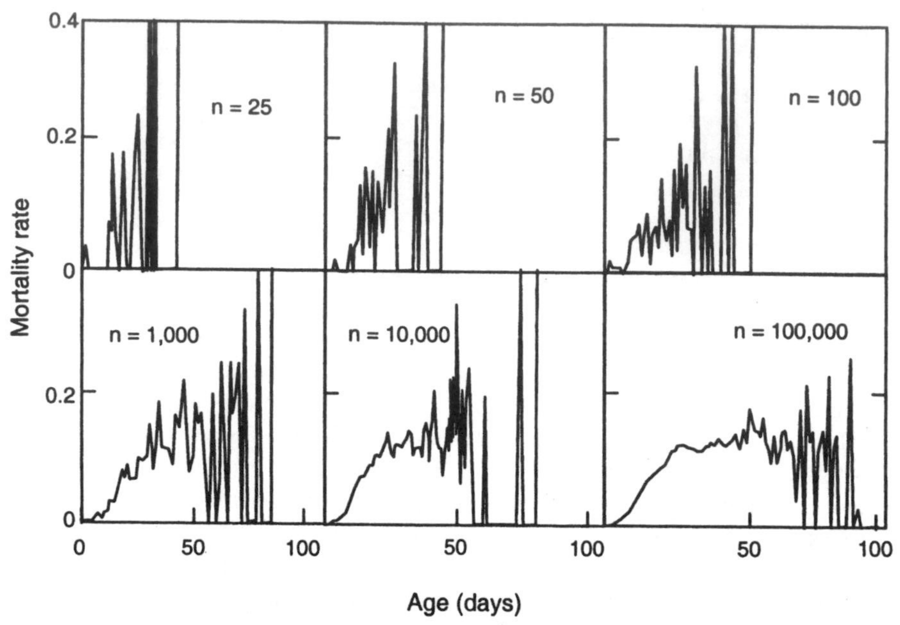

FIGURE 3.4. Age-specific mortality rates computed for cohorts of 6 different initial numbers from $n = 25$ to $n = 100{,}000$. Cohorts were created by randomly subsampling the 1.2 million medfly deaths observed in experiment 3. The results reveal the inadequacy of small cohort size for determining age-specific mortality rates even at young ages and show that large cohorts are needed to determine mortality patterns at advanced ages. Larger n values provide new information on mortality rates at progressively older ages (Carey et al. 1992).

at a survival curve; mortality curves are superior for this purpose. Survival curves are often plotted not because they are the best curve for studying mortality patterns but because they are fairly smooth and regular, even for small populations. In contrast, mortality curves tend to fluctuate erratically when population sizes are small. This problem can be alleviated by using larger populations or various techniques of smoothing.

3.4. Demographic Selection

Three explanations for the slowing of mortality at older ages were suggested following publication of the results of the large-scale life table study (Carey, Curtsinger, and Vaupel, 1993a, 1993b): (1) compositional effects—heterogeneity and selection as noted in the original paper, (2) density effects—the concept that conditions improve at older ages because of reduced crowding due to attrition (examined in chapter 5), and (3) changes at the level of the individual—physiological changes resulting from different experiences such as mating, reproduction, and/or diet (examined in chapter 6). Two papers were published showing how demographic heterogeneity arguments could be used to explain the leveling off of mortality at older ages (Kowald and Kirkwood 1993; Vaupel and Carey 1993). The model developed by Vaupel and Carey (1993) is presented in this section. The other two arguments used to explain the leveling off will be examined in chapters 5 and 6.

3.4.1. Background and Concept

As populations age, they become more selected because individuals with higher death rates will die out in greater numbers than those with lower death rates, thereby transforming the population into one consisting mostly of individuals with low death rates (Carey and Liedo 1995a; Rogers 1992; Vaupel and Carey 1993; Vaupel et al. 1979). The concept of subgroups endowed with different levels of frailty is known as demographic heterogeneity and the winnowing process as the cohort ages is referred to as demographic selection. The actuarial consequence of cohorts consisting of subcohorts each of which possesses a different level of frailty is that the mortality trajectory of the whole may depart substantially from Gompertz rates even though these each of the subcohorts is subject to Gompertz mortality rates.

A number of researchers argued that demographic heterogeneity accounted for the leveling off of the medfly mortality study by Carey, Liedo, Orozco, and Vaupel (1992) including Kowald and Kirkwood (1993), Hughes and Charlesworth (1994) and Brooks, Lithgow, and Johnson (1994). The male-female mortality crossover can also be explained by invoking demographic selection arguments. The possibility exists that female medflies at emergence were, on average, frailer than male medflies and there was also greater variance in female frailty; there may have been a relatively large proportion of frail females *and* of robust females compared with males (Carey, Liedo, Orozco, Tatar, and Vaupel 1995a; Vaupel and Carey 1993).

3.4.2. Explaining Mortality Patterns

Vaupel and Carey (1993) fitted the observed medfly mortality pattern presented in figure 3.1 with mixtures of increasing Gompertz curves. To model the age trajectory of medfly mortality, assume that

$$\mu_{x,z} = z\mu_0(x) \tag{3.1}$$

TABLE 3.2
Values of Frailty z and Proportions of Flies at Each Level of Frailty for the Gompertz Model

Frailty z	*Proportion of Initial Cohort*
3.7	0.41
0.75	0.38
0.17	0.13
0.03	0.046
0.0093	0.20
0.0020	0.0082
0.00036	0.0017
0.000074	0.00046
0.000011	0.00013
0.0000014	0.000053
0.000000058	0.000013
0.00000000073	0.000043

Source: Vaupel and Carey 1993.

where $\mu(x,z)$ denotes the hazard rate of individuals with fixed frailty z at exact age x and $\mu_0\,(x)$ is the baseline hazard function of the form

$$\mu_0\,(x) = 0.003e^{0.3x} \tag{3.2}$$

Experiments with different numbers of subpopulations suggested that 12 groups were sufficient to capture the observed pattern of medfly mortality using values of frailty z and proportions p of flies given in table 3.2. The modeling results are shown in figure 3.5. The two most frail subgroups (1 and 2) constituted 41% and 38% of the initial cohort, respectively, for a combined total of nearly 8 out of 10 flies of the initial cohort. Their life expectancies were 10.2 and 15.3 days, respectively. In contrast, the two most robust subgroups combined (11 and 12) constituted only around one out of 20,000 flies of the initial cohort, and their life expectancies were 86.2 and 88.1 days, respectively. Comparisons of the model parameters for the most frail to the most robust subgroup reveals that the frailty parameter varied by over a billion-fold ($z = 3.7$ vs. $z = 7.3 \times 10^{-10}$), the proportions of the initial cohort varied by nearly a thousand-fold (0.41 vs. 0.000043), and the life expectancies at birth (denoted e_0) varied by nine-fold ($e_0 = 10.2$ to $e_0 = 86$ days). As Vaupel and Carey (1993, p. 1667) state,

> Any change in a population statistic may be a result of individual or compositional change. The method of calculating frailty distributions provides an engine for generating concrete compositional counter examples to any direct attribution of observed changes in population hazard rates over age, time, or duration to corresponding changes in the characteristics of individuals.

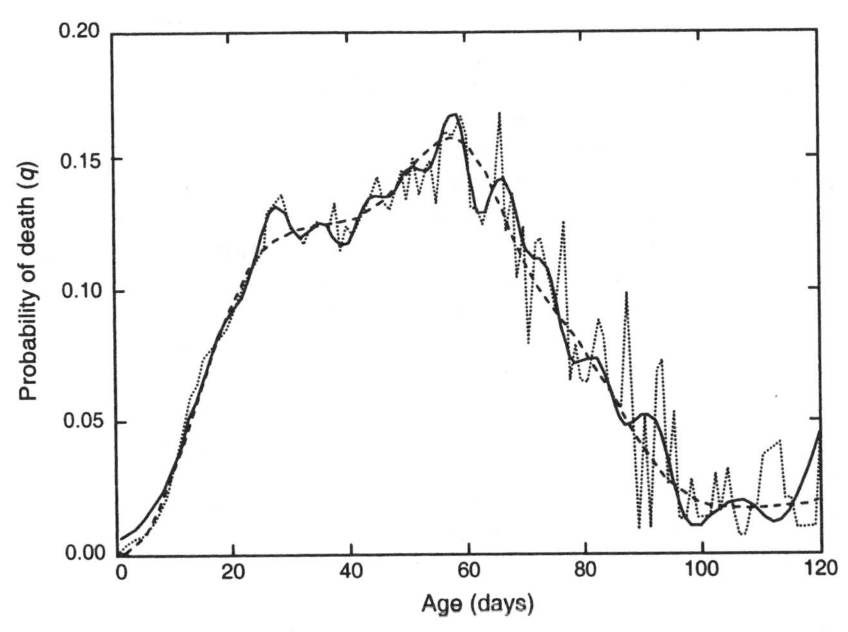

FIGURE 3.5. Results of frailty model with 12 subpopulations and with a composite cohort (Vaupel and Carey 1993).

3.4.3. Epistemological Concepts of Demographic Selection

The selective survival argument is a general concept that applies wherever variation and selection occur (Carey and Liedo 1995a; Vaupel and Yashin 1985). As Peters (1991) notes, the selectionist approach has been applied to competition among theories in epistemology, to business and economics, and to evolution by natural selection. However, an epistemological constraint of all selectionist arguments is that they are tautological (Peters 1991): i) the underlying argument is circular—the most robust (fittest), identified by their better survivorship, survive better; and ii) the explanations are not open to falsification because they accommodate every possible situation. For example, every pattern of age-specific mortality—increasing, leveling off, and decreasing—can be "explained" by invoking selectionist arguments. In other words, selection is a logical truth and not a scientific theory, per se (Peters 1991).

3.5. Sex Differentials

After the discovery and subsequent publication of mortality deceleration at older ages in the 1.2 million medfly cohort, one of the first questions that was addressed was the extent to which differences existed between the sexes in mortality, survival, and life expectancy and the inter-cage variation in these rates. Indeed, the data set consisted of approximately 600,000 flies of each sex and therefore questions about nuances in mortality and survival could be detected using the power of these numbers. In addition, it was felt that the big question about male-female life table differences could be answered: Do female generally outlive males? Whereas smaller studies were suggestive, this massive data set revealed that the answer to this question is not at all straightforward. Indeed, the paper from which much of the material in this chapter was taken was titled "A male-female longevity paradox in medfly cohorts" (Carey, Liedo, Orozco, Tatar, and Vaupel, 1995a).

3.5.1. Background

While women generally outlive men by a margin of 4 to 10 years throughout the industrialized world (Hazzard 1986, 1990; Hazzard and Applebaum-Bowden 1990; Holden 1987; Stolnitz 1957; Waldron 1992), a long-standing question in biology is whether this female advantage in longevity is a general characteristic of most nonhuman species as well. The scientific literature contains conflicting views. For example, Hazzard (1990) states, "The greater longevity of females than males appears to have a fundamental biological basis. Studies of comparative zoology suggest that greater female longevity is virtually universal." Hamilton and Mestler (1969) begin their paper with this statement: "Males tend to die at an earlier age than females in most species of animals for which data are available." Brody and Brock (1985) write that "there are basic and fundamental questions posed by the fact that female survival [advantage] seems to be one of the most pervasive findings within the animal kingdom." In contrast, the paper by Lints and coworkers (1983) made 218 comparisons between female and male life span in *Drosophila melanogaster* and concluded that mean life span of females exceeds that of males in only about half the cases. D.W.E. Smith (1989) notes that "while there is some evidence that adult populations of many animal species contain more females than males, most of these studies do not consider survival to an age approaching the potential limit for the species as implied by the word longevity." Gavrilov and Gavrilova (1991) state, "The hopes connected with the search for general biological mechanisms underlying these [sex] differences seem to be in vain, since, despite the wide-spread opinion to the contrary, the greater life span of females is not in itself a general biological regularity."

In examining sex-specific demographic data from a large-scale medfly life table study (experiment 3 in Carey, Liedo, Orozco, and Vaupel 1992), my coworkers and I discovered a paradox with respect to male-female life table traits—in 167 cohorts averaging 7,200 flies each, males usually possessed the higher life expectancy but females were usually the last to die. These patterns suggested that the underlying mortality schedules for the sexes were inconsistent with several long-held assumptions about the nature of sex mortality and longevity differences in the population biology, ecology, and gerontology literature (Charlesworth 1994; Charnov 1982; Fisher 1958; Hamilton 1948; D.W.E. Smith and Warner 1989; Trivers 1972)—that the sign of mortality differences is unchanging throughout the adult life course, that females are typically longer lived than males, and that sex ratio biases always favor one sex at all ages.

The demographic incongruency in medfly male-female longevity can only be explained as due to an underlying mortality crossover where age-specific death rates of one sex must be higher up to a particular age and lower thereafter (Coale and Kisker 1986; Manton et al. 1979; Manton and Stallard 1984; Petersen and Petersen 1986). Because a demographic relationship of the type described by the medfly data has not been previously documented in a nonhuman species, and, more generally, reliable data on age-specific mortality in most nonhuman species is relatively rare (Promislow 1991; Promislow and Harvey 1990), the specific objective of this research was to confirm the existence of mortality crossovers in the medfly cohorts. The broad importance of this research concerns two-sex models in demography (Caswell 1989; Goodman 1953, 1967; Gupta 1973; Keyfitz 1966; Pollak 1986, 1987; Schoen 1978), forecasting sex differentials in mortality (Carter and Lee 1992), and comparative studies on male-female aging (Comfort 1979; Greenwood and Adams 1987; Kitagawa and Hauser 1973; Lopez and Ruzicka 1983; Partridge 1986; Peterson 1975; Preston 1976; Tatar and Carey 1994a; Tatar and Carey 1994b). Indeed, Brody and Brock (1985) believe that sex differences alone provide one of the most promising areas of research into longevity available to science.

3.5.2. Mortality Trajectories

SEX AND AGE-SPECIFIC MORTALITY

The age-specific mortality rates for male and female medflies are given in figure 3.6. Female mortality increased more rapidly at young ages than did male mortality. However, at day 16 female mortality abruptly began to level off while male mortality continued to increase thus causing a mortality crossover at around 20 days. Male mortality peaked at the same time as female mortality and mortality in both sexes began declining at about the same age. Male and female mortality rates were similar at ages beyond 60 days.

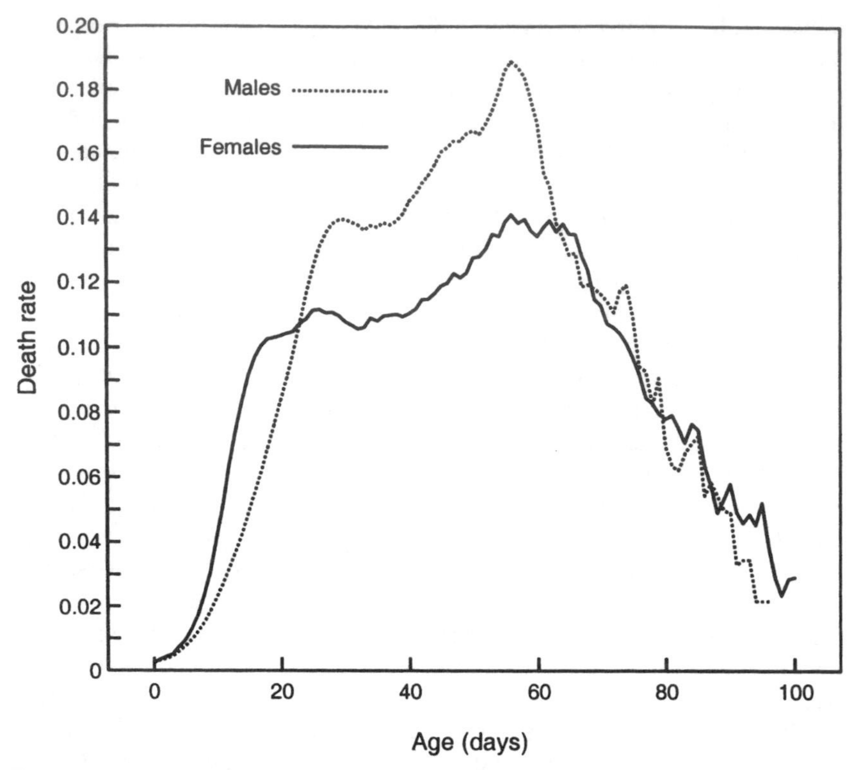

FIGURE 3.6. Smoothed male and female age-specific mortality rates from cohorts consisting of approximately 600,000 medflies of each sex. Curves were smoothed using a 7-day running mean (geometric) (Carey, Liedo, Orozco, Tatar, and Vaupel 1995a).

MORTALITY AND SURVIVAL RATIOS

The ratio of the age-specific male and female mortality schedules show both the relative differences in the mortality levels and the patterns of convergence, crossover, and divergence (figure 3.7). The sex mortality ratio indicates that the greatest relative difference between male and female mortality at the young ages occurred around day 16, when male rates differed from female rates by a factor of 0.7. After the mortality crossover, male mortality was higher than that for females by a factor of around 1.3 from 30 to 60 days. The effects of these mortality differences on the relative abundance of each sex were not offsetting since the male advantage occurred when the rates of both sexes were relatively low, whereas the female advantage occurred when the rates of both sexes were relatively high.

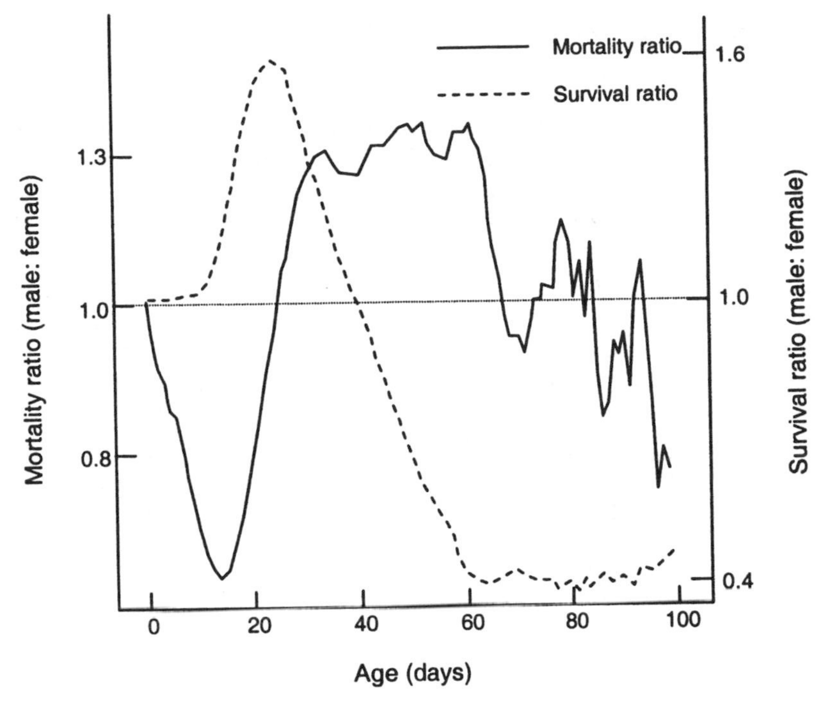

FIGURE 3.7. Sex mortality ratios for medflies (male to female age-specific mortality ratios) using the smoothed rates shown in figure 3.6 and the survival sex ratio (ratio of male to female survival schedules) (Carey et al. 1995a).

The numerical consequences of these mortality ratios on the relative abundance of each sex is indicated by the sex survival ratios—the ratio of male to female survival rates. At 25 days the males outnumbered females by 1.6-fold. However, by 40 days the number of each sex was equal and at older ages the number of males was only 0.4 that of females. These trends show that age-specific sex ratio is neither fixed nor biased toward only one sex at all ages. The relationship between sex mortality rates and relative abundance reveals that the sex survival ratio cannot be used to estimate sex mortality differentials. Three comparisons illustrate this point: (i) from ages 0 to 14 the sex survival ratio was constant and near unity but the proportional sex mortality differences were high and increasing; (ii) at 40 days the proportional mortality differences were again high (though reversed) and constant. However, the male-female ratio was near unity and decreasing; and (iii) at ages greater than 60 days the cohort was strongly female biased but the sex mortality differences were virtually non-existent.

3.5.3. Medfly Mortality in Solitary Confinement

A male-female mortality crossover and female bias at older ages was also evident in mortality data from the two experiments in which medflies were maintained in solitary confinement, as shown in figure 3.8. In both cases male mortality was less than female mortality until 18 to 22 days, when the mortality crossover occurred—female mortality was then lower than male mortality. The male to female ratio at older ages ranged from 0.7 to 0.8 at older ages for both experiments, as shown in the survival ratios. These findings for medflies that were held in uncrowded environments and were not allowed to mate or reproduce suggests that the male-female mortality crossover and female bias at older ages is due to differences in the basic biology of the sexes and is not unique to conditions for either mated flies or those maintained in groups. The combined effects of mating and density as observed in the mortality data for flies maintained in cages amplify the sex mortality differential but does not change its fundamental pattern.

3.5.4. Cohort Variability

Evidence of the occurrence of male-female mortality crossovers in most of the 167 cages is shown in figure 3.9. The clustering of points above the isometric (diagonal) line for mortality rates at 10 days reveals that female mortality exceeded that of males in nearly 90% of all cages at this early age. In contrast, the clustering of points below the isometric line for mortality at 30 days indicates that male mortality exceeded that for females in over 95% of all cages at this later age.

Further evidence of the widespread occurrence of mortality crossovers is given in figure 3.10, which shows sex-specific expectation of life at ages 0 and 30. The expectation of life at eclosion (age 0) for males exceeded that for females in over 95% of all cages, but life expectancy at day 30 for females exceeded that for males in over 90% of the cages. The net result of the crossover was to bias the cohort in favor of females at older ages. The last fly to die in a cage was four times more likely to be a female than a male (figure 3.11).

3.5.5. Sex Mortality Crossover

EXPLANATIONS

There are two possible explanations for the male-female mortality crossovers. The first explanation is that the mortality crossover could be an artifact of compositional change in the male and female subpopulations due to

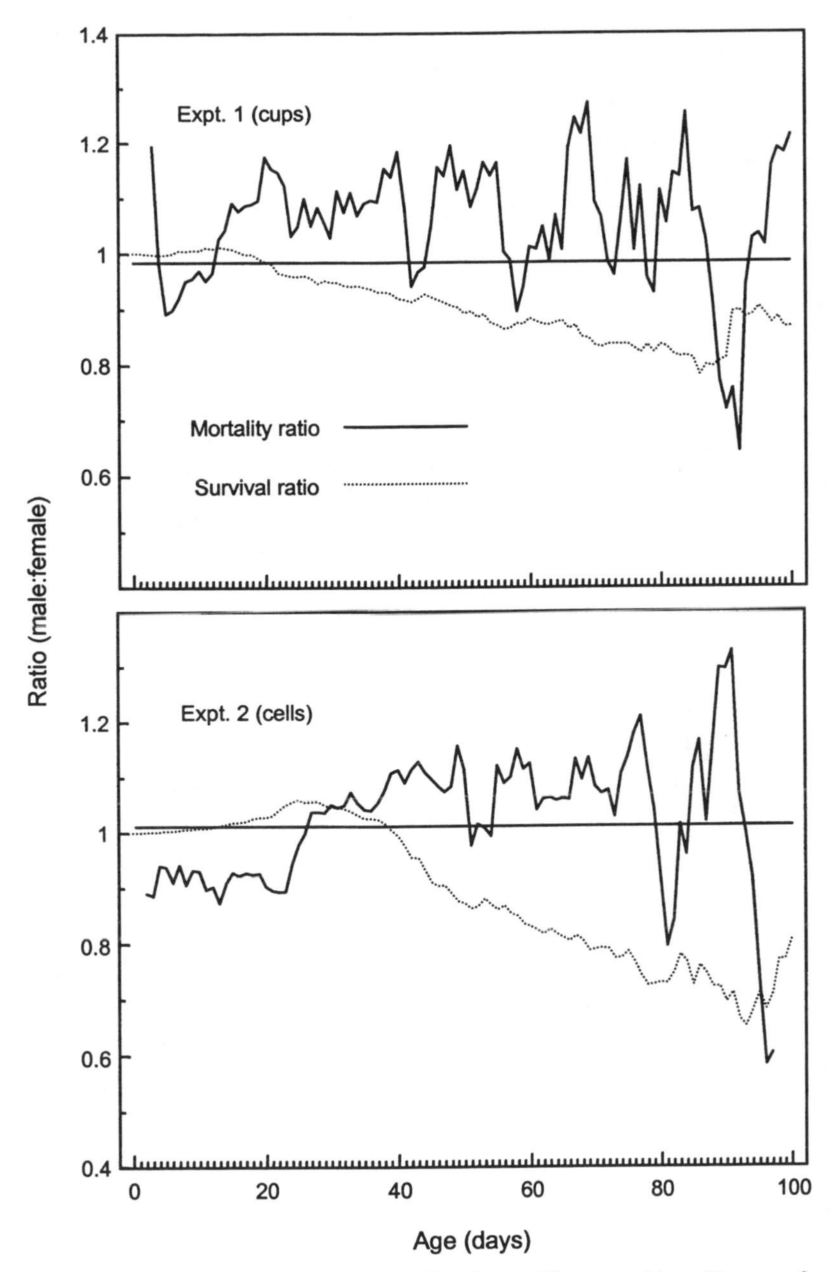

FIGURE 3.8. Mortality and survival sex ratios for medflies reared in solitary confinement. Approximately 21,000 individuals were used in experiment 1 (top) and approximately 27,000 individuals were used in experiment 2 (bottom) (Carey et al. 1995a).

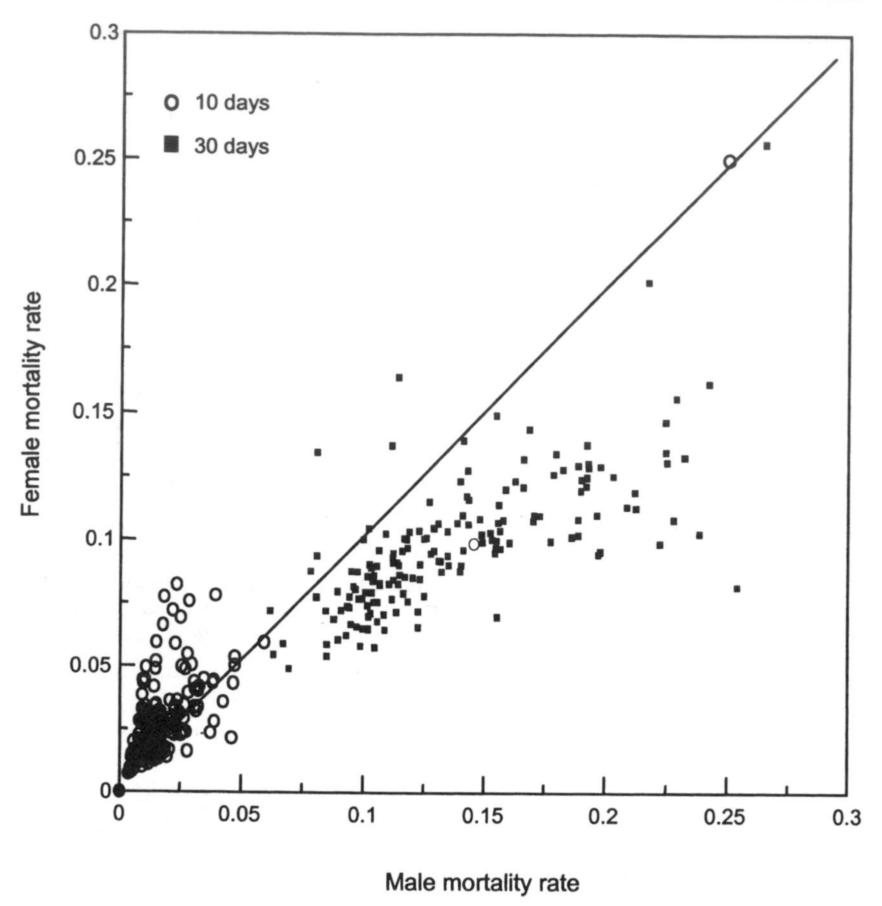

FIGURE 3.9. Male versus female mortality rates at 10 and 30 days for 167 medfly cohorts of approximately 3,600 individuals of each sex. The diagonal line is the isometric line, where male = female mortality. Thus a point above the line indicates that female mortality within the cage was higher at the specified age than was male mortality at that age. Sex differences in mortality among all cages at both 10 and 30 days are statistically significant ($P < .001$) (Carey et al. 1995a).

demographic heterogeneity (Vaupel and Carey 1993; Vaupel et al. 1979). As populations age, they become more selected because individuals with higher death rates will die out in greater numbers than those with lower death rates, thereby transforming the population into one consisting mostly of individuals with low death rates (Rogers 1992). This explanation is also referred to as the "cohort-inversion model," which is based on the concept that cohorts experiencing particularly hard or good times early in life will respond inversely later in life (Elo and Preston 1992; Hobcraft et al. 1982). Thus the

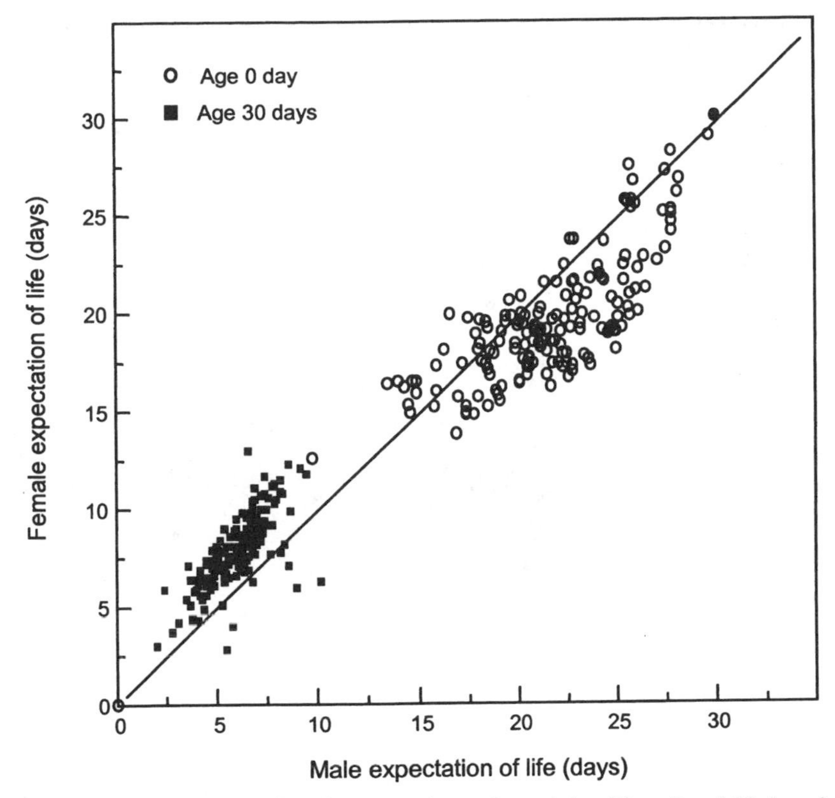

FIGURE 3.10. Male versus female expectations of remaining life at 0 and 30 days for 167 medfly cohorts of approximately 3,600 individuals of each sex. The diagonal line is the isometric line, where male = female expectation of life. Thus a point above the line indicates that female expectation of life at the specified age within the cage was higher than the male expectation of life at that age. Sex differences in expectations of life among all cages at 0 and 30 days are statistically significant ($P < .001$) (Carey et al. 1995a).

possibility exists that male-female mortality rates crossed because of heterogeneity at the cohort level. Perhaps females at emergence were, on average, frailer than males and there was also greater variance in female frailty; there may have been a relatively large proportion of frail females *and* of robust females compared with males.

A second explanation is that biological differences between males and females exist at the individual level and were manifested as differences in age-specific mortality. These sex-specific differences could include mating behavior (Matthews and Matthews 1978), physiology (Engelmann 1970), reproductive costs (Reznick 1985; Tatar and Carey 1994b), and hormonal ac-

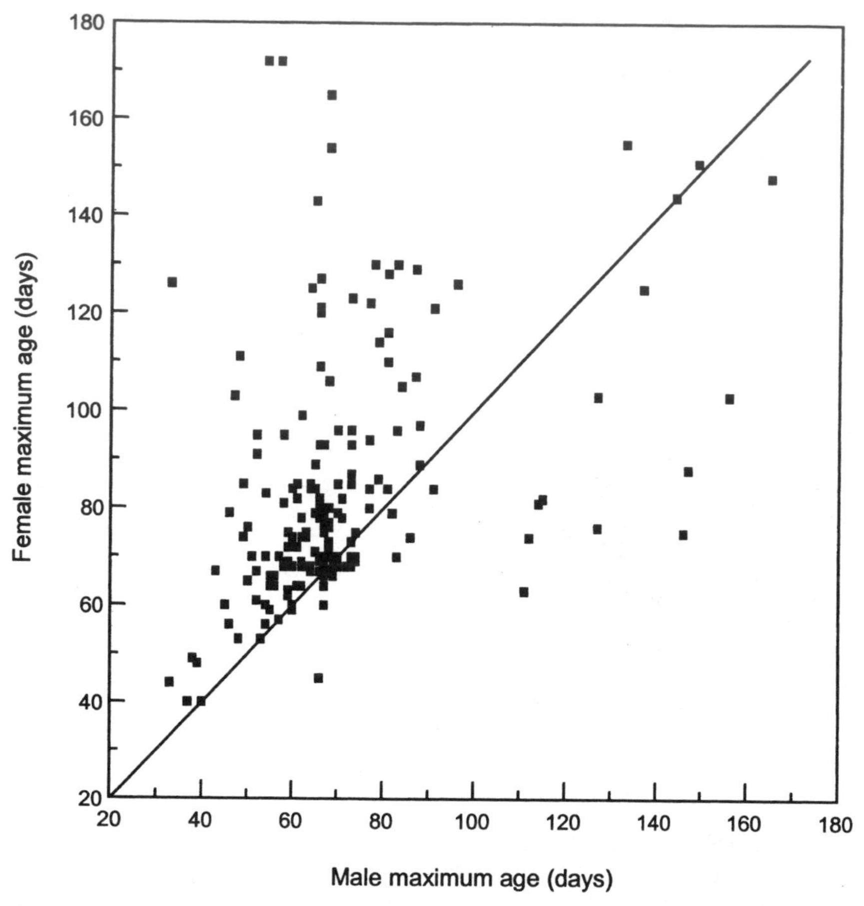

FIGURE 3.11. Ages of the last male and the last female to die in each of 167 medfly cohorts of approximately 3,600 individuals of each sex. The diagonal line is the isometric line, where male = female oldest age. Thus a point above the line indicates that the last female to die within the cage was older than the last male to die. Sex differences among all cages in oldest age attained are statistically significant ($P < .001$) (Carey et al. 1995a).

tivity (Engelmann 1968). A finding that suggests the existence of differences at the individual level is the male-female survival crossover at 40 days (figure 3.8). The reason for the importance of the survival ratio is that it represents the cumulative mortality experience of the population. If the cumulative mortality advantage is eliminated as is the case with the male and female survival shown in figure 3.7, then a mortality selection model cannot be a complete explanation of the mortality differentials (Manton and Stallard 1984).

Manton and Stallard (1984) believe that as a complement to the concept of crossover, attention should be paid to the concept of a peak mortality differential—the age at which the greatest proportional difference exists between the age-specific mortality rates for the two populations. They suggest that it is quite possible that an explanation of the peak differential may serve to explain mortality convergence and crossover. They state, "This is because once it is understood why the peak differential occurs at that point in the age range, the later convergence and crossover of the two mortality curves may turn out to be a natural consequence of the mortality dynamics involved in the explanation."

It is likely that the age patterns for the medfly sex mortality ratio are at least partly due to differences in both physiological and behavioral costs of reproduction (e.g., gonadal activity and mating) as reflected in two peaks in mortality differentials. The first peak in the sex mortality ratio occurred at about two weeks when mortality was lower for males than for females; egg production in females is greatest from 7 to 14 days (Vargas 1989). The second peak was flatter than the first and occurred at around 30 days. This was after the mortality crossover and extended through ages when male mortality exceeded female mortality; this period occurred when female egg production was low or nil (Carey, Yang, and Foote 1988) and male mating activity was reduced.

IMPLICATIONS

Discovery of a male-female mortality crossover has several implications in population biology, demography, and gerontology. One implication involves the specific question of male-female longevity differences. Without an awareness of the nature of the male-female mortality schedules and specifically the mortality crossover, sex differences in longevity could be interpreted in three ways: (1) males are longer lived if life expectancy at emergence is used as the longevity criteria; (2) females are longer lived if oldest age attained is the longevity criteria; and (3) that males and females are equally long lived if the cohort numbers were used as the longevity criteria to day 10 when the numbers of each sex were approximately equal or several days before and after day 40 when the sex ratio was at or near 50:50 due to the mortality crossover. Indeed, ambiguous and conflicting reports on longevity sex differentials are common in the literature (Clutton-Brock and Iason 1986; Ehrlich et al. 1984; Lints et al. 1983; D.W.E. Smith and Warner 1989).

A second implication is that the results cast doubt on the assumption that sex ratio is biased toward only one sex throughout the life course of most species. The consequence of the mortality crossover around day 20 and its persistence through day 60 was to shift the sex ratio from a male bias at young ages to a female bias at middle ages. Females were the last to die, not because female mortality was lower at the older ages, but because mortality rates of both sexes were low and mortality differentials were small at advanced ages. This aspect is important because it sheds light on the dynamics of sex biasing. For example, the sex ratio at birth in humans is 1.05 (male: female) in most developed countries such as the United States. However, by 30 years of age the sex ratio is 50:50 due to excess male mortality. By age 85 females outnumber males by 3-to-1 (Taeuber and Rosenwaike 1992).

A third implication is that sex ratio cannot be used as a proxy for sex mortality differentials for two reasons. The first involves survival differences, which not only lag behind mortality differences, but may result from mortality patterns that are exceedingly complex, as was shown in figure 3.6. This complexity is not revealed in survival differences because survival is cumulative. The second reason is that growing populations contain smaller fractions of their total membership in the older age classes than do stationary or declining ones. Therefore, a population in which a male-female mortality crossover exists may be biased toward one sex when it is increasing but biased toward the other sex when it is decreasing (Keyfitz 1985). Thus shifts in population sex ratio may reflect changes in growth rate and not changes in relative male-female mortality patterns.

A fourth implication concerns the population biology of species when two sexes are considered. For example, Werren and Charnov (1978) demonstrated that there are exceptions to the argument by R. A. Fisher (1958) that differential mortality between sexes will have no effect on the equilibrium 1:1 sex ratio. Specifically they showed that because most populations in nature are not at a demographic equilibrium (stability) selection can favor genes that result in the temporary overproduction of one or the other sex. Another example of an implication in population biology of differential mortality between sexes is the finding of Caswell and Weeks (1986) who demonstrated that complex bifurcation patterns may occur in populations if male and female survival differences exist and there is interstage competition for mates. This finding is important because, as Caswell (1989) notes, selection on the primary sex ratio when there is no such equilibrium adult sex ratio is an unsolved question.

A final implication is that discovery of the mortality crossover challenges long-held views on aging research including the nature of senescence and the use of certain life table parameters. For example, the mortality crossover makes it impossible to neatly classify two populations according to any of a number of demographic metrics widely used in aging studies. If the relative rate of change in mortality with age is used to compare senescence rates between two populations, then senescence for males relative to females is lower from ages 0 to 20 days, higher from 21 to 60 days, and the same from 60 days onward. If life expectancy differences at eclosion are used as the criteria for differences in aging rates, then female medflies age more rapidly than males. However, the majority of the 2.5-day gap in male-female life expectancies at eclosion can be explained as due to sex mortality differences at the relatively youthful ages between 11 and 20 days (Carey 1993).

That the complex dynamics of sex mortality differentials in nonhuman species has previously been unrecognized is of little surprise. This is because the vast majority of life table studies on nonhuman species are based on relatively small numbers of individuals, because determining sex-specific mortality and sex ratio in the field is exceedingly difficult, and because most life history studies have been concerned historically with survival rather than mortality differences. As Ehrlich and his coworkers stated (1984), "Thus a seemingly simple thing like sex ratio is, in detail, quite complex both to define and to estimate."

It is doubtful that sex mortality differentials observed in the laboratory for any species including the medfly would be similar to those in the field. Unlike populations in the field, which usually consist of individuals in a variety of different ages, the laboratory medflies used in the experiments described here were maintained as same-aged cohorts within each cage at densities far higher than would ever be experienced in nature. Consequently extrapolation of specific findings such as crossover age to field situations is probably not valid. However, the general finding that male-female mortality rates may crossover under some circumstances is important in a broader context and raises the likely possibility that sex-specific differences in life tables for many species may be far more complicated than previously realized. Indeed, whether males or females live longer may be equivocal in many species. However, as Zuk (1990) notes, generalization of specific results on male-female mortality differences from one species, which is adapted to a particular set of circumstances and to other species, is probably risky. I believe that future biological research focusing on causal mechanisms under lying convergence, cross over and divergence of male-female mortality rates with age will be more important to understanding gender differences in aging than will a continuing quest to demonstrate the universality of a female longevity advantage.

3.5.6. A General Framework

Despite the ambiguity of much of the data in the nonhuman biological literature, explanations for differences in male-female mortality have been framed around the assumption that males of most species experience higher mortality rates than females. The two most common explanations for why males (putatively) have shorter life spans (D.W.E. Smith and Warner 1989; Trivers 1972) are based on (i) behavioral aspects where males of many species are at higher risk due to different life history requirements than females, such as mate finding and territory defense; and (ii) the chromosomal hypothesis where, it is suggested, females have an advantage because in most species females are the homogametic sex (XX) whereas males are the heterogametic sex (XY). It is argued that having two X chromosomes is advantageous because the X chromosome is three times the size of the Y chromosome and contains far more expressed genetic information, most of which is for functions and molecules unrelated to the female geneotype (Greenwood and Adams 1987; Montagu 1974; D.W.E. Smith and Warner 1989; Zuk 1990).

In contrast to the ambiguity of male-female mortality differences in the biological literature, human data on sex mortality differentials in the demographic literature is unequivocal—females experience lower mortality than males at virtually all ages in almost every contemporary society (Hazzard 1986; Keyfitz and Flieger 1990). Two explanations have been proposed to account for the female longevity advantage in humans (Wingard 1984). The first is a biological explanation that women are biologically "more fit" than men. This explanation includes the chromosomal hypothesis as well as arguments related to the protective effects of female sex hormones such as estrogen or to the deleterious effect of testosterone in males (Alexander and Stimson 1988; Hazzard 1990; Hazzard and Applebaum-Bowden 1990). The second explanation for why women outlive men concerns social, lifestyle, and environmental factors—men behave in ways more damaging to health, for example, through smoking, alcohol consumption, occupation hazards, and violence. It has been suggested that the cigarette-smoking sex differential may account for over half of the sex differential in longevity in the United States (Hazzard 1986).

In light of reports in both the demographic and the biological literature on sex mortality differentials as well as the results of the current study on the medfly, the underlying mechanisms for sex mortality differentials can be grouped into three interrelated and coevolved categories. The first category is *constitutional endowment*, which includes all structural, physiological, endocrinological, and immunological factors affecting the ability of individuals of each sex to resist disease, stress, physical challenge, and deterioration. This category is concerned with overall "fitness" and includes the direct and indirect effects of chromosomal differences between males and females. The second category concerns factors associated with sex-specific *reproductive biology*, including the effects of male and female hormones, gonadal development, and production of eggs or offspring. This group of factors is concerned with *processes* typically classified as *costs of reproduction* (Reznick 1985): for example, virgin insects typically exhibit lower mortality rates than individuals that mate and reproduce (Carey, Krainacker, and Varga 1986; Partridge 1986; Partridge and Farquhar 1981; Partridge and Fowler 1992; Tatar et al. 1994). The third category of factors that influence the sex mortality differential is *behavioral predispositions*. These include behavioral traits evolved to maintain territories as well as the "high risk–high stakes" strategy of males of many species for locating, competing for, and defending mates (Zuk 1990).

Factors interact within and between categories. For example, gender-related exposure to parasites will be affected by differences in male and female behaviors (Bundy 1988; Tinsley 1989). Once infected the immunological response (endowment) of each sex will modulate survival of parasites, which, in turn, is influenced to a large degree by sex hormones (Alexander and Stimson 1988). The interaction of the three sex-specific factors—endowment, reproductive biology, and behavior—determine the overall *susceptibility to death* for each sex, which, when filtered through environmental, biological, and other factors, produces a *probability of death*. The two concepts—susceptibility and probability—are not equivalent (Kannisto 1991). This is because mortality often runs counter to constitutional frailty due to behavioral factors. As Kannisto (1991) notes, boys die of accidents more frequently than girls, not because boys are more frail but because they take greater risks; reproducing females often experience higher death rates than males of the same age due to the high cost of offspring production and not due to differences in frailty, per se. In both cases, however, the sex differentials in risk-taking or in costs often diminish with age. Consequently differences in frailty or endowment may account for most of the sex mortality differential at older ages whereas differences in both behavior and reproductive biology between the sexes may account for the largest proportion of the sex mortality differential at younger ages, ceteris paribus. The combination of all factors will ultimately determine the overall *sex survival differential* at advanced ages because survival is cumulative; differential mortality *at* young ages will affect the relative survival *to* older ages and thus influence the sex bias at advanced ages.

3.6. Summary

The results of the large-scale life table study of the medfly were important not only because they set the stage for subsequent medfly experiments but because: (1) they provided unequivocal evidence that the Gompertz model was not universal, which, in turn, forced a reexamination of the concepts of both senescence and life span limits; (2) the scale of the study with 167 cohorts (cages) totaling 1.2 million deaths opened up new possibilities for analysis of nonhuman actuarial data including sex-mortality differences, mortality ratios, smoothing techniques, and heterogeneity; (3) disaggregating the mortality schedule into the male and female components revealed the distinctiveness of the sex-mortality differentials (the sex-specific mortality responses were so different that it suggested from the very beginning of the analysis that there was no way that the age-specific mortality of the two sexes would likely be similar under any circumstances); (4) it followed from this finding that the question of which sex lives longer became less relevant than the more fundamental question involving the nature and origin of mortality differences between the sexes (in other words, the findings suggested that the sex-specific response to different environments could shed light not simply on gender differences in life table traits, but also on fundamental aspects of aging); (5) given the scale of the study, the age-specific mortality for the 600,000 female medfly cohort provided a baseline from which all other mortality trajectories were compared (indeed, later studies examined many of the traits of this schedule, such as the surge [shoulder] in mortality at young ages, the overall peak, and the decline at late ages); and (6) the data revealed that it was impossible to state unequivocally that either males or females lived longer because (i) although longevity can be characterized in a number of different ways, one measure of longevity (life expectancy) often favored one sex while another measure favored the other (record life span), (ii) considerable between-cohort variation existed for a given longevity measure, and (iii) relative longevity for the two sexes was conditional on the environment in which they were maintained or the treatment to which they were subjected.

4

Reproduction and Behavior

BECAUSE of the central importance of both reproduction and behavior in life history theory, in this section I present the findings of studies that were designed to examine the relationship between medfly longevity and female reproduction and mating status (i.e., virgin vs. nonvirgin) as well as a unique type of behavior exhibited by aging medfly males. The first section describes a graphical method for visualizing the relationship of individual-level reproduction in females and their longevity that was developed expressly for graphical analysis of the large-scale data on reproduction. The technique provides insights into both broad patterns and subtle changes in reproduction throughout the life course and relates these changes to female longevity. The same data are used in the analyses presented in the next section but with a more rigorous statistical emphasis. The next section describes the results of a study on the mortality dynamics of medfly females maintained in either same-sex (virgin) or mixed-sex (mated) cages. The final subsection contains the results of a study that was originally designed to study the relationship of mating behavior (sexual calling) and longevity in over 200 individual males but in which a new and unique behavior was observed in males that were approaching death—supine behavior that corresponds to the upside-down position of older flies within the cage.

4.1. Reproduction and Longevity: Visualizing Linkages

4.1.1. Graphic Concept—Event History Diagrams

Data sets on the longevity and age-specific reproduction of individuals, particularly those on insects, are often both large and detailed. For example, Partridge and Fowler (1992) monitored the 2-day reproductive rates in 430 individual *Drosophila melanogaster* females throughout their average 20–25 day lifetime. This effort produced records of 430 individual life spans and approximately 5,000 fertility records classified by individual and age. Although methods for analyzing this type of data using life history approaches (Roff 1992), life table concepts (Chiang 1984), and demographic techniques (Carey 1993; Carey, Yang, and Foote 1988) are relatively straightforward in application, there are no simple methods available to display clearly, precisely, and efficiently all of the reproductive data. Consequently techniques for the analysis of costly data on the reproductive rates of individuals are limited even though the importance of understanding reproduction and its relationship to mortality is widely acknowledged by ecologists (Ricklefs 1979), evolutionary biologists (Reznick 1985), and gerontologists (Finch 1990).

A simple graphical tool was developed by Carey, Liedo, Müller, Wang, and Vaupel 1998b) referred to as an event history diagram, which creates visual displays of cohort survival combined with the longevity and age patterns of reproduction for individuals. My objective in this section is to describe this graphical technique using medfly data as an example application and demonstrate how it provides new perspectives and insights into large data sets on individual reproduction. Although the specific focus and examples are on the medfly, the technique is general and thus can be applied to any measurements on a continuous or discrete covariate for any species.

4.1.2. Application to Medfly Reproduction

The graphics integrating individual reproductive data with cohort survival are based on three concepts: (1) the life course of a single individual female depicted as a horizontal line, the length of which is proportional to her longevity; (2) age segments of the lines shaded according to the number of eggs laid; and (3) the individuals rank-ordered from shortest- to longest-lived so that when the lines are plotted, they create a cohort survival (l_x) schedule depicting the number of individuals alive at each age. These concepts were used for organizing and graphing the reproductive data on 1,000 individual female medflies. Data were entered into a computer spreadsheet (Microsoft Excel) and arranged by age (columns) and individual (rows). The data on individuals were then rank-ordered and plotted using the software graphics program DeltaGraph.

4.1.3. Insights from Graphs

The graphs of the medfly data integrating longevity, survival and reproduction are shown in figure 4.1. The concept of the dark gray coding scheme for the figures is a progression from highest egg-laying days in figure 4.1A (i.e., days in which > 40 eggs laid are coded black), to intermediate egg-laying days in 4.1B (i.e., days in which 21–40 eggs laid are coded black), to lowest egg-laying days in 4.1C (days in which 1–20 eggs laid are coded black). The graphical emphasis in 4.1D is on the zero-egg-laying days. Since the average individual in the 1,000-female cohort lived 35.6 days and laid 759.3 eggs, each figure portrays approximately 35,600 numbers representing the distribution of 759,300 eggs. Because the graphs are constructed from original rather than smoothed or curve-fitted numbers, they allow the data to "speak for themselves." This is important because abrupt changes in the level of reproduction between adjacent age classes (e.g., 0 eggs laid at one age followed by 50 eggs laid at the next) appear in these graphs that would not appear in figures designed to show, for example, "zones" of high and low reproduction. Thus subtle patterns and nuances can be detected by carefully studying the graphs.

A broad pattern that is immediately evident in figure 4.1 is the sigmoidal shape of the cohort survival schedule showing a gentle decrease from 0 to around 20 days, during which the first 100 flies died, followed by a more rapid drop from days 20 through 50, during which around 800 flies died, and ending in a long tail at the oldest ages when the remaining 100 flies died. This pattern is a manifestation of an underlying mortality schedule in female medflies that accelerates at young and middle ages and decelerates at older ages (Carey et al. 1995a; Carey, Liedo, Orozco, and Vaupel 1992; Müller et al. 1997b).

A number of specific aspects of these figures merit comment. First, there is a weak correlation between longevity and age of first reproduction. This is apparent from the absence of any distinct trends between the two variables portrayed in figure 4.1 (left-most bands depicting pre-reproductive ages versus life spans depicted by survival schedule) but is more explicit statistically with the correlation coefficient computed as $r^2 = 0.259$. The graphs provide a visual image of the quantitative relationship between the variables.

Second, although there are high egg-production days throughout the cohort and across all ages as shown in figure 4.1A, most are concentrated in a "reproductive window" spanning ages 5 through 25. This 20-day band of high egg-production appears in nearly all longevity levels and thus implies that the correlation between early reproduction and longevity is also weak. Indeed the correlation coefficient for longevity and cumulative reproductive for the first 15 days of life was $r^2 = 0.031$. High egg-production days were noticeably absent from flies living beyond 60 days. The eggs laid on high-production laying days (i.e. black-coded days in figure 4.1A) accounted for approximately 60% of all eggs laid in the cohort.

Third, the distribution of the days in the cohort in which intermediate levels of eggs were laid is shown in figure 4.1B. This figure reveals a uniform scatter of intermediate egg-laying days across virtually the entire cohort from 20 to 50 days. In other words, moderately high rates of egg laying persisted into older ages for most of the longer-lived individuals. The eggs laid on the intermediate-production days (i.e., black-coded days in figure 4.1B) accounted for approximately 30% of all eggs laid in the cohort.

Fourth, low levels of egg production were evident (figure 4.1C): (i) during the first few days of oviposition and the last several days of life for each fly; (ii) at the oldest ages (see survival tail); and (iii) throughout the cohort as shown in the more-or-less even scatter of low egg-production days at all

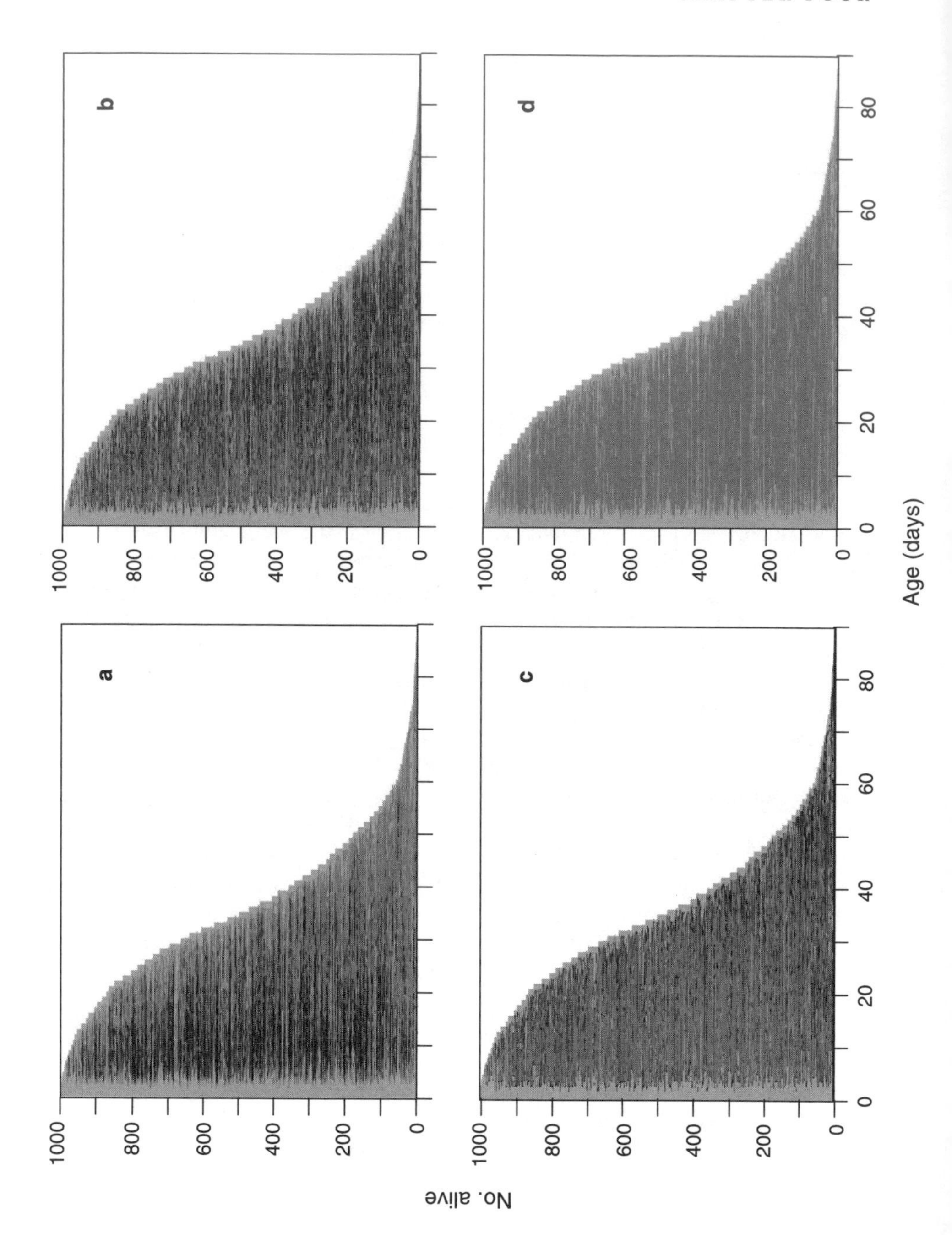
a
b
c
d
No. alive
Age (days)
1000
800
600
400
200
0
0
20
40
60
80

ages across individuals. This reflects the variation in daily egg laying by individual flies. The eggs laid on low-production laying days (i.e., black-coded days in figure 4.1C) accounted for approximately 10% of all eggs laid in the cohort.

Fifth, the distribution of days in which no eggs were laid is presented in figure 4.1D. Of the 35,600 fly-days in the cohort, a total of around 14,500 or 40% were days in which no eggs were laid. These zero-production days can be classified into three types: (i) the immature periods experienced by all flies prior to laying their first eggs (these accounted for roughly 50% of all zero-egg days); (ii) zero-production days resulting from lack of egg laying in 64 sterile flies (these "infertile fly-days" accounted for about 10% of the zero-egg days in the cohort); (iii) those due to day-to-day variation in egg laying by fertile flies. These accounted for about 40% of the zero-egg days and include post-reproductive periods experienced by many flies, occasional periods when individual flies did not lay for several days, and a clustering of zero-egg days for the longest-lived flies in the tail.

Sixth, egg-laying patterns in the oldest flies (> 60 days) consisted of a mixture of many days of no egg laying (figure 4.1D) and days of low egg production (figure 4.1C). This pattern occurred during the ages when mortality rates were high but at a plateau. This lack of reproductive activity sheds light on the period of leveling off at older ages and suggests that this phenomenon may be partly due to a reduction in the overall reproductive activity of individual flies rather than exclusively to demographic selection (Vaupel and Carey 1993).

4.1.4. Implications

There are several reasons why longitudinal data on individuals is preferred over data that is grouped or cross-sectional (Vaupel and Carey 1993) and

---

FIGURE 4.1. Graphic depicting age-specific cohort survival and lifetime reproduction for 1,000 individual Mediterranean fruit fly females. Each horizontal line portrays the longevity of a single individual and the shades designate the level of reproduction for each age. 4.1A. (top left corner) Codes: light gray = zero eggs; medium gray = 1–40 eggs; black = > 40 eggs (i.e., black highlights ages of *highest* levels of reproduction). Note that the light gray vertical band on the left depicts the distribution of ages of first reproduction. 4.1B. (top right corner) Codes: black = 21–40 eggs, medium gray = 1–20 or > 40 eggs (i.e., black highlights ages of *intermediate* levels of reproduction). 4.1C. (lower left corner) Codes: black = 1–20 eggs, medium gray = > 20 eggs (i.e., black highlights ages of *lowest* levels of reproduction). 4.1D. (lower right corner) Codes: medium gray = > 0 eggs (i.e., black coding absent; emphasis on *qualitative aspects* of reproduction) (Carey, Liedo, Müller, Wang, and Vaupel 1998b).

thus why graphical techniques that help visualize individual-level data are important. The first reason is that as a study-population ages it becomes more and more selected due to attrition (Vaupel et al. 1979). By 40 days the fly population available for study represents less than half of the original cohort. Averages derived from measurements in young females are thus based on observations of flies some of which did not live to age 40, while data from older flies obviously represents a select population that has survived. Another reason that individual-level data are important is that they provide insight into the between-fly variation in egg laying and thus reveal compositional influences on the cohort average. For example, individual-level data show whether a decrease in cohort reproduction with age is due to an increase in the fraction of females that lay zero eggs or due to an overall decrease in the level of egg laying by each individual (Carey, Yang, Foote 1988). A third reason that individual-level data are important is that if periods of more intensive egg laying vary from fly to fly, this intra-individual variation can be wiped out in the process of averaging across individuals. For instance, the shape of a peak of egg-laying in the averaged or cross-sectional egg laying display may not resemble any of the peaks observed in individuals' egg laying behavior. A final reason that reproductive data on individuals are important is that the data allow between-fly comparisons to be made in lifetime levels of reproduction and, in turn, on the long-term trajectories of reproduction in each individual over a specified period. In particular, they provide important insights into the reproductive age patterns of flies by comparing high versus low lifetime reproductive rates, early versus late ages of first reproduction, and short versus long lifetimes (Carey, Liedo, Müller, Wang, and Chiou 1998a).

My broad objective in this section is to demonstrate how the graphics technique could provide visual displays of the data on individual medfly reproduction and longevity and thereby enhance the depth and scope of the analysis. The concept underlying the technique is that longitudinal data on demographic events such as reproduction can be portrayed by shading the data on individuals and ordering them according to any number of life history criteria. The cohort survival schedule emerges when the data are plotted for individuals rank-ordered from shortest- to longest-lived. However, other important relationships can be visually displayed using this technique, for example, by rank-ordering the individual-level data by lifetime reproduction or by age of maturity, each of which provide different patterns and shaded gradients and, hence, different insights.

The characteristics of the graphs presented in this chapter are consistent with ideals concerning the visual display of quantitative information outlined by Tufte (1983)—that graphical displays should show the data, induce the viewer to think about the substance rather than about methodology, present many numbers in a small space, make large data sets coherent, encourage comparisons, reveal the data at several levels of detail from a broad overview to the fine structure, and be closely integrated with the statistical and verbal descriptions of the data set. My coworkers and I believe that this technique is capable of complementing the conventional methods of demographic and actuarial analyses, shedding new light on individual variation in age-specific reproduction, encouraging data gathering on the life history traits of individuals, helping to visualize the quantitative nature and degree of statistical correlation, and providing a more sensitive tool for revealing nuances in life history data.

4.2. Relationship of Reproduction and Mortality

Despite the vast literature on insect reproduction and nearly universal agreement among ecologists (Ricklefs 1979), gerontologists (Finch 1990), and evolutionary biologists (Charlesworth 1994) regarding the importance of understanding insect age-specific reproduction, surprisingly little is known about the demographic details of the age patterns of reproduction over the lifetime of even a single insect-species. This paradox exists because experiments on reproduction in basic biodemographic research have historically not been designed to determine differences in reproductive age patterns, per se, but rather to parameterize models (Birch 1948; Caswell 1989), to assess mortality costs (Calow 1970; Partridge and Harvey 1985; Reznick 1985), to document the relationship with adult size (Roff 1992), to examine birth-death trade-offs in life history theory (Roff 1992), and to test hypotheses in the evolutionary biology of aging (Rose 1991; Rose and Charlesworth 1980). The problem is further compounded by the tendency of researchers to publish summary measures of reproduction (e.g., gross/net reproductive rates) rather than to include the age-specific fertility data for cohorts or for individual females.

One of the broad consequences of this lack of understanding of the details of age-specific reproduction in insects is that a wide variety of basic biodemographic questions remain either partially or completely unanswered. Of particular interest are questions associated with aging and senescence and their relationship with patterns of reproduction, such as (1) What is the interrelationship of the *trajectories* of age-specific fecundity and mortality? (2) What are the differences and similarities in the age *patterns* of reproduction among females with substantially different levels of lifetime reproduction? (3) How does mortality change with respect to progressively increasing levels of egg laying (so-called parity progression)? (4) Do *all* individual females that lay the greatest number of eggs when young, lay fewer eggs at older ages *and* experience reduced life spans? and (5) What is the relationship between individual longevity and age-specific reproduction over an entire cohort? Answers to these and related questions will not only provide a deeper understanding of the relationship between reproductive and actuarial components of life histories, but will also enhance the sensitivity and breadth of analyses concerned with reproductive data. My broad objective in this section is to report the results of a study of daily reproduction in 1,000 individual medfly females including the patterns of reproductive timing and amplitude and the association of these traits with individual longevity (Carey, Liedo, Müller, Wang, and Chiou 1998a).

4.2.1. Empirical Framework

Medfly pairs were caged in 6.5 × 6.5 × 12 cm plastic bottles kept horizontally, in which water was provided through a cotton-stoppered vial inserted through a hole in the back of the cage and adult food was placed in a small bottle cap and replaced as needed. As an oviposition devise, the lid of each cage (bottle) was replaced by a fine mesh through which females would lay their eggs, which, in turn, would fall into a dish lined with a damp, black cloth. These oviposition dishes were collected daily for egg counting. Males were replaced with same-aged, virgin males if the male died before the female. A total of 34 successive cohorts of 10, 25, 50 or 100 pairs were set up at irregular intervals, and daily egg production and age at death were recorded from a total of 1,000 females over a 3-year period.

4.2.2. Basic Birth and Death Rates

A summary of the birth and death rates by 30-day period for the 1,000-medfly female cohort is given in figure 4.2. Slightly over a third of all deaths occurred during the first month, whereas nearly 60% of all medfly females died during their second month. Only about 5% of all flies lived beyond two months. Expectation of life was over 3-fold greater during the first month than the last month and daily mortality was around 6-fold greater in the last month that in the first month. The small difference between the average daily mortality for the second (i.e., 7.7%) and third months (i.e., 9.1%) suggests that mortality decelerated at older ages as was observed in previous studies on this species (Carey, Liedo, Orozco, and Vaupel 1992).

Of the 936 females that produced eggs, 927, or over 99%, matured during the first month and the remaining flies matured during the second month. A total of 81.0, 18.3, and 0.7% of all eggs were laid during the first, second, and third periods, respectively. Average fecundity was over 3-fold greater during the first 30-day period compared with the last 30-day period. Eggs were laid on average once every 1.2 days during the first 30-day period but

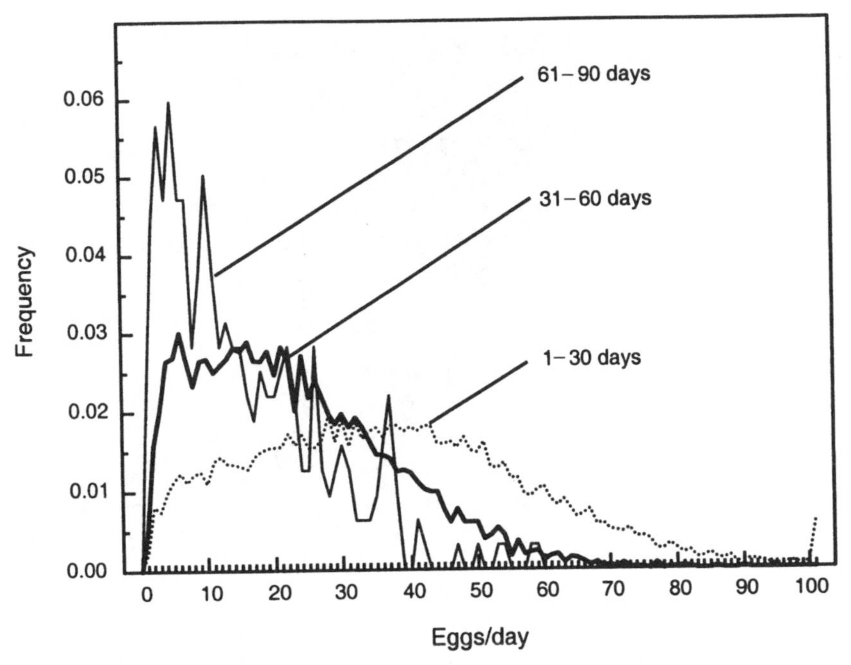

FIGURE 4.2. Frequency distribution of the level of daily egg laying (> 0) in 1,000 medfly females over 3 age periods relative to the total number of oviposition-days within the period. The number of fly-days in which at least one egg was laid for each of the 1-month periods was 16,387 and 5,878, and 318 for ages 0–30, 31–60, and 61–90, respectively. The average number of eggs/day for each of the age periods was 37.3, 22.8, and 15.5 for 0–30, 31–60, and 61–90, respectively (Carey, Liedo, Müller, Wang, and Chiou 1998a).

an average of only once every two days for the 55 females that lived beyond 60 days. The distributions of the daily number of eggs laid (> 0 eggs) for the three periods (see figure 4.2) revealed that the range and average number of eggs laid per day decreased in the successive months with around 37, 23, and 15.5 eggs/day laid on average for the first, second, and third months, respectively.

Cumulative reproduction of individual females over the first 30 days was not correlated with either subsequent reproduction (see figure 4.3) or future longevity (figure 4.4). Even though the distribution of eggs laid during the first 30 days ranged from nearly 2,000 eggs to zero, the egg production level of an individual female cannot be used to predict how many more eggs she will lay or how long she will likely live. The r-values for the correlation coefficients were less than 0.10 for both relationships. The lack of correlation between longevity and reproduction is consistent with the findings of

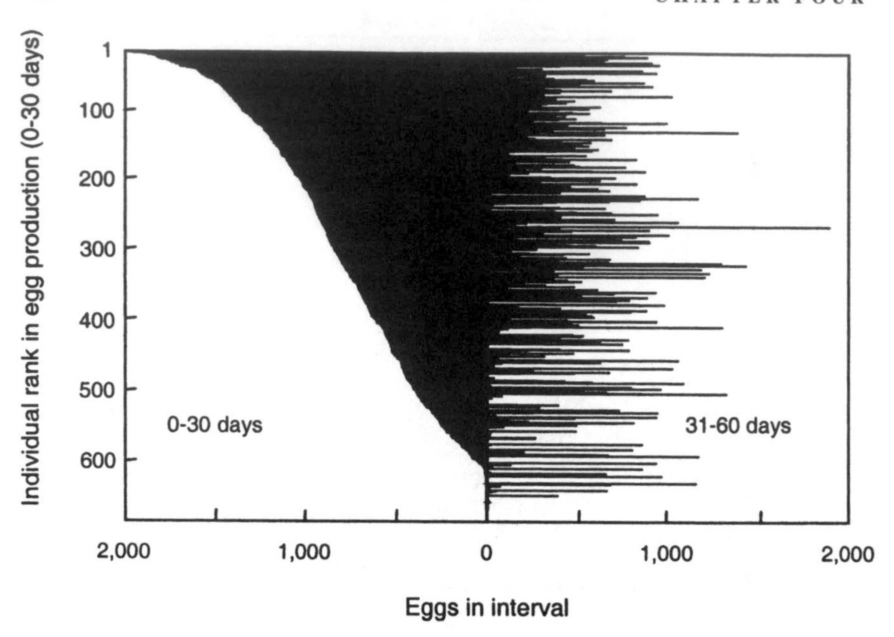

FIGURE 4.3. Relationship between cumulative egg production in female medflies through 30 days and the total number of eggs each subsequently produced. Individuals were rank-ordered according to cumulative egg laying over the first 30 days (Carey et al. 1998a).

Aigaki and Ohba (1984a) on *Drosophila virilis*, which reported that reproductive activity and life span showed a nearly normal distribution in the genetically homogeneous population but that reproductive activity did not correlate simply with life span in individual flies. They argued that death due to aging and age-associated changes in reproductive activity could be considered as under the control of different physiological processes.

Although the low statistical correlation between longevity and lifetime reproduction suggests that long life is not a necessary condition for high lifetime reproduction, the analysis obscures an important aspect of the demographic relationship. Change-point analysis revealed that there is a positive relationship between life span and lifetime reproduction but only up to 51 days (figure 4.5). In other words, lifetime reproduction increased with longevity but there was no reproductive gain due to added longevity for the 150 females that lived past 51 days. However, of the 10 females that produced the greatest number of eggs in their lifetime, the life span of 9 was greater than 50 days or two weeks greater than that of the average female. Also females in the first quartile of life spans (average longevity = 56.1 days) produced an average of nearly 1,100 eggs/female, whereas females in the

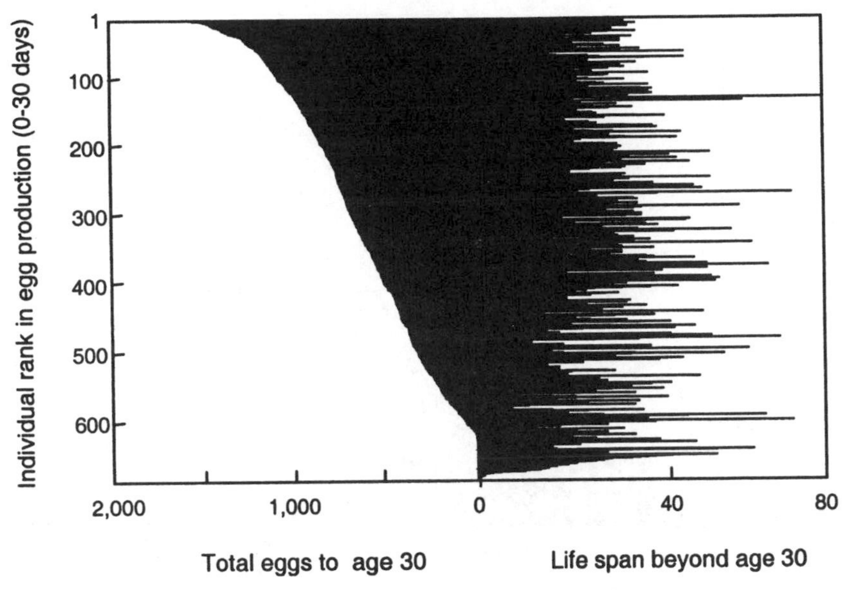

FIGURE 4.4. Relationship between cumulative egg production in female medflies through 30 days and remaining life span. Individuals were rank-ordered according to cumulative egg laying over the first 30 days (Carey et al. 1998a).

last quartile (average longevity = 18.2 days) produced an average of only 300 eggs/female. The importance of these particular findings is that they help to clarify the complex relationship between longevity and age-specific fecundity. That is, females that exhibit high levels of lifetime reproduction must necessarily live long enough to lay eggs over a sustained period. However, this relationship becomes progressively weaker at older ages because: (i) egg laying decreases at older ages thus reducing the rate at which lifetime totals accumulate; and (ii) inasmuch as there is no cost of reproduction, some long-lived females continue to lay eggs well into their advanced ages while others lay few or no eggs at the oldest ages. This increases the variance in lifetime reproduction of long-lived flies and therefore reduces its correlation with longevity.

The smoothed age trajectories of mortality in the cohort for both zero-egg layers and fertile females are given in figure 4.6. Daily mortality for fertile (egg-laying) females was less than 1% for the first 10 days and then increased to around 6% by 30 days when 642 individuals remained alive. Mortality then leveled off for 10 days, after which time it increased to over 10% by 52 days. At 60 days only 55-of-1,000 individuals remained alive with mortality fluctuating between 7 and 12% for the remainder of the life of the cohort. Daily mortality for zero-egg layers exhibited two surges—one

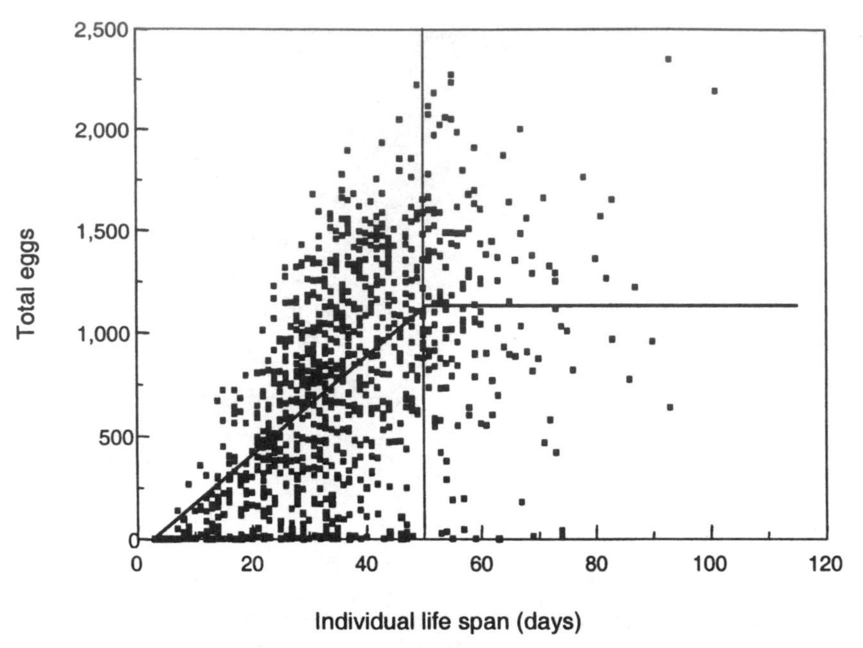

FIGURE 4.5. Lifetime reproduction vs. life span in 1,000 medfly females. The correlation coefficient for the regression of all points is $r = 0.5512$ and $p \approx 0.00$. The diagonal line depicts the results of change-point analysis, which shows the significant positive slope (= 26.09) before day 51 and the negligible slope (= 2.03) after day 51 (Carey et al. 1998a).

from 0 through 20 days and another from 20 through 40 days. The consequences of these two surges with respect to the frequency distribution of deaths are described in the next section.

4.2.3. Bimodal Distribution of Deaths for Infertile Females

The age distribution of deaths for the 64 females that laid no eggs in their lifetime exhibited two separate peaks, as shown in figure 4.7. This timing of the peaks in this bimodal distribution suggests that there are two "types" of zero-egg-laying individuals. The first type, which corresponds to the distribution of deaths around the left-hand peak, are those that can be characterized as "generally defective," inasmuch as they died young without laying eggs. The second type, corresponding to the distribution of deaths around the right-hand peak, correspond to females that were only "reproductively defective" but normal in other respects. This conclusion is based on the similarity in shape and timing of this group relative to the deaths within the remaining

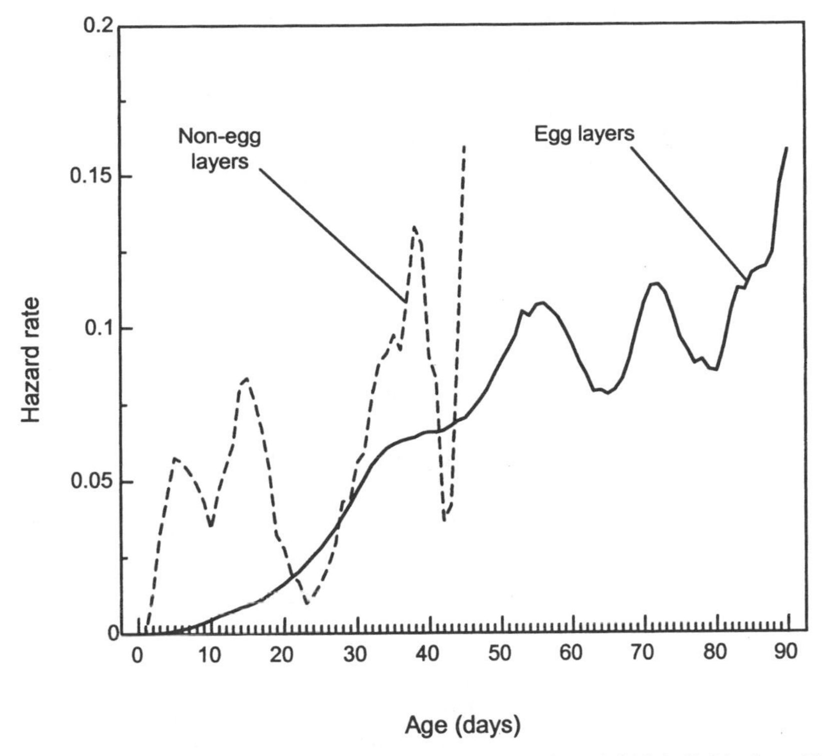

FIGURE 4.6. Trajectories of age-specific mortality based on 1,000 individual medfly females—64 zero-egg-laying females (dashed line) and 936 egg-laying females (solid line) (Carey et al. 1998a).

cohort of egg layers. The importance of this finding is that it decouples longevity and egg laying and thus underscores the earlier finding that there was no cost of reproduction—the distribution of deaths for the "reproductively defective" was the same as that for the reproductively "nondefective."

4.2.4. Age Patterns of Lifetime Reproduction

Analysis of the relationship of life span and age-specific reproduction revealed the following (Carey, Liedo, Müller, Wang, and Chiou 1998a). First, a high density reproductive "window" existed between 7 and 20 days for flies that lived between 22 and 55 days. This distinct "window" was absent in both short-lived (< 20 days) and long-lived flies (> 60 days) but for different reasons. Many of the short-lived flies died before they were old enough to produce high levels of eggs. In contrast, the long-lived flies may have

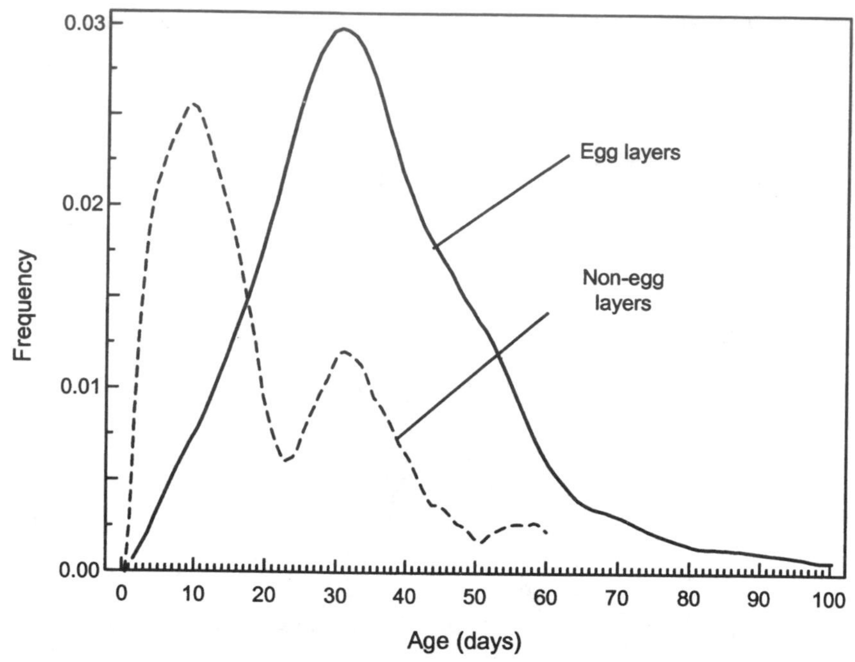

FIGURE 4.7. Distribution of deaths in females that laid no eggs (dashed line) and fertile females (solid lines). Note the bimodal distribution of deaths in zero-egg-laying females and the correspondence in timing of the right peak for zero-egg layers and the single peak for egg layers (Carey et al. 1998a).

experienced low mortality risk because they did not undergo an intense period of egg laying when they were young.

Second, flies did not experience high levels of egg laying at the end of their lives, regardless of whether they were short- or long-lived. This suggests that the vitality of an individual fly as reflected in its daily egg production is linked to its current robustness. That flies seldom died when they were at their peak egg production is evident from the graphical analysis (Carey, Liedo, Müller, Wang, and Vaupel 1998b).

Third, long-lived flies ($>$ 70 days) experience several peaks of egg laying every three weeks beginning around 20 days and extending through 70 days. This may reflect a reproductive cycle in which females undergo a burst of egg laying followed by a recovery period. This pattern may be more conducive to long life than is a single intense reproductive burst followed by a steady decline as was observed in shorter-lived flies.

One of the central questions in this study of reproduction in the medfly concerns whether qualitative differences in the lifetime patterns of egg production for flies that produced the greatest number of eggs versus flies that

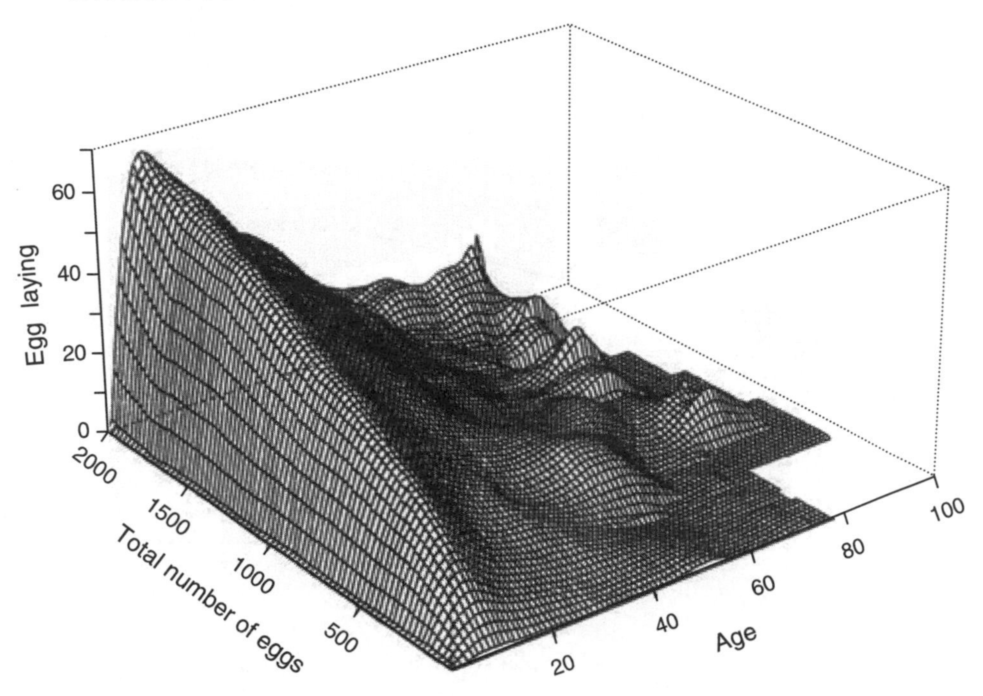

FIGURE 4.8. Smoothed 3 D plot showing the relationship between the age-specific reproduction (i.e., egg/female/day) of females rank-ordered from high (left-most) to low (right-most) lifetime reproduction. Plot is based on egg-laying data for 1,000 females (see text) (Carey et al. 1998a).

produced many fewer eggs in their lifetime. The answer to this question is revealed in the surface chart in figure 4.8 showing the relationship between age-specific reproduction and lifetime (total) number of eggs, and in figure 4.9 showing cross sections of the chart in figure 4.8 at 500, 1,000, 1,500, and 2,000 eggs per female. These figure reveal that females that lay the greater numbers of eggs in their lifetime scale the level of egg production during a peak period at around 10 days. Indeed, there was no difference in the timing of the peak production among females at 1-of-4 lifetime reproductive levels. Peak egg production for females that produced only 500 eggs in their lifetime was around 25 eggs/day, whereas peak egg production for females that produced 2,000 eggs in their lifetime was over 60 eggs/day. Age-specific levels of egg production was greater for high lifetime egg producers than for low lifetime egg producers but the overall age patterns were nearly identical through age 60.

Previous examination of the unsmoothed data by Carey et al. (1998b) revealed that, even though the daily fecundity of the average young female

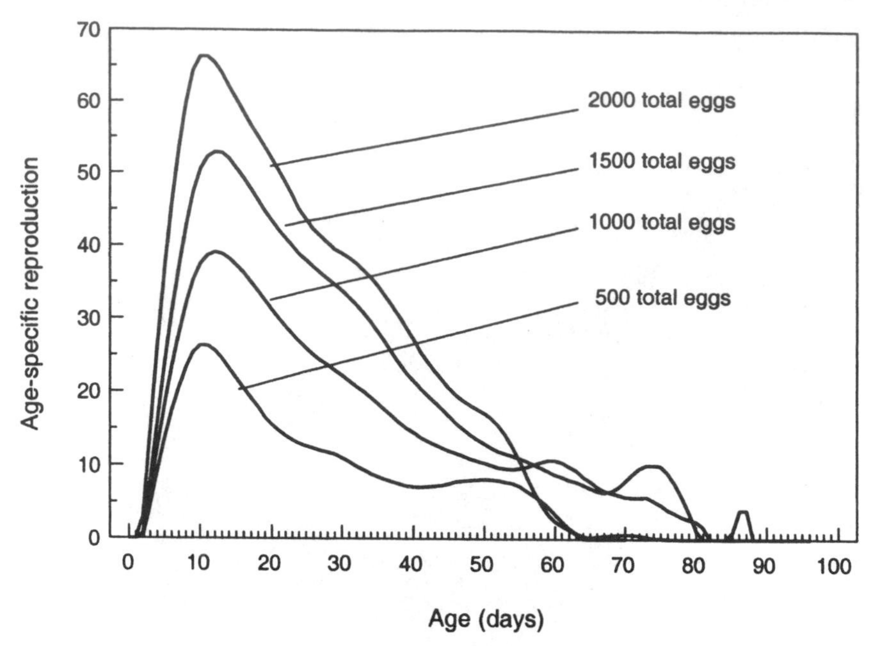

FIGURE 4.9. Cross section of surface plot shown in figure 4.8 for the Mediterranean fruit fly at 4 levels of lifetime egg production—500, 1,000, 1,500 and 2,000 eggs/female. See caption for figure 4.8 (Carey et al. 1998a).

was far higher than the fecundity of the average old female, a great deal of variation existed in day-to-day reproductive levels. In other words, young flies often interspersed days of high egg production with a few days of low or zero-egg production. Similarly, old flies sometimes produced a large number of eggs on a given day even though their longer-term production at the older ages had been low up to that age. This variation in daily egg laying probably reflects a natural cycling of "recovery" or "resting" periods after bouts of heavy egg laying. Knowledge of this daily variation in reproductive levels and in egg-laying intervals may be as important to characterizing the medfly's life history as is knowledge of the age-specific pattern of reproduction in the cohort.

4.2.5. Implications

Conventional life history theory (Roff 1992) holds that because egg laying is stressful and requires a major expenditure of energy, females that are reproductively active at young ages should be more frail and less fecund at older ages than are females that were reproductively less active at young ages.

Based on this reasoning, this weakening effect should be manifested as either decreased fecundity at later ages or as increased mortality, i.e., demographic cost of reproduction (G. Bell and Koufopanou 1986).

There are two possible explanations for why no cost of reproduction occurred in this study. The first explanation is based on the argument presented by Reznick (1985) that the data were obtained from nonmanipulative studies and therefore essentially represent naturally occurring variation in reproductive effort. There was no cost because there was no "demand" (e.g., food, host, or mate deprivation) and thus the measure is based on phenotypic correlation rather than experimental manipulation, genetic correlation, or responses to selection.

The second explanation for the absence of a demographic cost of reproduction is based on the argument by Tuomi et al. (1983), who noted that demographic theory represents only a specific line of reasoning and a specific approach to life history evolution. That is, the concept of reproductive effort implies a direct trade-off—an increase in reproductive effort increases current reproductive output at the expense of survival due to reduced somatic investment. Tuomi et al. (1983) point out that the postulates of the trade-off hypothesis do not necessarily hold in an allocation system where increased resource input (e.g., more food) uncouples somatic costs from the direct influences of reproductive effort. Reproduction can take place only partially or not at all at the expense of somatic investment. Also reproductive individuals can adapt to resist somatic costs without any survival costs.

One of the most surprising aspects of this study was that, despite the large number of published papers on age-specific reproduction in fruit flies and other groups of insects, my colleagues and I discovered several properties of age-specific reproduction that had not previously been documented or that were poorly understood, including (1) the complete lack of correlation between early reproduction and subsequent fecundity and longevity; (2) the complex relationship between lifetime reproduction and longevity—direct correlation up to day 51 but no gain in lifetime reproduction with additional increases in life span thereafter; (3) the bimodal distribution of deaths of infertile females suggesting that there are two "types" of zero-egg layers—"generally defective" and "reproductively defective"; (4) the life history strategy of increasing lifetime reproduction by scaling the amplitude of the peak rather than by shifting the overall age pattern; and (5) the complex relationship between mortality and cumulative reproduction (parity progression).

We believe that there were three aspects of the study that accounted for these new insights, all of which could easily be incorporated into the design and analytical framework of similar studies on insect reproduction in the future. First, the emphasis on the reproductive history of *individual females* made it possible to examine variation at multiple levels—individuals at single ages, over age periods (young vs. old), and over entire lifetimes. The composite patterns of individuals could then be related to cohort (i.e., mortality) and subcohort (i.e., levels of lifetime reproductive) properties. Second, the *study scale* (1,000 females) allowed the patterns of reproduction over all age groups to be examined including those at the most advanced ages. The scale also provided sufficient numbers for the actuarial properties of the cohort to be measured and related to phases of reproduction. Third, the statistical *smoothing and graphical techniques* brought to bear on the large data set of individual reproductive histories provided insights that were not possible using more conventional demographic methods.

4.3. Cost of Virginity

The concept of mortality cost of mating and reproduction in insects is fundamental to ecology, evolution, population biology, and gerontology because it provides a framework for life history trade-offs between reproduction and longevity (G. Bell and Koufopanou 1986; Reznick 1985). Altering longevity by manipulating reproductive activity has been used to study the evolution of aging and thus provides material for the study of its genetic mechanisms. Comparisons of male and female differences in mortality also shed light on the effects of sex-specific reproductive processes on life span (Carey and Liedo 1995a, 1995b).

Virgin insects are widely used in experiments concerned with the mortality effects of mating and reproduction since virginity is generally considered "cost free"; survival to older ages is usually greater for unmated than for mated individuals (Bilewicz 1953; J. M. Smith 1958). An implicit assumption of virtually all studies concerned with cost of mating and reproduction is that, because survival *to* older ages is higher in virgins than in nonvirgins, mortality *at* older ages is greater in non-virgins. Because of the central importance of the reproductive and mating cost concept to theories of aging (Kirkwood and Rose 1991; Rose and Charlesworth 1980) and for understanding male-female mortality differentials (Hazzard 1986, 1990; Hazzard and Applebaum-Bowden 1990), my colleagues and I conducted experiments to determine whether the effects of mating in females were beneficial at older ages relative to virginity, as reflected in mortality rates of virgin and nonvirgin cohorts.

4.3.1. Operational Framework

For the experiments, around 1,500 individuals of each sex were placed in one cage for monitoring death rates of females with access to males, and

TABLE 4.1
Number of Female Mediterranean Fruit Flies Alive at Age x (N_x) and Fraction of Initial Number Surviving to Age x (l_x) for Cohorts Maintained in Either All-Female or Mixed-Sex Cages

	Mixed-Sex		*All-Female*	
Age	N_x	l_x	N_x	l_x
0	30,003	1.0000	35,162	1.0000
10	18,110	0.6036	22,394	0.6369
20	6,837	0.2279	11,505	0.3272
30	2,425	0.0808	3,494	0.0994
40	757	0.0252	856	0.0243
50	236	0.0079	212	0.0060
60	55	0.0018	52	0.0015
70	4	0.0001	11	0.0003
80	0	0.0000	0	0.0000

Source: Carey et al. 2002.

3,000 females were placed in another cage for monitoring death rates of females denied access to mates. This design was replicated 19 times to obtain mortality data on approximately 35,000 females maintained in either all-female (virgin) cages and 30,000 females maintained in mixed-sex (nonvirgin) cages. Cohorts that totaled over 30,000 flies were used to ensure that mortality rates at both young and old ages would be based on large numbers. Minor contamination of several all-female cages was discovered by the technicians while counting dead flies. A total of eight male Mediterranean fruit flies were found in cohorts intended to be all females. No adjustment in the analyses was made inasmuch as the number of male "contaminants" was extremely small relative to the total number of females (8 out of 35,162 or approximately 1 male in 4,400 females).

4.3.2. Sex-specific Survival of Virgins

The greater survival rate of virgin females relative to nonvirgins was consistent with findings on other insect species (G. Bell and Koufopanou 1986) as presented in table 4.1. For example, the survival rates to day 20 for females maintained in mixed-sex and all-female cages were around 23 and 33%, respectively. The expectations of life at eclosion for females maintained in mixed-sex and all-female cages were 14.7 and 16.1 days, respectively. Thus the life expectancy difference between virgins and nonvirgins was 1.3 days for females. The leveling off of mortality at older ages is consistent with the

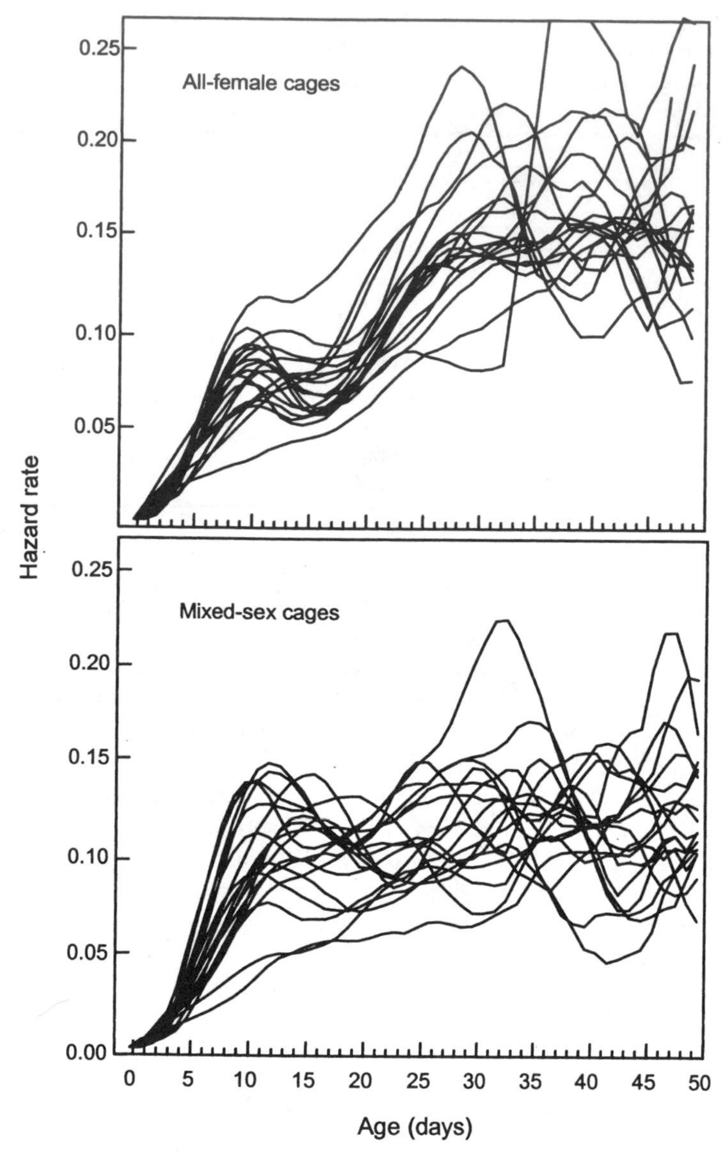

FIGURE 4.10. Smoothed hazard rates for 33 female Mediterranean fruit fly cohorts maintained in either all-female or mixed-sex cages. Each curve is based on deaths in approximately 1,000 flies (Carey et al. 1998a).

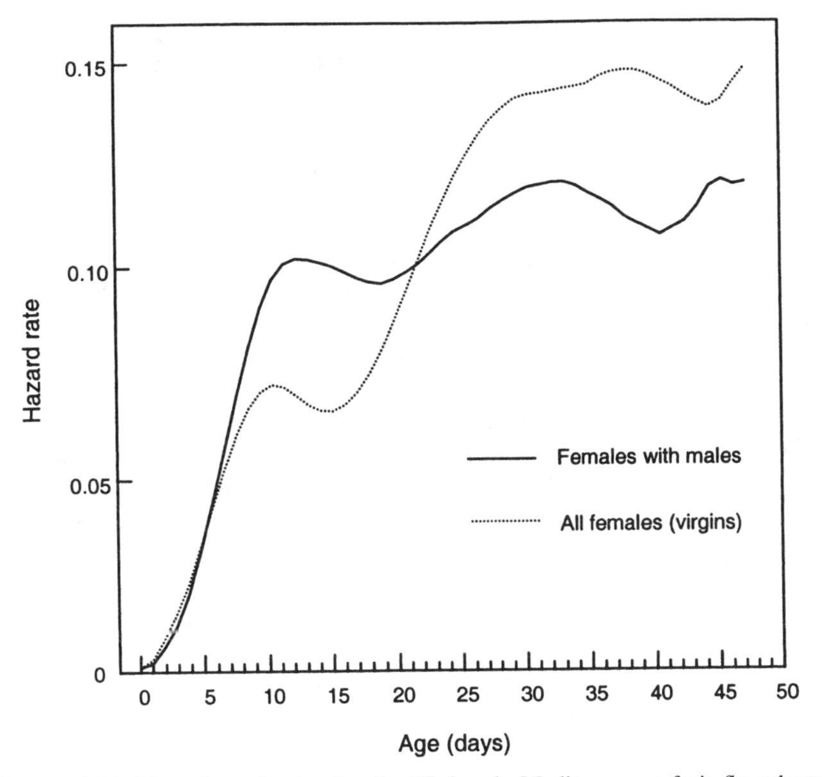

FIGURE 4.11. Mean hazard rates for the 33 female Mediterranean fruit fly cohorts shown in figure 4.10. Each curve is based on deaths in a total of over 33,000 individuals (Carey et al. 1998a).

findings of a previous study on 1.2 million Mediterranean fruit flies (Carey, Liedo, Orozco, and Vaupel 1992).

The trajectories of age-specific mortality maintained under the two different mating regimes are shown for all 19 cohorts in figure 4.10 and as a composite in figure 4.11. Mortality rates for virgin females were slightly higher than for nonvirgins at young ages. However, these mortality differences had little impact on life expectancy differences because the absolute levels of mortality were low at the young ages. Mortality in virgin females increased to about 0.08 after 7 days, declined to about 0.06, and then increased again to about 0.13, where it remained relatively steady. Mortality in nonvirgin females increased to 0.10 after about 8 days and then began leveling off. This early leveling off of nonvirgins and the continual increase of mortality in virgins caused the two schedules to cross over at about 3 weeks so that the mortality of virgin females was 1–4% higher than mortality in nonvirgins at older ages. Note that both virgins and mated flies show a peak of mortality around day 10, indicating that both have a vulnerable period around this age (Müller et al. 1997), but at a much higher level of mortality for mated flies (probably reflecting "cost of mating").

4.3.3. Mortality Crossovers

The mortality crossovers between cohorts of virgins and nonvirgins at older ages for each sex reversed the life expectancy advantage. Female life expectancy advantage favored virgins only until day 16, at which time it switched to nonvirgins. At 50 days, the expectation of life for both virgin and nonvirgin females was nearly identical. The death rates from 0 to 20 days of mated females were substantially higher than for unmated females, but from 21 to 45 days the death rates were considerably lower in mated females than in virgin females. χ^2 tests for differences in 5-day death rates between females maintained in mixed-sex cages and females in all-female cages. The differences in death rates were highly significant ($P < .001$) for all ages except at 45–50 days. The effects of the mortality crossover on relative survival of 30-day-old females are shown in table 4.2. For example, over 30% and 9% of females maintained in mixed-sex cages (nonvirgins) survived to 40 and 50 days, respectively, whereas around 25% and 6% of females maintained in all-female cages (virgins) survived to these respective ages. This is a reversal of relative survival of these two groups of females compared to young ages. Statistical comparisons of mean differences for virgin and mated females at selected ages are given in table 4.3. All comparisons are highly significant, with the first two (0 and 10 days) showing

TABLE 4.2
Number of Female Mediterranean Fruit Flies Alive at Age x (for $x \geq 30$) and Survival (l_x) Normalized Using Number Alive at 30 Days for Cohorts Maintained in Either All-Female or Mixed Cages

	Mixed-Sex		*All-Female*	
Age	N_x	l_x	N_x	l_x
30	2,425	1.0000	3,494	1.0000
40	757	0.3122	856	0.2450
50	236	0.0973	212	0.0607
60	55	0.0227	52	0.0149
70	4	0.0016	11	0.0031
80	0	0.0000	0	0.0000

Note: The results are the same as in table 4.1 except survival (l_x) is normalized using the number alive at 30 days (N_{30}) (Carey et al. 2002).

TABLE 4.3
Table of Pair-wise Comparisons for Log-transformed Remaining Lifetimes for Groups of Virgin and Mated Medflies

Group	$T > 0$		$T > 10$		$T > 20$		$T > 30$	
Statistic	V_0	M_0	V_{10}	M_{10}	V_{20}	M_{20}	V_{30}	M_{30}
Mean	16.57	15.16	12.08	10.31	8.70	9.74	7.54	8.86
Variance	104.753	94.271	77.511	86.237	52.838	72.639	44.469	62.637
Std	10.235	9.709	8.804	9.286	7.269	8.523	6.669	7.914
t-test	12.095*		19.6169		−5.3684*		−4.7989*	
df	64982.95		37827.36		13778.4		5005.71	
P	$< 10^{-5}$		$< 10^{-5}$		$< 10^{-5}$		$< 10^{-5}$	

Note: The table contains the mean and standard deviation on the original (time in days) scale for remaining lifetimes, and contains the t-test results for log-transformed remaining lifetimes (test statistics and p-values). The groups are defined as V = virgins and M = mated. The subscript after V and M denotes the remaining lifetime after age x (e.g., V_{20} is life expectancy at day 20 for virgins). (Carey et al. 2002).

*The t-test is based on the transformed data (log).

longer remaining lifetimes for the virgins, and the latter two (20 and 30 days) showing longer remaining lifetimes for the mated flies.

4.3.4. Implications

Higher survival rates of virgins relative to nonvirgins have been reported in other insect studies, including those on *Drosophila* (Fowler and Partridge 1989; Malick and Kidwell 1966; J. M. Smith 1958), the housefly, *Musca domestica* (Ragland and Sohal 1973), the bruchiid beetle, *Callosobruchus maculatus* (Tatar et al. 1993), the tribolium beetle, *Tribolium castaneum* (Mertz 1975; Sonleitner 1961), grasshoppers, *Melanoplus spp.* (Dean 1981), the bug *Dysdercus fasciatus* (Clarke and Sardesai 1959), the leaf miner, *Agromyza frontella* (Quiring and McNeil 1984), the apple maggot fly *Rhagoletis pomonella* (Roitberg 1989), and the moth, *Ephestia kuhniella* (Norris 1933). However, a study of *D. virilis* (Aigaki and Ohba 1984b) revealed that mating status has a reverse effect on survival of female and male flies—unmated females lived around 14% longer than mated females, whereas unmated males lived about 12% shorter than mated males. A study on the backswimmer, *Corixa punctata*, by Calow (1977) and another study on *D. subobscura* by Lamb (1964) reported no survival differences between virgin and mated individuals. No studies on insects could be located in which mortality rates of virgins and nonvirgins were examined at older ages. However, the findings are similar to those on *D. melanogaster* in which the cost of mating to the individual is an immediate but short-term increase in mortality at the time of mating (Partridge 1986; Partridge and Andrews 1985). The cumulative effect is to reduce the fraction of individuals that attain older ages.

There are several explanations for why the mortality rate of female Mediterranean fruit flies maintained in all-female cages crossed over the mortality rate of females Mediterranean fruit flies maintained in cages with males. One explanation is that the changing mortality pattern in each cohort was due to heterogeneity at the cohort level (Vaupel and Carey 1993). That is, in cages where females were maintained with males, the most frail females may have died at young ages due to mating costs, and thus the most robust subgroup of females (i.e., those with the lowest death rates) survived to the older ages (Vaupel et al. 1979).

A second explanation is that the differences in mortality patterns were an artifact of the particular environmental conditions. For example, it is possible that the longevity advantage would favor virgins over mated females throughout their lives if both mated and unmated females were given a protein source for food. For example, my colleagues and I found that the male-female mortality crossover occurred later and was less pronounced in cohorts provided with protein than in cohorts that were protein-deprived (Müller et al. 1997b).

A third explanation for the relative changes in the mortality schedules of both treatments is that age-changes in the reproductive biology and behavior of individuals alter their age-specific vulnerability and, in turn, the age-specific mortality trajectory. For example, it is possible that the short-term survival advantage of virgins may eventually disappear due to costs associated with the constant arousal and subsequent thwarting of the sex drive (Morris 1955), to the possible absence of protective hormones elicited by mating or reproduction (Engelmann 1970), or to the elicitation of costly behaviors associated with the opposite sex (Daly 1978; see also Arita and Kaneshiro 1983; Shelly and Whittier 1995; Whittier et al. 1994). For example, when Mediterranean fruit fly females are maintained as virgins several days beyond the optimum sexual maturation period, females begin to mimic the courtship actions of the male such as "pseudomale" behavior in Mediterranean fruit fly females (Morris, 1955; Arita and Kaneshiro 1983).

A fourth explanation is that the findings are consistent with a protective effect of reproduction that comes to bear at later ages, as opposed to a fairly immediate cost of mating. For example, increased throughput of eggs with many eggs present in early oogenic stages may have a protective effect, so that in particular the flies that remain active in egg laying and reproduction are protected in older ages. This is consistent with earlier findings by Müller and coworkers (Müller et al. 2001) that an individual's mortality is associated with the time-dynamics of the egg-laying trajectory in medflies.

The results of this study, demonstrating that virginity bears a relative cost at older ages, suggest that there may not be a single identifiable mortality "cause" (Preston et al. 1972) in the cost of mating. This concept challenges the conventional implicit view that (i) reproductive activity in most insects increases mortality (i.e., cost of reproduction) over *all* age groups and (ii) this increase is due to a specific cause. For example, Chapman and coworkers (1995) reported that the cause of death in mated female *D. melonogaster* was the result of the transfer of male accessory-gland main-cell products. The findings suggest that the effects on mortality of mating in particular and of reproduction in general are probably quite complex and result from multiple, interacting causes.

4.4. Supine Behavior—a Predictor of Time-to-death

A biomarker of aging is a behavioral or biological parameter of an organism that either alone or in some multivariate composite will better predict functional capacity and/or mortality risk at some late age than will chronological age (Markowska and Breckler 1999). Behavior biomarkers of aging are important because (1) behavior changes with aging and thus behavior itself can be used as an index of aging; and (2) any intervention to alter the chronological course of aging must be assessed behaviorally since quality of life as well as longevity must be considered in the evaluation of any intervention (Baker and Sprott 1988). Because of their rapid generation time and low cost relative to mammalian models such as rodents or primates, understanding behavioral biomarkers in fruit flies has important implications regarding research designed to understand morbidity dynamics, behavioral neuroethology, and gerontology, and regarding the interpretation of longevity extension in model organisms.

4.4.1. Background

While monitoring the behavior of male Mediterranean fruit flies (*Ceratitis capitata*) throughout their lives, Papadopoulos and coworkers (2002 [forthcoming]) discovered a behavioral trait that is unique to older, geriatric flies in general but especially to individuals that are gradually approaching death. They termed this trait *supine behavior* in accordance with the upside-down position of the temporarily-immobile flies. Supine males lie on their backs at the bottom of their cage appearing dead or moribund. But these flies are very much alive and moderately robust as becomes evident when they right themselves either spontaneously or after gentle prodding and initiate walking, eating, and wing-fanning behaviors, some of which are indistinguishable from that of normal flies.

Inasmuch as any upside-down adult insect is dysfunctional, it follows that the health of medfly males that was observed spending progressively greater amounts of time in the supine position was both poor and declining relative to the health of flies capable of remaining upright. Because virtually nothing is known about the nature of chronic conditions (illnesses, infirmity, dementia) in insects, because no viable animal model systems exist for studying morbidity dynamics (e.g., determinants, age of onset, persistence, association with mortality), and because of the importance of understanding morbidity incidence and prevalence in human populations (Crimmins 2001; Crimmins et al. 1994, 1996, 1989), experimental studies were initiated to address and quantify a number of questions regarding supine behavior in medfly males: When do flies first begin to exhibit supine behavior? Is it sporadic or progressive? What is the age and frequency distribution of this behavior for both cohorts and individuals? Is supine behavior "reversible" or is it a behavioral marker of impending death?

4.4.2. Empirical and Statistical Framework

Supine behavior was studied in 203 male medflies that were derived from a wild-caught colony (which was established from infested hosts collected in Thessaloniki, Greece, in Fall 1999) by placing individual 1-day-old males in $5 \times 12 \times 7.5$ cm plastic transparent cages with adult food (protein hydrolysate-sugar mixture) and water and maintaining them throughout their lives at 25 ± 2 °C, $65 \pm 5\%$ relative humidity, and 14:10 light:dark cycle with photophase starting at 06:00. Supine and other behaviors (calling, walking, resting, feeding) were recorded for each fly in 10-minute intervals over a 2-hour period from 1200 to 1400 h each day until it died. Thus the daily supine number for each fly could range from 0 to 12 (i.e., 2-hour period at 6 observations per hour).

The model using daily supine level $X(t)$ recorded as age t as the time-dependent covariate is given by $\lambda(t) = \lambda_0(t) \exp\{\beta X(t)\}$, where $\lambda(t)$ and $\lambda_0(t)$ denote the hazard rate and baseline hazard rate, respectively. The null hypothesis H_0: $\beta = 0$ was tested and rejected it with $P < 0.0000$. The estimated value of $\beta = 0.23357$. The numbers provided are for models that have just one predictor. Fitting a model that has both the onset and level of supine behavior as predictors indicates that both are highly significant, with mortality risk ratios of 17.59 and 1.18 for onset and level of supine behavior, respectively.

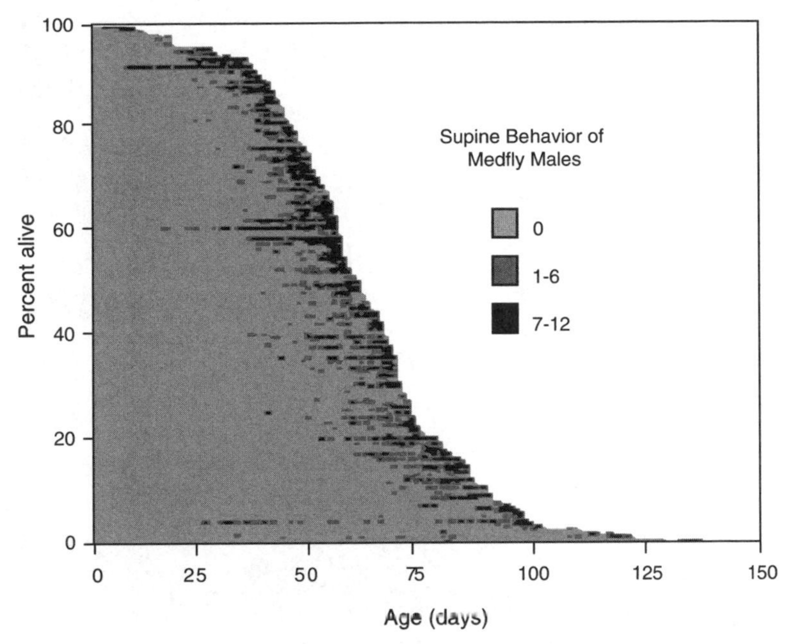

FIGURE 4.12. Event history charts for supine behavior in 203 male medflies relative to cohort survival. Each individual is represented by a horizontal "line" proportional to his life span. Each day an individual displayed supine behavior it is coded according to the number exhibited— light gray indicates no supine behavior observed, medium gray indicates 1–6 supine observations, and black indicates 7–12 supine observations (Papadopoulos et al. 2002 [forthcoming]).

4.4.3. Results

An event-history chart (Carey, Liedo, Müller, Wang, and Vaupel 1998b) showing the age-patterns of the supine behavior for each of the 203 males (figure 4.12) reveals a distinct association of individual life span and both the age of onset and the intensity of this behavior. The band depicting supine behavior shows that in many flies it begins to occur about 2–3 weeks prior to their death and therefore the period of its occurrence closely follows the cohort survival (l_x) schedule. Supine behavior seldom occurred in very young flies (< 25 days) but frequently occurred in flies that were over 50 days old—ages at which mortality rates began to increase substantially. The event history chart (figure 4.12) reveals four general properties of the medfly supine behavior: (1) *persistent*—occurs on subsequent days after onset; (2) *progressive*—intensity increases with age; (3) *predictive*—onset and inten-

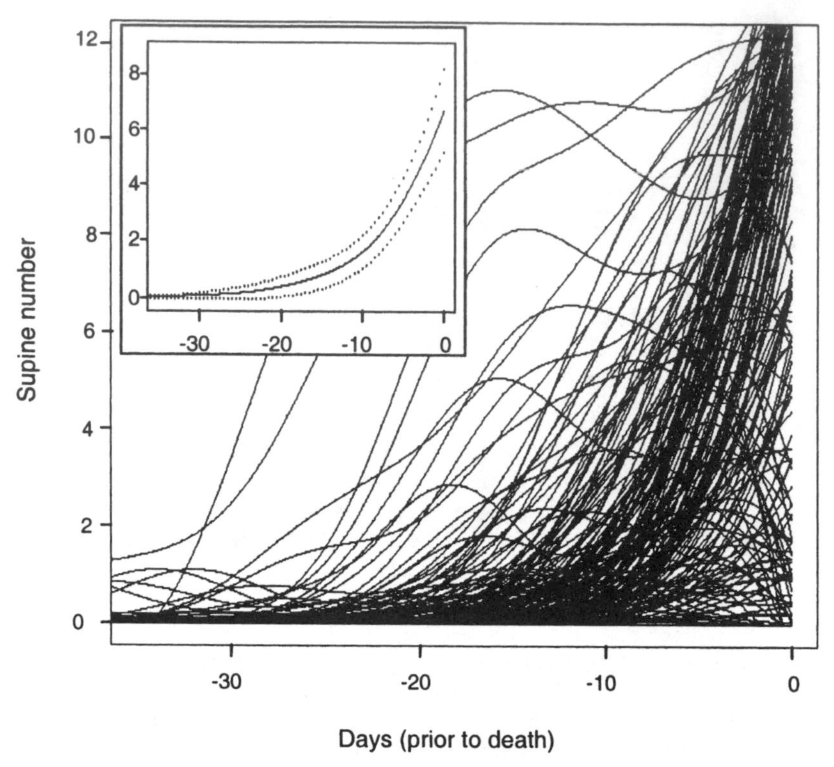

FIGURE 4.13. Smoothed plots for the supine number recorded in the 2-hour observation period of each of 203 males aligned relative to time of death. Inset: mean supine number for the male medfly cohort. Dashed lines indicate ± 2 SD (Papadopoulos et al. 2002).

sity is a strong indicator of impending death; and (4) *universal*—nearly all male medflies exhibit this behavior regardless of their age of death.

Smoothed plots of the relationships between onset and age progression of supine behavior relative to the time of death for each of the 203 male medflies (figure 4.13) reveals the heavy clustering of supine numbers near the time of death. Most of the supine behavior began between 10–15 days prior to death with the highest concentrations occurring 5–7 days prior to death. Coxproportional hazards regression models (Klein and Moeschberger 1997) were fitted with time-varying predictors to the data shown in figure 4.13 to test for the significance of both onset (presupine-to-supine) and level of supine behavior with respect to mortality (hazard) rate. Using first day of supine onset, we defined the time-dependent covariate as $z(t) = 0$ if the fly is in presupine status and $z(t) = 1$ when the fly has reached supine status. The onset of the supine behavior is a highly significant predictor of mortality

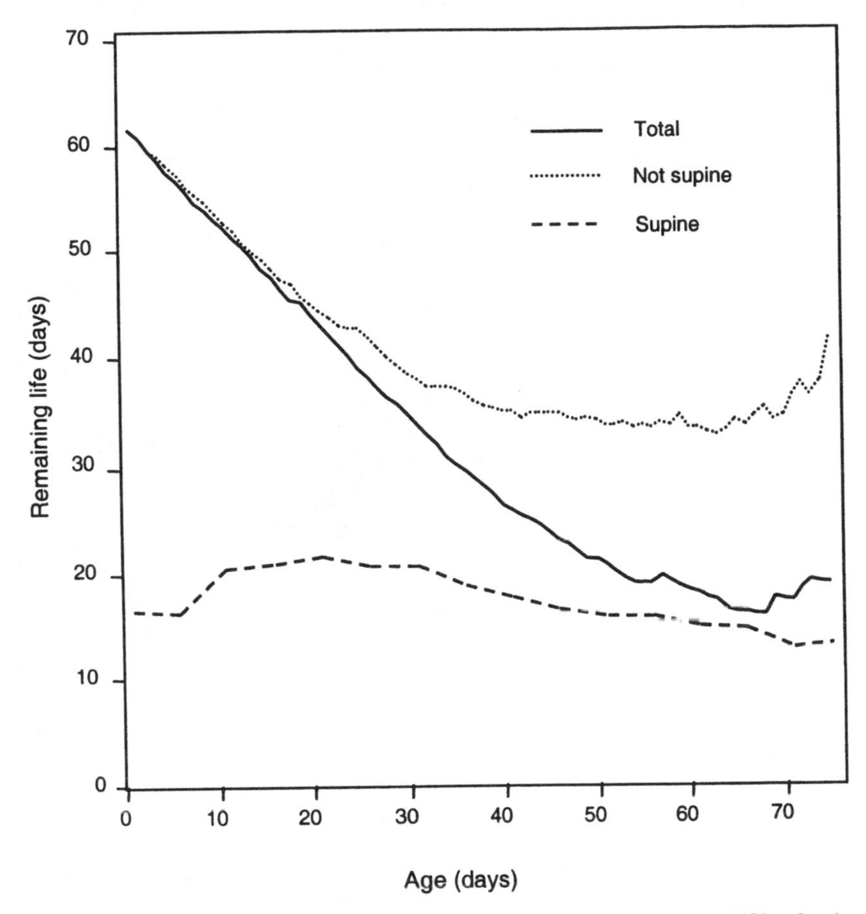

FIGURE 4.14. Schedules of remaining life expectancy (days) for male medflies for the cohort as a whole (center line) as well as for males both prior to (top line) and after (bottom line) the onset of supine behavior (Papadopoulos et al. 2002).

($p < 0.0001$) and increases mortality risk considerably more than the level of supine behavior. The model showed that a fly's mortality risk increased by 39.5-fold once it began experiencing supine behavior. Level of supine behavior is also a significant predictor of mortality, and for every increase in supine level by 1.0 the hazard rate increases by 26.3%.

The impact on mortality risk of the onset of supine behavior is illustrated in figure 4.14, which shows the remaining life expectancy at each age for all flies (middle curve) as well as flies disaggregated according to whether they had exhibited supine behavior (lower curve) or not (upper curve). Three aspects of this figure merit special comment. *First*, striking differences occurred in the remaining life of young flies (< 20 days) that had not exhibited

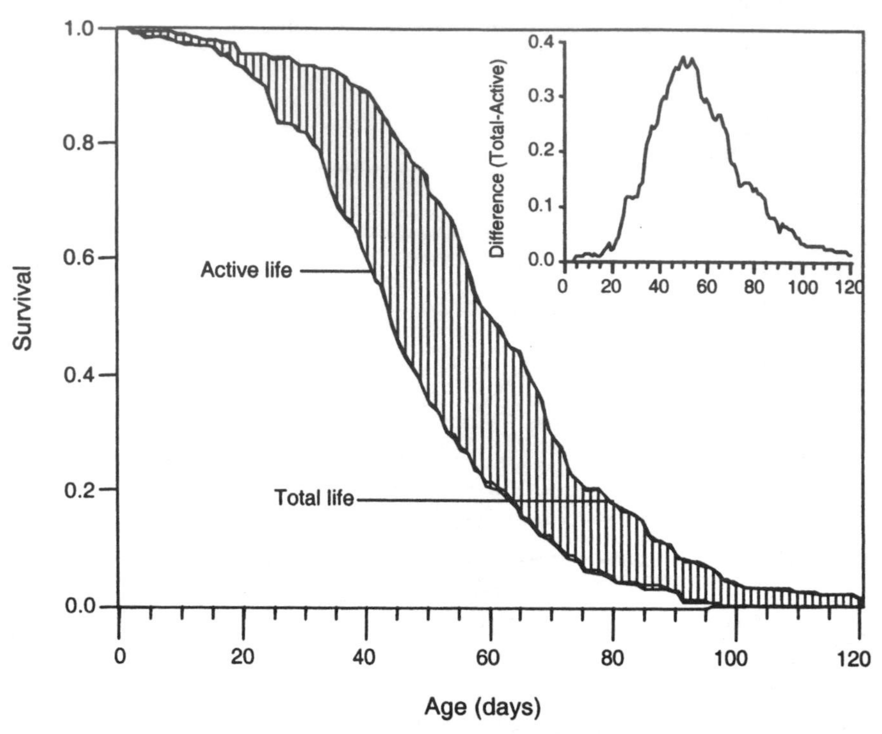

FIGURE 4.15. Schedules of active life and total survival for the male medfly cohort. The age at which active life for an individual ended was defined as the age at which supine behavior was first observed. Inactive life was thus defined as all life-days beyond this onset age. Inset shows the within-cohort distribution of life-days that the average fly experienced in the inactive stage, computed as the difference between total and active life (shaded area in main figure) (Papadopoulos et al. 2002).

supine behavior and young flies that were experiencing supine behavior—over 50–60 days of life remaining for the former (nonsupine) but only 20 days or less for the latter (supine). *Second*, the remaining life expectancy of a nonsupine fly 25–70 days of age remains constant at about 35 life-days. However, once it enters the supine state its life expectancy immediately drops by 20 days to approximately 15 days. *Third*, the remaining life of flies that initiated the supine behavior is also remarkably constant, ranging from slightly over 20 days for flies that are 10–30 days old to around 16–18 days for very young flies and flies older than 30 days.

We used the age at which a male medfly first exhibited supine behavior to demarcate the end of the "active" period of a male fly's life to construct a cohort schedule of "active life" (Katz et al. 1983; Manton and Land 2000). This schedule, combined with the conventional survival (l_x) schedule, was used to estimate the age distribution and prevalence of inactive life by taking the difference at each age between the inactive and total life schedules (figure 4.15). Total and active life expectancies were computed as 61.7 and 45.6 days, respectively and thus the average male experienced 16.1 days or 26.1% of its 61.7-day lifetime partially in the supine state. The distribution of inactive life for the cohort varied with age (figure 4.15 inset) with the greatest prevalence occurring between 30 and 50 days and the least occurring at the younger and older ages. This measure of inactive life provides information, other than death, about the dimensions of health and morbidity for a fly and can be used as a model system for actuarial analysis as a health-status measure.

4.4.4. Implications

The current findings are important for several reasons. First, the use of supine behavior as a proxy for health status and impending death in a medfly cohort provides insights into questions regarding morbidity dynamics (Fries 1980; Fries et al. 1993). The constancy of life expectancy once flies entered the supine state regardless of age of entry suggest that virtually all individuals undergo a transition period of declining health prior to death; that morbidity is a natural stage of the aging process and its compression may have limits. Thus time-to-supine may serve as an additional endpoint of interest in different experiments concerned with health and vitality (e.g., feeding, mating, reproduction).

Second, supine behavior as a proxy for deteriorating health and fitness may be important in ecological contexts since it is morbidity that determines vulnerability and thus the age trajectory of mortality. This may explain why the cause-of-death structure in nature is often only weakly related to the dynamics of a population. If increased vulnerability due to aging effectively represents the underlying "cause" of death in nature, then it follows that the immediate cause of death (predator) may be capricious and thus have little relevance to population regulation.

Third, the observation that flies commonly spend part of each day lying upside down over a period of several days or weeks prior to dying, raises not only the technical issue regarding the possibility that the recorded age of death is incorrect, but a related conceptual one as well—age patterns of mortality and morbidity can become confounded when unhealthy (supine) flies that are capable of living many more days are mistaken for dead ones. This does not present a serious analytical problem when comparing rates of actuarial aging if the schedules of active and total life are approximately parallel and/or if the incidence of supine flies being mistaken for dead flies is random. However, determination of mortality patterns can be substantially affected if the mistakes in the determination of age of death are nonrandom or if the length of the morbidity (inactive) period is shorter at young and old ages.

Fourth, if most individuals experience a stage of morbidity prior to their death, then this raises questions concerning whether the increases in longevity reported in model organisms through genetic and/or environmental manipulations are due to extensions in the "active" or the "inactive" life-span segment. This is a particularly relevant question concerning the results of aging studies on the nematode, *Caenorhabditis elegans*—longevity extension in mutant strains nematodes is achieved exclusively by extending their post-reproductive period, which includes the ages when their physical activity and feeding is practically nil (Apfeld and Kenyon 1999; Hsin and Kenyon 1999; T. E. Johnson 1990). This suggests that longevity extension in nematodes is achieved largely by extending the inactive rather than active life span.

Fifth, if the underlying cause(s) of the supine behavior are neurological and central in origin as opposed to declines in the structure and function of the peripheral nervous system or of the skeletal musculature, then medflies and presumably other insect species that exhibit this behavior could be used as models for studying the onset and progression of neurological diseases, dementia, and brain aging and, in turn, for elucidating the discourse between individual behavior and the neural structures that underlie these behaviors (Neuweiler 1999; D. L. Price et al. 1999; Scheibel 1999). The results may especially pertinent to emerging fields of evolutionary neurobiology and behavioral neuroethology (Francis 1995) as they relate to behavioral aspects of gerontology.

In general, the results reported in this section on supine behavior in male medflies open up new opportunities for research on age-specific changes in behavior and the relevance of these changes to aging and longevity. In particular, the findings raise the possibility of identifying a suite of behaviors comparable to the activities of daily living (ADLs) in humans—behaviors that could be used as disability indices for studying the antecedents of the diseases and infirmities that occur in the end-phase of life in model organisms.

4.5. Summary

Inasmuch as changes in mortality for any species are not independent of changes in other life history traits, it follows that baseline studies on reproduction and behavior will provide important background and context for interpreting results of studies where conditions alter these life history traits. One of the main technical results presented in this chapter was the introduction of the event history chart as a powerful method for visualizing the relationship of longevity and the intensity of any age-specific event such as reproduction. This new technique was used in the first section to capture in a single graph a wide variety of age patterns related to the reproduction of individual females. One of the strengths of the graphic technique is that it visualizes in one picture the age relationship among 5 or 6 reproductive traits including age of first reproduction; egg-laying intensity at young, medium, and old ages; general rate of change in egg laying; post-reproductive period, and cohort survival. This is not possible using conventional X-Y plots or surface charts.

The main biological and demographic results presented in this chapter on reproduction include (1) the fine details of medfly egg laying in individual females, including the frequency of daily egg laying, daily and lifetime reproduction, and rate of change with age; (2) cohort data including schedule of age-specific reproduction, disaggregation of cohort into two subgroups classified according to egg-laying pattern; (3) relationship between early egg laying and both life span and egg-laying at older ages; and (4) the age-specific mortality pattern of virgin versus nonvirgin females, which shows that although virgins experience lower mortality at young ages than nonvirgins, their cohort mortality rates are higher at older ages, hence the cost of virginity. Findings of a behavioral study on the relationship of what was termed "supine behavior" in male medflies to their time-of-death was reported in the last section. The key result was that the onset of supine behavior increased mortality by nearly 40-fold relative to flies that had not exhibited this behavior. The implications of using this model system for understanding morbidity dynamics were discussed.

5

Mortality Dynamics of Density

DENSITY is a part of all controlled experiments involving animals in much the same way as temperature is a part—neither factor can be excluded. All captive environments restrict movement of individuals within an area that can be characterized by volume or surface area and thus every experiment will necessarily involve individuals per unit area or volume. After publication of the results from the large-scale medfly study (Carey, Liedo, Orozco, and Vaupel 1992) some scientists expressed concerns that the slowing of mortality was due to the lower densities at older ages, which resulted from the daily attrition (Nusbaum et al. 1993; Olshansky et al. 1993). In this section I describe the results of studies that were conducted on density effects showing that a reduction in density at older ages did not account for the slowing of mortality. The majority of the text and results presented in this chapter was taken from the paper by Carey and Liedo (1995b) with literature updates from Curtsinger (1995).

5.1. Background

Most animals are widely spaced in nature due to behavioral patterns evolved to maintain territories (Tanner 1966). Thus when individuals are cohoused with conspecifics as is often the case in biological studies, stress is increased because behaviors associated with territoriality must be modified (Price 1984; Ranter and Boice 1975). This heightened stress, combined with an increase in the incidence of physical damage due to fighting and accidents in animals confined in close quarters, usually causes a decrease in longevity. The effects of crowding on longevity and physical injury is well documented in many insects including *Drosophila* (Graves 1993; Mueller et al. 1991; Pearl et al. 1927; Pearl and Parker 1922; Prout and McChesney 1985), the milkweed bug, *Oncopeltus fasciatus* (Dingle 1968), the cotton stainer, *Dysdercus fasciatus* (Dingle 1968), the corn earworm, *Heliothis zea* (Jones et al. 1975), the dermestid beetle, *Trogoderma creutz* (Davis 1945), the honey bee, *Apis mellifera* (Smaragdova 1930), and the house fly, *Musca domestica* (Barber and Starnes 1949; Finch 1990; Patterson 1957; Ragland and Sohal 1973, 1975; Rockstein 1957; Rockstein et al. 1981; Rockstein and Lieberman 1959). However, the effects of density on the underlying demographic determinant of the longevity differences—age-specific mortality—are virtually unknown.

For example, it is not known whether crowding reduces life expectancy in cohorts by altering the overall pattern of the mortality schedule, by increasing the slope of the mortality schedule over particular age groups, or by increasing the level of mortality at each age. Nor is it known whether different levels of density have qualitatively different effects on mortality patterns of subgroups such as males vs. females or short- vs. long-lived strains.

Understanding the mortality dynamics of density—how different numbers of animals per cage affects the levels and patterns of cohort mortality—is important for several reasons. The first reason concerns the use of parameters derived from mortality models for inter- and intraspecific comparisons. Books by Comfort (1979), Finch (1990) and Gavrilov and Gavrilova (1991) all contain life table information gathered from the literature on scores of species and include estimates of parameters derived from the Gompertz or the Gompertz-Makeham mortality models (Gompertz 1825; Makeham 1967), such as the initial mortality rate, the exponential coefficient (age-dependent component), and the Makeham constant (age-independent mortality). Because the majority of life tables cited in these books were constructed from data on confined cohorts, the use of the parameters for comparing the mortality properties of groups, species, or strains is questionable if initial density affects any aspect of the mortality schedule. The same concept is relevant to selection experiments designed to isolate "density tolerant" strains of *Drosophila* (Mueller et al. 1991). It is not known, for example, whether longer life spans in strains that have been selected to live longer under crowded conditions are due to a delay in the onset of senescence, to a decrease in the exponential rate of aging (age-dependent component), or to a uniform reduction in mortality at each age (age-independent component).

A second reason that understanding the effects of density on age-specific mortality is important concerns gender differences in longevity. If the stress and wear-and-tear caused by crowding affects the mortality rate in one sex more than the other, then the outcome of many experiments on sex-specific mortality differences (Hamilton 1948) may be an artifact of density effects. It is conceivable that life expectancy in some species maintained under high densities could favor one sex but favor the opposite sex if maintained under low densities; thus, which sex lives longer is conditional on environmental parameters.

A third reason for understanding the importance of the effects of density on age-specific mortality concerns findings in previous studies on fruit flies (Carey et al. 1992; Curtsinger et al. 1992): mortality rates level off and decline at the most advanced ages. This interpretation was challenged by others (Carey et al. 1993a; Carey et al. 1993b; Kowald and Kirkwood 1993; Nusbaum et al. 1993; Robine and Ritchie 1993) who argued that the leveling off and decline was likely an artifact of density effects—mortality decreased due to declining densities with age. The interpretation by Carey, Liedo, Orozco, and Vaupel (1992) and Curtsinger and coworkers (1992) is wrong if this explanation can account for the deceleration of mortality at older ages. However, if density effects do not alter the overall pattern of mortality at older ages, then the interpretation stands and the leveling off and decline of mortality at advanced ages will have to be seriously addressed by gerontologists, evolutionary biologists, entomologists, demographers, and others.

Because information on the effects of density on mortality is almost nonexistent and because the mortality schedule is fundamental to life table studies on aging and senescence (Gavrilov and Gavrilova 1991; Manton and Stallard 1984), the broad objective of this chapter is to report the findings on the effects of different levels of crowding on age-specific mortality in the Mediterranean fruit fly using the large-scale experimental system described in Carey and coworkers (1992). The specific objective in the study was to test the hypothesis that the overall pattern of mortality in the medfly and particularly the leveling off and decline at older ages is independent of initial, current, and cumulative densities. Although the medfly is used as a model, the findings have implications that pertain to any life table and aging study of confined organisms.

5.2. Operational Framework

5.2.1. Types of Density Effects

Density effects can be separated into three types (figure 5.1): (i) initial density—the starting number of individuals in each cohort; (ii) cumulative density—the cumulative number of surviving flies in a cage up to day x; and (iii) current density—the average number of flies alive in the age interval x to $x + 1$. Initial density is often used as a proxy for the other two types of densities. However, initial density is a constant in that it does not involve the concept of fly-days, or duration of exposure. Cumulative density is the sum total of fly-days and current density is the total number of flies in a cage at a specified time. Initial density is constant and cumulative density increases with age so neither factor can account directly for a decline in mortality at older ages. The three types of densities are not independent.

5.2.2. Experimental Details

Two sets of data were used in the overall study, both of which were gathered at the Moscamed medfly mass rearing facility. The first set of data were those gathered for experiment 3 described in Carey, Liedo, Orozco, and Vaupel (1992). Data for a total of over 1.2 million medflies was gathered from 167 cages containing an average of about 7,200 flies each. Densities varied from the target density because of variability in the number of pupae technicians placed in a cage and in the proportion of pupae that successfully

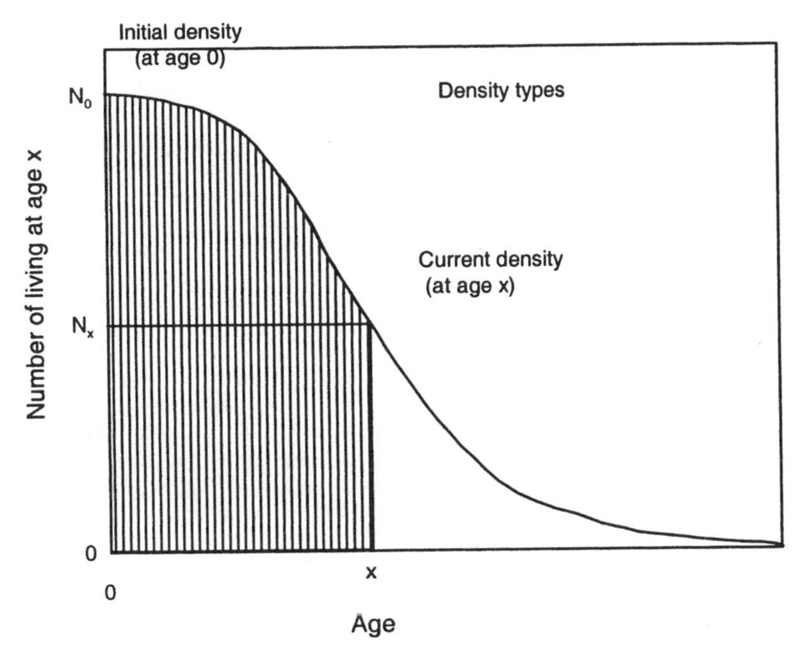

FIGURE 5.1. Survival in a hypothetical cohort showing 3 types of densities: (i) initial density, which is the number of individuals at age 0; (ii) current density, which is the number of individuals alive at age x; and (iii) cumulative density, which is the total number of fly-days from day 0 through age x (Carey, Liedo, and Vaupel 1995b).

emerged and survived to the beginning of the first day. The technicians placed approximately the same volume of pupae in each cage but depending on the size of the pupae, two equal volumes can contain different numbers.

The second set of data was gathered using an experiment explicitly designed to measure density effects on medfly mortality. On the same day and using the same batch of pupae, 1 cage with about 10,000 flies, 2 cages with about 5,000 flies each and 4 cages with about 2,500 flies each were set up. This procedure was repeated 7 times for a total of 49 cages maintained under the same environmental conditions as in experiments 3 described in Carey and coworkers (1992). Mortality data was obtained on a total of 214,735 flies or around 70,000 flies for each of the three treatments. Because no larval food (diet) was available in cohort cages, it was not possible for eggs laid by females to develop into second generation adults and thus contaminate the cohort with younger aged flies.

5.2.3. Age-Density Models

Since density declines with age in life table experiments where dead individuals are not replaced, the observed age-trajectory of medfly mortality could be an artifact of the shifting balance between an age-effect and a density effect. Let $m(x,N)$ be the death rate at age x in a cage with N flies. Then the two propositions underlying the hypothesis are

A. For cages at the same current density, death rates increase with age, i.e.,

$$\partial m(m, N)/\partial x > 0. \tag{5.1}$$

B. For flies at the same age, death rates increase with density, i.e., $\partial m(m, N) / \partial N > 0$. Since density N declines with age, i.e., $dN(x) / dx < 0$, death rates can either increase or decrease with age:

$$\frac{dm(x, N(x))}{dx} = \left.\frac{\partial m(x, N)}{\partial x}\right|_{N=N(x)} + \frac{dN(x)}{dx}\left.\frac{\partial m(x, N)}{\partial N}\right|_{N=N(x)}. \tag{5.2}$$

The first term in this expression can be interpreted as the age-effect and the second term as the density-effect. If at younger ages the age-effect outweighs the density-effect but at older ages the density-effect outweighs the age-effect, mortality will rise and then fall.

To simultaneously control for density and differential mortality, estimations were made for the coefficients (separately for each day) $m_0(x)$, $a(x)$, $b(x)$, and $\delta(x)$ of the model

$$\begin{aligned} \ln m_i(x) = \ln m_0(x) + a(x)\ln N_i(x) + b(x)\ln(H_i(x)/\overline{H}(x)) \\ + I_i \ln\delta(x) + \varepsilon(x), \end{aligned} \tag{5.3}$$

where $m_i(x)$ is the central death rate at age x in cage i, $N_i(x)$ is the number of surviving flies in cage i at the start of day x, $H_i(x)$ is the cumulative hazard, $\overline{H}$ is the cumulative hazard for all cages combined, I_i equals one for the cages in the density experiment and zero for the cages in the original 1.2 million medfly experiment, and $\varepsilon_i(x)$ is the error. The coefficient $m_0(x)$ is the estimated baseline death rate controlling for density and differential mortality. The central death rate is given by

$$m_i(x) = \frac{N_i(x) - N_i(x + 1)}{(N_i(x) + N_i(x + 1))/2}. \tag{5.4}$$

The cumulative hazard can be calculated by $H_i(x) = -\ln(N_i(x) / N_i(0))$. Note that the ratio $H_i(x)/\overline{H}(x)$ can be interpreted as the average relative risk of mortality in cage i up to age x: if death rates in cage i are z times the average for all cages, then $H_i/\overline{H} = z$. The coefficients were estimated using the least-squares criterion in a multiple regression analysis. In some cages at some ages, especially advanced ages, there were no deaths. To avoid values of zero for $m_i(x)$ and to smooth the erratic death rates observed in cages with few survivors, the average death rate in the interval $x \pm k$ for all cages was substituted for $m_i(x)$ with no deaths or with fewer than 10 survivors, using the formula

$$m_i(x) = \frac{N_i(x - k) - N_i(x + k)}{\sum_{j=-k}^{k} (N_i(x + j) + N_i(x + j + 1))/2}. \tag{5.5}$$

The smallest value of k was used such that there was at least one death in the interval and at least 10 fly-days of exposure. In the regression analysis, cages were excluded with no surviving flies on day x. Regression equations were estimated with the additional term $\theta(x)ln(m_i(x - 1)/\hat{m}_i(x - 1))$, added after the initial day to account for serial correlation between the estimation error on successive days: $\hat{m}_i(x - 1)$ is the value estimated by the regression equation for the previous day.

Daily age-specific death rates at specified densities was computed over a wide range of densities from one fly per cage to 5,000 flies per cage for (i) the density experiment and (ii) the density experiment and the original 1.2 million medfly experiment combined. Values were estimated for successive 10-day intervals and were plotted at the midpoint of the relevant interval. Within each interval, death rates for a specified density were based on death rates in the cages that achieved this density on some day in the interval, the death rate used being the rate on the day the density was achieved. For the interval running from the start of day x to the start of day $x + 10$ and for density D, the death rate m was calculated by

$$m = \frac{\sum_{k=0}^{9} I(N_i(x + k))[N_i(x + k) - N_i(x + k + 1)]}{\sum_{k=0}^{9} I(N_i(x + k))[(N_i(x + k) + N_i(x + k + 1))/2]}, \tag{5.6}$$

where $I(N_i(x + k)) = 1$ if $N_i(x + k) \geq D$ and $N_i(x + k + 1) \leq D$ and $I(N_i(x + k)) = 0$ otherwise, where $N_i(x)$ is the number of surviving flies in cage i at the start of day x. Values of m were plotted only if there were at least 50 days of exposure at the specified density and age interval, i.e., the denominator of the formula had to exceed 50.

5.3. Mortality Dynamics

5.3.1. Density Effects on Survival and Life Expectancy

A demographic summary by sex for the three density treatments is presented in table 5.1. Several aspects of this table merit comment. *First*, the 2- to 4-fold differences in number of flies per cage at the beginning of the experiments were reduced to around 1.5 to 2.5-fold differences by 20 days due to

TABLE 5.1
Summary of Results for Male and Female Medflies Maintained in Mixed Cages at Three Initial Densities—2,500, 5,000, and 10,000 Flies Per Cage

	Males			*Females*		
Parameter at Age x	*2,500*	*5,000*	*10,000*	*2,500*	*5,000*	*10,000*
Number per Cage[a]						
x = 0	1279	2568	5189	1253	2577	5068
	(292.9)	(512.7)	(826.3)	(329.0)	(490.1)	(948.2)
20	560	997	1568	409	615	854
	(206.5)	(543.3)	(815.5)	(151.9)	(113.0)	(201.6)
40	18	29	26	26	25	20
	(13.3)	(28.2)	(29.1)	(15.7)	(13.6)	(12.4)
Survival to Age x (l_x)[b]						
x = 20	.4380	.3882	.3022	.3263	.2385	.1685
	(0.110)	(0.152)	(0.123)	(0.112)	(0.080)	(0.061)
40	.0140	.0113	.0050	.0206	.0096	.0039
	(0.009)	(0.009)	(0.005)	(0.014)	(0.006)	(0.003)
Expectation of Life (e_x)[c]						
x = 0	19.2	18.3	16.6	17.5	15.7	14.2
	(1.92)	(2.67)	(2.14)	(2.31)	(1.42)	(1.01)
20	6.6	6.4	5.5	8.2	6.8	6.0
	(0.95)	(1.36)	(0.88)	(1.25)	(1.10)	(0.66)
40	5.9	6.0	5.4	7.7	5.9	5.7
	(2.56)	(1.65)	(1.76)	(2.46)	(1.22)	(1.47)

Note: Values are per cage averages (SD) using 28 cages of approximately 2,500 flies, 14 cages of approximately 5,000 flies, and 7 cages of approximately 10,000 flies (Carey, Liedo, and Vaupel 1995b).
[a]Number by sex; sum of numbers of both sexes gives average cage densities at specified ages.
[b]Fraction of the original cohort surviving to age x.
[c]Number of days remaining to the average individual alive at age x.

the higher mortality in the higher density cages. The relative differences in numbers due to different initial densities diminished with time and were virtually nonexistent by around 4 to 5 weeks. *Second*, the fraction of the original cohort surviving to 20 and 40 days was inversely related to initial number. For example, there was about 5-fold greater survival of females to 40 days in cages starting with 2,500 individuals relative to survival of females to this age in cages starting with 10,000 individuals. *Third*, the sex and density trends were also reflected in life expectancies at different ages. For example, life expectancies at eclosion (day 0) for both males and females were less in higher density cages than in lower density ones. Also, male life expectancy at eclosion was 16% greater in low density cages than in high density cages, whereas female life expectancy at eclosion was 23%

greater in the low versus high density cages. These findings are similar to those of Dingle (1968), who reported that *Oncopeltus* and *Dysdercus* females experienced proportionately higher mortality than males at high versus low densities. However, Rockstein et al. (1981) reported the opposite effect of density on male-female differences in life expectancy in *Musca domestica*. Anderson (1961) examined the effects of density on sex ratio in a wide range of species but reported little that was general. *Fourth*, a male-female mortality crossover was evident at all three densities and is consistent with findings from analysis of male/female data from original life table study on 1.2 million medflies (Carey, Liedo, Orozco, Tatar, and Vaupel 1995a; Carey, Liedo, Orozco, and Vaupel 1992). For example, expectation of life was greater for males than for females at age 0 but less for males than for females at 40 days.

5.3.2. Effects of Initial Density on Mortality

Comparisons of the average mortality ratios among the three crowding experiments are given in table 5.2 for each sex. Relative differences in mortality between high- and medium-density cages were similar for males and females, averaging around 1.20-fold greater in the higher-density cages. However, male mortality averaged only 1.10-fold higher in medium-density cages relative to low-density cages, whereas female mortality averaged 1.19-fold higher. In general, each doubling of initial density increased mortality for the first 40 days by 10 to 20% which, in turn, decreased life expectancy at age 0 by about 5 to 10%.

TABLE 5.2
Average Mortality Ratio (SD) by Sex for 3 Density Experiments from Day 0 through 40

Mortality Ratio	*Males*	*Females*
High-to-Medium	1.20 (0.130)	1.19 (0.131)
Medium-to-Low	1.10 (0.120)	1.19 (0.106)
High-to-Low	1.32 (0.209)	1.42 (0.186)

Note: High, medum and low densities denote 10,000, 5,000 and 2,500 flies/cage, respectively. For example, the average male mortality at ages 0 through 40 days was 1.20-fold greater in the high density cages (10,000 flies/cage) than in the medium density cages (5,000 flies/cage) (Carey et al. 1995b).

The smoothed age-specific mortality schedules for all density treatments for both sexes are presented in figure 5.2, which shows the uniformity of the three sex-specific mortality patterns corresponding to each treatment. In all treatments male mortality increased monotonically to around day 20, abruptly leveled off from days 20 through 40 and decreased thereafter. Female mortality increased to around day 18 at which time it continued to increase at a slower rate for two more weeks. At around day 35 female mortality rates began to decrease for all three density treatments. In short, mortality differences among the three treatments were quantitative and not qualitative; the shapes of all the schedules were similar to the mortality schedules observed in the 1.2 million medfly cohort (Carey 1997; Carey et al. 1992). The leveling off of mortality observed in this study was also observed in *Drosophila* by Curtsinger et al. (1992), Fukui et al. (1993) and Clark and Guadalupe (1995) and in bean beetles by Tatar et al. (1994), and Tatar and Carey (1994a, 1994b). Similar patterns of rising, leveling, and then decreasing death rates were also found in the life table study by Krainacker (Krainacker et al. 1987) based on around 2,000 medflies maintained in cages of 25 to 50 pairs.

5.3.3. *Effects of Initial Density on Mortality Patterns*

The rate of change at each age for the sex-specific mortality curves are presented in figure 5.3. These plots reveal the following relationships between density and mortality: First, the geometric rate of change in mortality rates declines to unity for cohorts from all density treatments and for both sexes. This supports the hypothesis that slowing of mortality at older ages is independent of initial density. Second, the ages at which the rate of change in mortality reaches unity are similar among the three treatments (i.e., the three initial densities)—approximately 27 days for males and 20 days for females. That the convergence age is independent of the initial number suggests that the slowing of mortality at these older ages is *not* conditional on decreases in density. Third, the average number of individuals within a cage differs by several fold (lower inset in figure 5.3) at the respective ages of the convergence of mortality rate of change to 1.0. For example, at 20 days when the rate of change in female mortality converged to 1.0 the densities in the three treatments ranged from 2,500 to 1,000 flies/cage. This implies that no "threshold density" exists such as a critical per-fly spacing requirement (Carpenter 1958; Connolly 1968; Luckinbill and Clare 1986).

The relationship shown in figure 5.3 also serves as a test of the hypothesis that demographic heterogeneity accounts for the leveling off of mortality rates, as was suggested by Kowald and Kirkwood (1993), Hughes and Charlesworth (1994), and Brooks et al. (1994). This hypothesis assumes that the

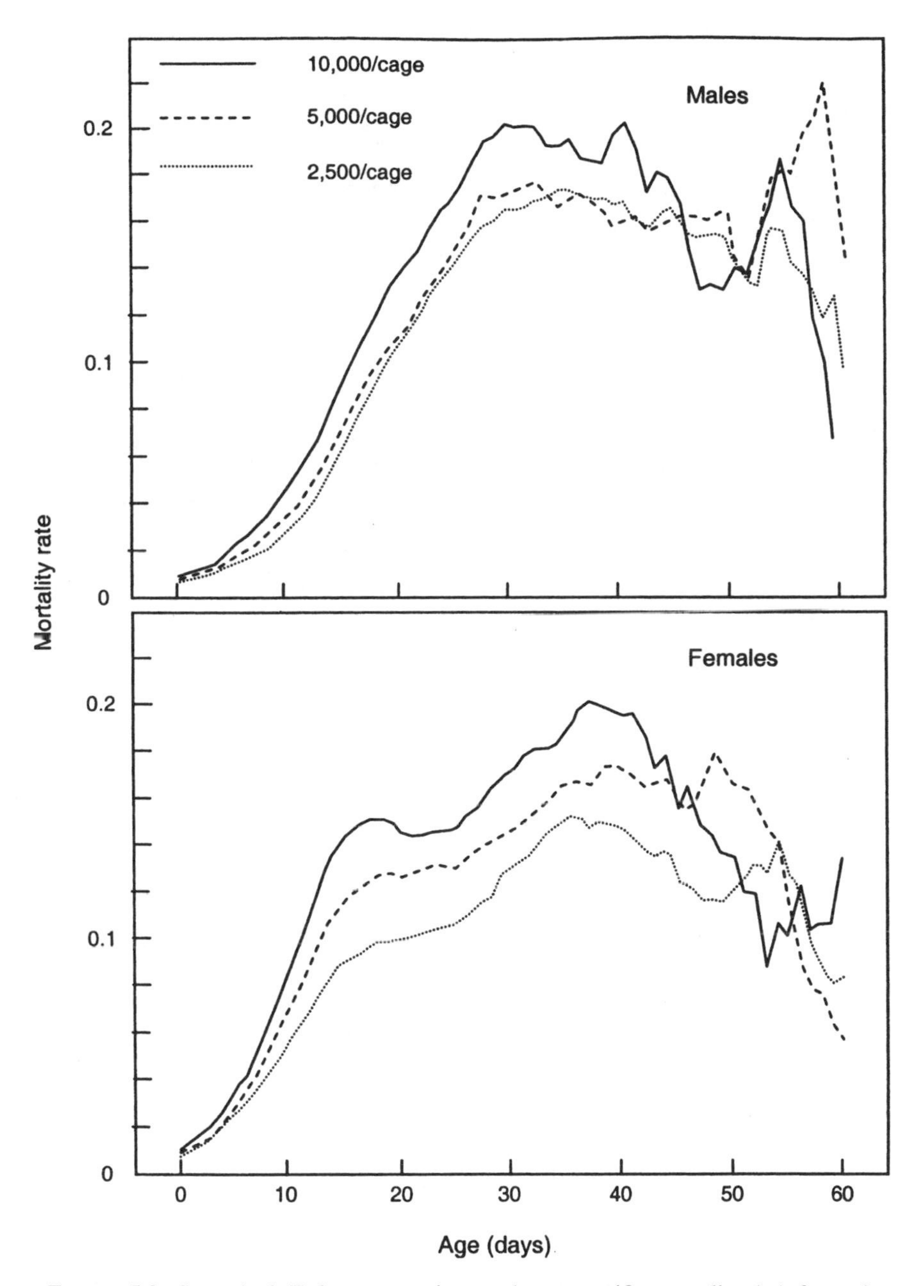

FIGURE 5.2. Smoothed (7-day geometric mean) age-specific mortality (q_x) for male (top) and female (bottom) medflies at 3 different initial densities. Each curve is based on an initial number of approximately 35,000 individuals (each sex). Age-specific mortality is defined as the fraction of flies alive on day x that die in the interval x to $x + 1$ (Carey et al. 1995b).

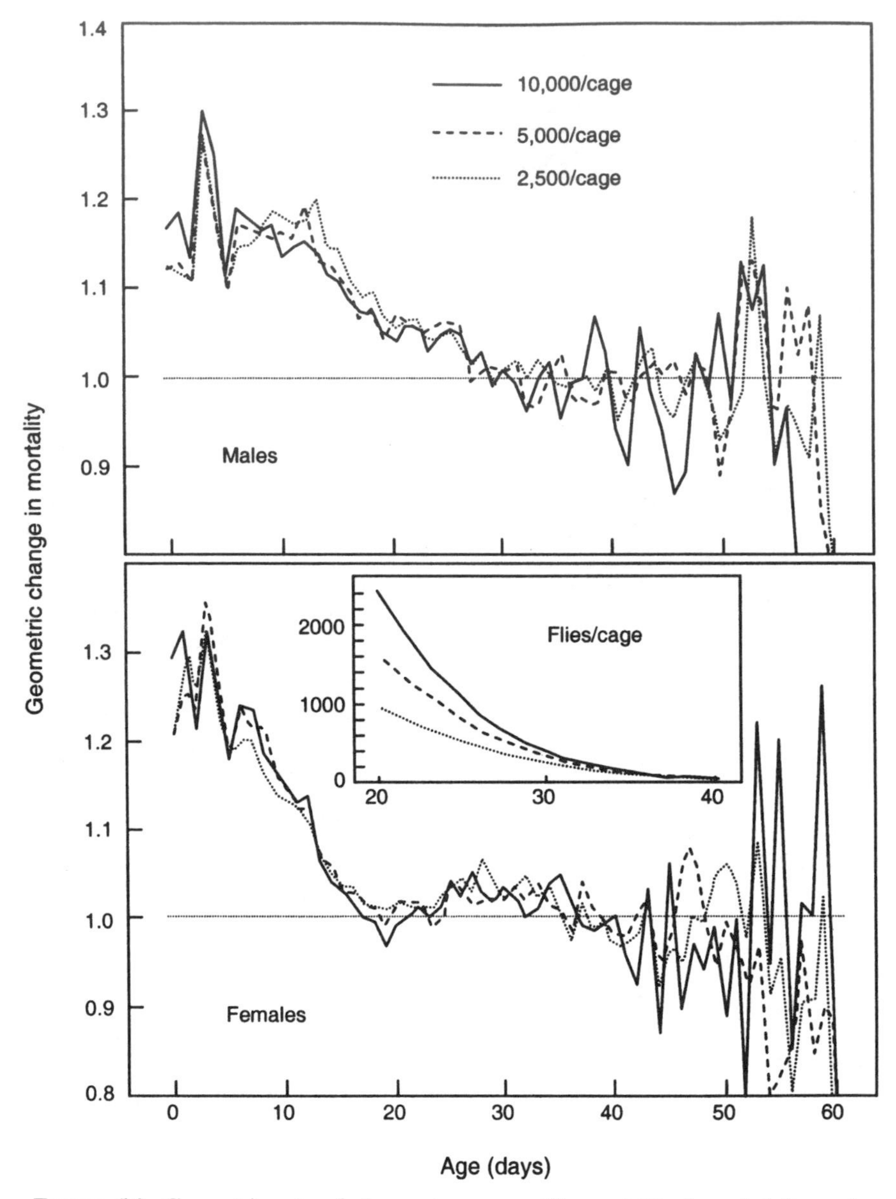

FIGURE 5.3. Geometric rate of change in age-specific mortality (q_{x+1}/q_x) for male (top) and female (bottom) medflies at 3 different initial densities. The inset shows the average total number of individuals in each cage from days 20 through 40 for the three treatments (Carey et al. 1995b).

cohort consists of subcohorts with different levels of frailty but that all subcohorts exhibit Gompertz mortality rates. Therefore, it is argued, as the cohort ages it becomes more selected because the subgroups with higher death rates die out leaving the more robust subgroups with lower death rates (Kowald and Kirkwood 1993; Vaupel and Carey 1993; Vaupel et al. 1979). Thus it is suggested that the differential mortality among subgroups that exhibit Gompertz mortality patterns creates the non-Gompertzian pattern of leveling off in the whole cohort. A test of this hypothesis became possible when it was observed that changes in initial densities scaled mortality uniformly across all ages. The concept for testing the hypothesis is this: if mortality is increased uniformly across all subgroups then, if heterogeneity accounts for the departure of the mortality pattern from the Gompertz (exponential), leveling off should occur at younger ages when total mortality is high rather than when total mortality is low. This shift in timing would occur because survivorship decreases more rapidly at high levels of mortality than at low ones and thus individuals in the most robust subgroups should be the predominant mortality type at an earlier age. However, this was not the case with the medflies as is evident in figure 5.2 and figure 5.3; the timing of the deceleration was independent of the level of mortality that resulted from different initial densities. Therefore it can be concluded that heterogeneity does not explain the leveling off of mortality in the medfly cohorts. This finding does not rule out the existence of demographic selection. Rather it suggests that the effects of changes at the level of the individual (e.g., reproductive physiology) supercede the effects of demographic heterogeneity and selection; in short, the leveling off of mortality is not an artifact of changes in cohort composition.

5.3.4. Correlation of Density and Mortality

There was a high correlation between initial density and cumulative density (figure 5.4), which suggests that initial density can be used as a proxy for cumulative density. The correlation coefficients for both cumulative density and current density versus mortality at each age for days 0 through 40 are plotted in figure 5.5. This figure shows (1) a moderately high correlation for females between mortality and density (both cumulative and current) for 0 through 14 days but a low correlation between mortality and the two density measures for males at all ages; and (2) that the correlation coefficients for mortality and current density are virtually identical with those for cumulative density at young ages (<10 days) but diverge at older ages. This overall pattern is due to the direct age dependence of the two types of densities.

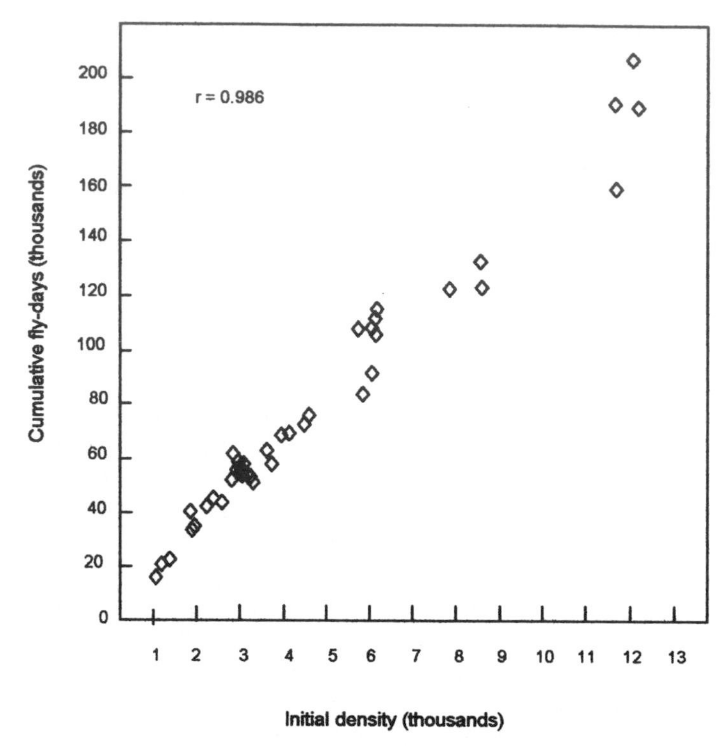

FIGURE 5.4. Relationship between initial density and cumulative density (fly-days summed through day 40) for medfly density experiments (Carey et al. 1995b).

5.3.5. Density Effects on Sex Survival Ratios

Two sets of age-specific survival ratios for the two sexes were examined in the density studies. The first was the ratio of the survival schedule computed for each sex in the lowest-density experiment (2,500 flies/cage) to the survival schedule of the respective sex in the highest-density experiment (10,000 flies/cage), the results of which are given in figure 5.6. These results show that the long-term effects of density on female survival are much greater than the effects on males and continue to increase for nearly 80 days. By 40 days the lower-density cages contain an average of 6-fold more of their initial number of female flies than the higher-density cages. In contrast, at this same time the lower-density cages contain an average of 3-fold more of their initial number of male flies than the higher-density cages. This reinforces the findings reported earlier that high initial density has a greater effect on female mortality than on male mortality.

The second set of survival ratios examined were the male:female survival

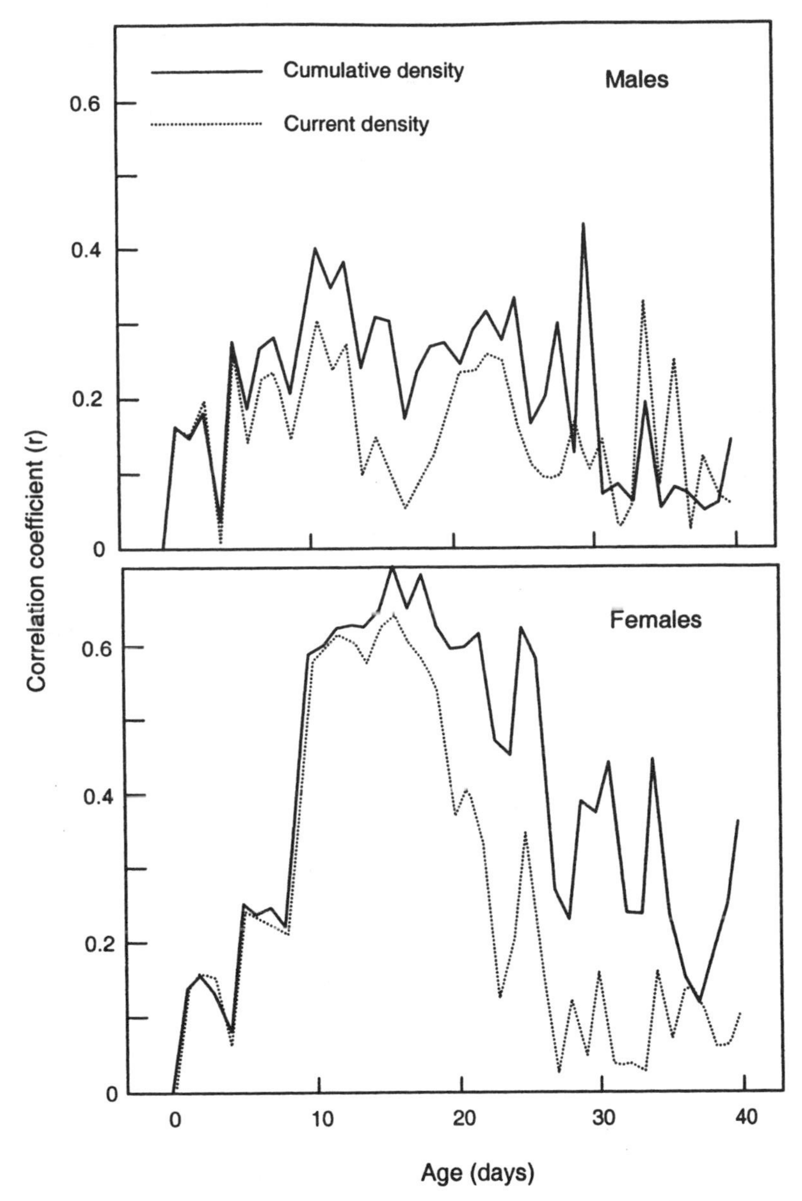

FIGURE 5.5. Correlation coefficients for cumulative and current density vs. mortality at each age for male (top) and female (bottom) medflies. Results are based on 7 cages with an average density of about 10,000 medflies, 14 cages with an average density of about 5,000 medflies, and 28 cages with an average density of about 2,500 medflies (Carey et al. 1995b).

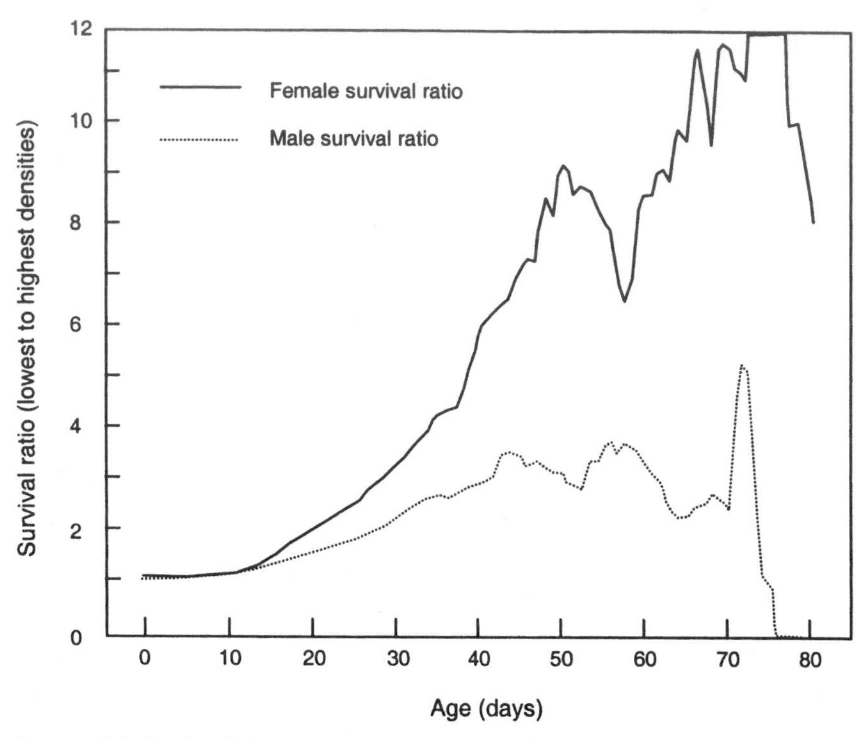

FIGURE 5.6. Ratio of the sex-specific survival schedules for the low- (2,500 flies/cage) vs. the high- (10,000 flies/cage) density experiments. For example, there were 9-fold more female survivors in the low-density trials than in the high-density trials at day 50, whereas there were 3-fold more male survivors in the low-versus high-density trials on day 40 (Carey et al. 1995b).

ratios for each of the three densities, the results of which are shown in figure 5.7. This figure shows two important patterns. First, the increasing male: female survival ratio to 20 days followed by the decreasing survival ratio for at least the next 10 days in all cohorts reveals the male:female mortality crossover. The relative abundance of male medflies could not increase and decrease without a mortality crossover—male mortality lower at young ages and female mortality lower at older ages. Second, the higher-density cages amplify the relative differences favoring males. Because female die off at a much faster rate than do males at the young ages, the relative advantage of females after the mortality crossover does not offset the sex bias created by the differentials during the first three weeks. Therefore the survival schedules do not cross over until nearly 60 days. In short, the quantitative differences in sex-specific mortality accounted for the qualitative difference in sex bias at the older ages; for the medfly, sex bias at older ages is partly attributable to the effects of crowding.

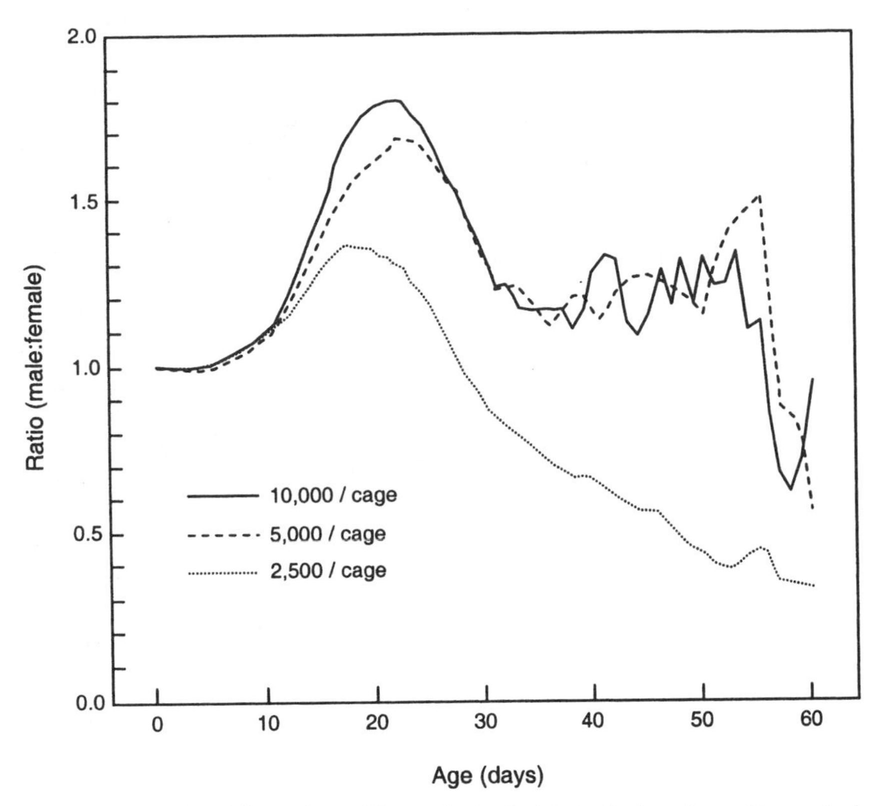

FIGURE 5.7. Ratio of the male medfly survival schedule to the female medfly survival schedule for each of the 3 initial cage densities (Carey et al. 1995b).

5.3.6. Equivalent Current Densities

The underlying concept for the analysis of "equivalent current densities" is that each cohort will progressively decline through all numerical levels ranging from the initial number to zero (extinction) as its members die off. Because of differences in starting numbers and of mortality among cages, there will exist variation among cages in the day at which a particular numerical level is attained. A central death rate can be computed for each of the 216 cages (cohorts) at specified numerical levels (e.g., 5000, 4000, 3000, 2000, 1000, and so forth), which will, in turn, be distributed over a range of age classes. This will yield a series of mortality schedules at "equivalent current densities," as shown in figures 5.8A and 5.9B. With density held constant, death rates tend to rise at younger ages, stay approximately level at middle ages, and fall at older ages. For example, the central death rate in cages with 1 to 5 flies at 50 days was 0.3 or greater, whereas the central death rate in

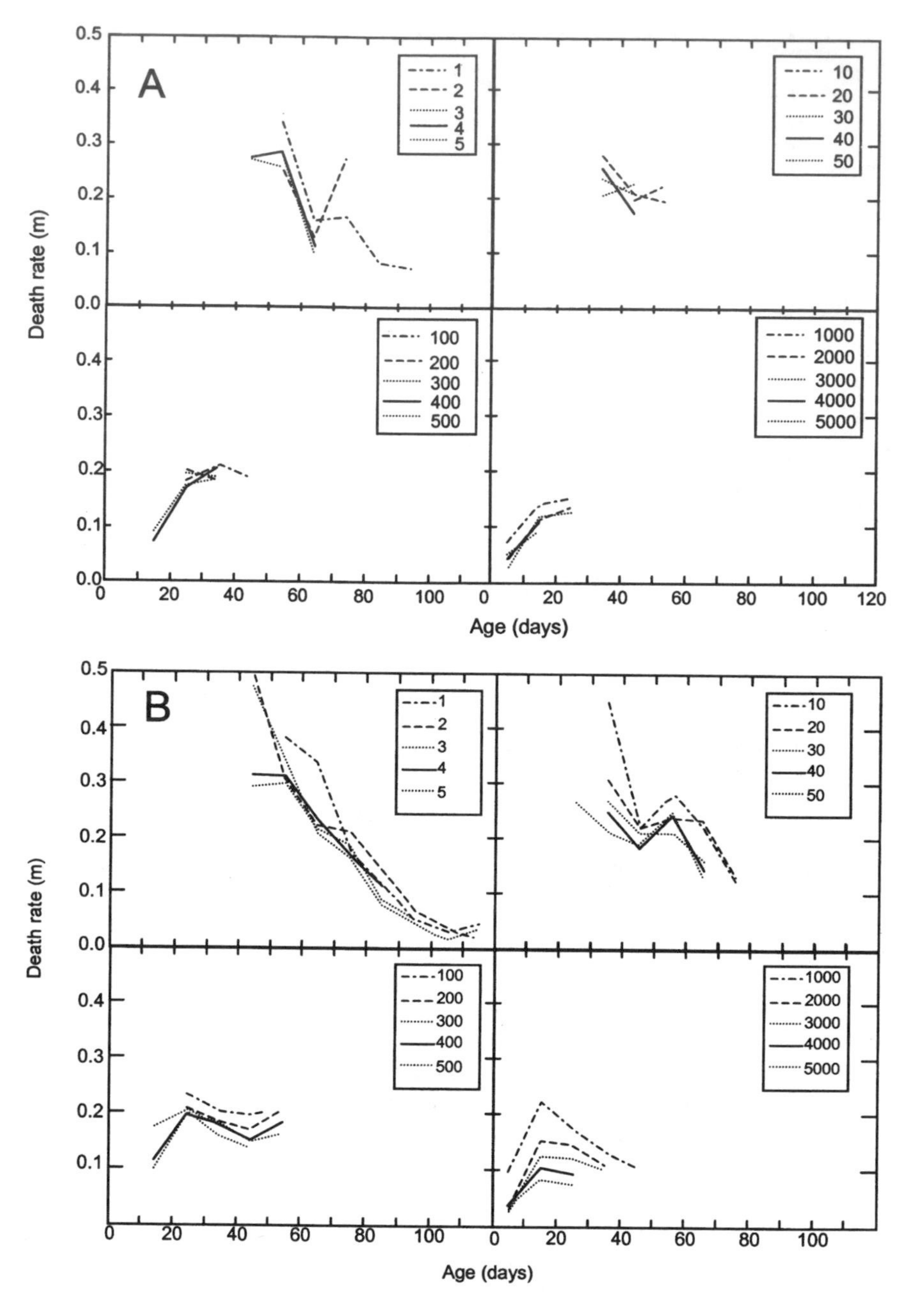

FIGURE 5.8A, B. Daily age-specific death rates at specified densities from one fly per cage to 5,000 flies per cage for (A) the density experiment and (B) the density experiment and original 1.2 million medfly experiment combined. Along the lines plotted—i.e., holding density constant—death rates tend to increase at younger ages, roughly level off at middle ages, and decline at older ages. Values were estimated for successive 10-day intervals and are plotted at the midpoint of the relevant interval (Carey et al. 1995b).

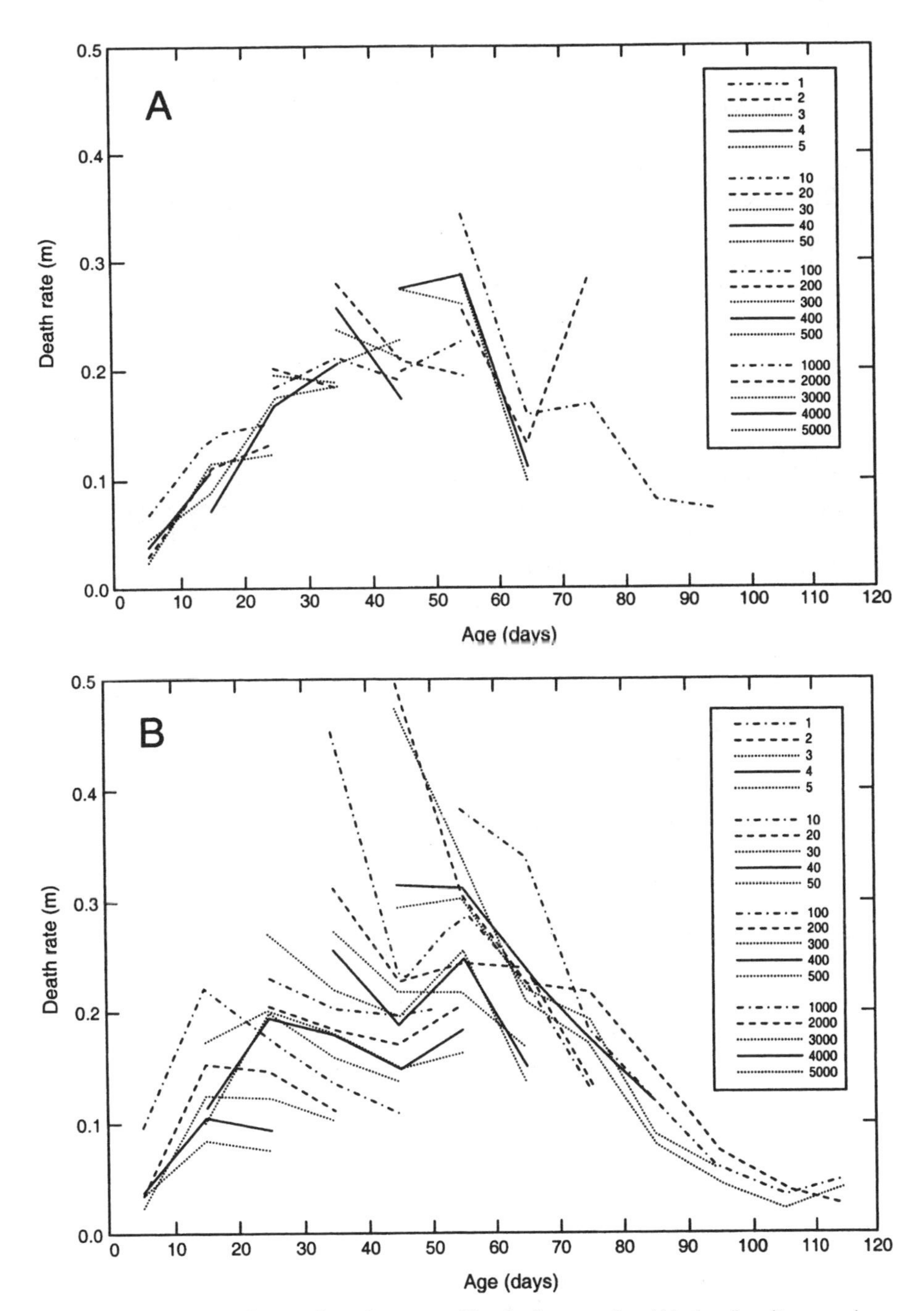

FIGURE 5.9A, B. Composite of age-specific death rates for (A) the density experiment and (B) the density experiment and original 1.2 million medfly experiment combined (Carey et al. 1995b).

cages with 1 to 5 flies 30 to 60 days later was 0.1 or less. Thus cages with equivalent current densities exhibited different death rates at different ages. At most ages, death rates are roughly the same at different densities, with a tendency for death rates to be lower at higher densities.

The data underlying figures 5.8A, B and 5.9A, B permit 151 paired comparisons of cohorts of the same age but different current densities. For example, in the age interval from 10 through 19 days, death rates were available for seven densities—400, 500, 1,000, 2,000, 3,000, 4,000, and 5,000 flies per cage. This permitted paired comparisons of whether mortality rates were higher or lower at the higher densities. For all age intervals combined, a total of 151 paired comparisons could be made. In 74% of the cages, death rates were lower in the cage with the higher density, a highly significant results ($P < 10^{-8}$). The data from the 49 cages in the density experiment were combined with data from the 167 cages in the original 1.2 million medfly experiment (table 5.3). The results are similarly inconsistent with proposition A (for cages at the same current density, death rates increase with age) and with proposition B (for flies at the same age, death rates increase with density).

The unexpected negative density-effect observed in the experiments may be explained as due to two possibilities. The first possibility concerns cohort heterogeneity. Some cages may have contained disproportionate number of robust flies; environmental conditions in some cages may have been especially salubrious. "Good" cages would tend to experience relatively low mortality and thus reach a particular density at a later age than bad cages. Consistent with this hypothesis, mortality in the 10 days following day x is positively correlated with mortality up to day x, at younger ages. However, at the middle and older ages when the decline in death rates occurs, the correlation is weak and insignificant (table 5.4).

The second possibility is that there exists a complex relationship in cohort numbers (current densities) and the effects of the autocorrelation between earlier and later mortality. It may be that high initial density raised mortality so much at young ages that the high initial-density cases turn into low current-density cases at later ages. In other words, because of the autocorrelation between earlier and later mortality, current density starts becoming inversely related to mortality.

5.3.7. Regression Model

The coefficient m_0—the estimated baseline death rate controlling for density and differential mortality—rises from day 5 until about day 35, then is roughly level until about day 45, and falls thereafter. The estimated value of a, the density coefficient, hovers around zero and is insignificant ($P > .05$) at nearly all ages. The value of b, the coefficient of differential mortality, is

TABLE 5.3

Summary Statistics for Analysis of Medfly Cage Density Effects on Mortality

x	Density Experiment						Original Experiment					
	Cages	*Min*	*LQ*	*Med*	*UQ*	*Max*	*Cages*	*Min*	*LQ*	*Med*	*UQ*	*Max*
1	49	1054	2799	3157	5991	12,029	167	1770	6437	7518	8208	9708
10	49	692	2407	2605	4917	9755	167	800	5799	6924	7636	9357
20	49	350	930	1182	1517	3853	167	205	2430	3229	4498	7111
30	49	66	205	270	340	850	167	36	443	773	1541	3266
40	49	10	27	45	57	157	165	3	81	170	427	1265
50	49	1	5	8	13	30	162	1	17	35	92	402
60	42	1	1	2	3	14	152	1	4	7	16	99
70	30	1	1	1	2	6	111	1	1	3	5	34
80	13	1	1	1	1	2	77	1	1	3	3	15
90	7	1	1	1	1	1	54	1	1	1	2	8
100	3	1	1	1	1	1	41	1	1	1	2	5

Note: Number of nonextinct cages and minimum, lower quartile (LQ), median, upper quartile (UQ), and maximum number of surviving flies at the start of day x in nonextinct cages in the density experiment ($n = 210{,}000$) and the original 1.2 million medfly experiment (Carey, Liedo, and Vaupel 1995b).

TABLE 5.4
Correlation between the Number of Deaths before and after Age x for 2 Medfly Experiments Alone and Combined

Age x	*Density Experiment*	*Original Experiment*	*Combined Experiments*
10	**.40** ± .12	**.52** ± .06	**.52** ± .05
20	**.80** ± .05	**.76** ± .03	**.77** ± .03
30	**.40** ± .12	**.69** ± .04	**.70** ± .03
40	**.48** ± .11	**.28** ± .07	**.28** ± .06
50	.08 ± .14	−**.17** ± .08	−.13 ± .07
60	−.01 ± .15	−.07 ± .08	−.11 ± .07
70	−.24 ± .17	−.11 ± .09	−.13 ± .08
80	−.16 ± .27	.07 ± .11	.04 ± .11
90	**	−.03 ± .14	.01 ± .13

Note: Correlation between proportion dying before start of day x and proportion of those surviving that die in the subsequent 10-day period, for the density experiment, the original 1.2 million medfly experiment, and both experiments combined. Standard errors are given after the ± symbol. Correlations that are significant (P < .05) are in bold face (Carey et al. 1995b).
**fewer than 10 cages had any survivors.

consistently positive, averaging roughly 0.6 up to day 70, and is significant (P < .05) at most ages up to day 40: cages with high past-mortality tended to have high current-mortality. The value of δ, which measures excess mortality in the density experiment relative to the original 1.2 million medfly experiment, is positive and significant (P < .05) during the first week, hovering around a value of 0.4, but is close to zero and insignificant (P > .05) thereafter. The value of 0.4 implies that death rates during the first week tended to be about 40% higher in the density experiment than in the original experiment. Mortality differentials of this magnitude, which are not unusual in replications of insect life table experiments, are probably attributable to subtle differences in pupal quality and in such environmental factors as temperature, humidity, and food. For the model containing the additional term $\theta(x)\ln(m_i(x-1)/\hat{m}_i(x-1))$, estimated value of $m_0(x)$, $a(x)$, $b(x)$ and $\delta(x)$ were very close to the values estimated in the simpler model. The coefficient $\theta(x)$ was consistently and significantly positive, varying around a value of about 0.4 up to about day 55 and a value of about 0.8 thereafter. Since persistent cage-specific factors affecting mortality are captured by $b(x)$, the positive values of $\theta(x)$ suggest that there are shorter-term factors that elevate mortality on adjacent days. In addition to doing the analysis for single days, regression coefficients for 10-day periods were estimated, using the formula for $m_i(x)$ for *Eqn*(4) with $k = 5$. The results are similar to those reported above. The results of the test of the hypothesis that the coefficients $m_0(x)$, $a(x)$, $b(x)$, and $\delta(x)$ were constant after day 10 was that only the hypothesis for $m_0(x)$ (P < .05) could be rejected. Estimated constant values for a, b, and δ are $-.03$, .46, and $-.04$; only the value of b is significant (P < .05).

5.4. Implications

5.4.1. Medfly Mortality Patterns at Older Ages

One of the most significant results of this study is that the leveling off and decline in medfly mortality at older ages cannot be explained as a simple artifact of decreasing current density. The decline in medfly death rates occurred when densities were very low—from day 60 to day 100, fewer than 10 flies were typically alive in cages initially holding thousands of flies. Consequently other factors associated with heterogeneity among flies or cages are likely to be more important than declines in density in explaining the decline in mortality. Furthermore even controlling for heterogeneity, age-specific mortality may decelerate at older ages. As Kowald and Kirkwood (1993) note, "old flies lead quieter lives." There may not be a simple link between activity levels and mortality: humans, for instance, tend to slow down with age but death rates continue to rise. Finch (1990) reviews a variety of developmental and post-maturational influences on senescence.

A second important result of this study is that changes in the level of medfly crowding has a quantitative but not a qualitative effect on the age-specific mortality schedule. Virtually all aspects of the overall mortality patterns as well as relative differences in mortality and longevity between males and females were independent of initial density including: (i) age at which the leveling off of mortality occurred; (ii) geometric rates of change in mortality with age; (iii) relative differences in male and female longevity and mortality rates including mortality crossovers; and (iv) mortality decline at advanced ages. These results suggest that increased crowding amplifies intrinsic age- and sex-specific patterns of vulnerability that are modulated by changes in age patterns of individuals' reproductive biology. Crowding affects the rate of dying in a cohort but not the rate of aging as measured by changes in the slope of mortality rates, inflection points, direction, and male-female differentials.

These main results have three important implications. The first implication is that it is probably impossible, even in theory, to eliminate crowding effects from an experiment. No evidence was found of an "optimal" density (Pearl et al. 1927), although the range of densities was limited to only three. Even if an optimal density for medflies did exist and was known, the problem of interpreting density effects on mortality and longevity would still be present. Density should be viewed simply as an environmental continuum like temperature, which can be adjusted and standardized but never eliminated. The fact of being in a captive, enclosed environment itself creates a density effect that is impossible to remove because it is an integral component of the controlled experiment. In general, understanding specific density effects may be less important for questions involving relative differences in life expectancy between two subgroups or treatments than for questions involving absolute differences.

The second implication is that life expectancy differences cannot be used as a proxy for changes in the mortality dynamics (Carey, Liedo, Orozco, and Vaupel 1992). The current analysis demonstrates that life expectancy differences caused by variation in densities reveal little about the nature of the mortality differences. Expectation of life or survival curves are summary measures and shed little light on deeper demographic and biological differences.

The third implication regards selection experiments. If the mortality differences observed between medfly cohorts reared at different densities are similar to the underlying mortality differences in *Drosophila* density selection experiments (Graves 1993; Mueller et al. 1993), then selection may be acting on traits that affect the mortality level and not on those that affect the mortality pattern. Understanding how selection acts on mortality in density experiments—whether on age-dependent or age-independent components—will provide more meaningful information than will knowledge of the consequences of mortality changes as reflected in summary measures of life expectancy and survival.

In general, the results show that the arguments by Graves (1993) and Mueller et al. (1993) that density-effects account for the leveling off of mortality at older ages in medflies are unfounded. Curtsinger (1995) also argued that the Graves and Mueller density experiments on *Drosophila* were unpersuasive because (i) the biology of the medfly is quite different than the biology of *Drosophila* and direct extrapolations from one species to another is questionable at best; (ii) densities studied by Graves and Mueller were up to 120 times higher than in the medfly experiments (*Drosophila* at high densities were packed like a "swarm of bees" and thus it is not surprising that mortality was high at high densities); (iii) Graves and Mueller ignored the leveling off of medfly mortality at older ages in cohorts of individuals maintained in solitary confinement; and (iv) the Graves and Mueller *Drosophila* density experiments were quite small; two lines were studied at two densities with an average of about 275 flies per treatment. In contrast, the medfly studies consisted of 1.2 million individuals from the original study of caged medflies and 210,000 individuals maintained at 1-of-3 densities.

The density studies suggest that the overriding consideration in an experiment where controlling for density effects is important is first to ensure the consistency of initial densities among treatments and replicates. The biological effects of initial density may supercede the effects of the other types of density because the effects of the initial density occur earlier than the others.

Young, maturing flies are likely to be more vulnerable to the effects of crowding than older ones. Indeed, Pearl et al. (1927) found that the most marked effect of density of population was produced early in life. This "early, long-term impact" concept of the effects of initial density may be the biological reason that higher initial densities simply increased medfly mortality uniformly over all ages rather than altered the overall pattern and that the biological effects of both current and cumulative density on mortality are of secondary importance.

5.4.2. Density in Human Context

Density effects are virtually never considered in the context of human aging as is evident by the absence of this topic in almost all books on aging, senescence, and life span, including those by Comfort (1979), Lamb (1977), Finch (1990), and Gavrilov and Gavrilova (1991). This lack of consideration of density effects in gerontology is surprising since it has long been known that crowding has a profound influence on factors that affect mortality rates including heightened incidence of social pathologies (Calhoun 1962) and social subordination (Christian 1970) in animal populations, and increases in infection rates (Galle et al. 1972), stress (Milgram 1970), and violence (Hawley 1972) in human populations. Indeed, the foundation of the seminal essay by Malthus (1799) on human populations concerns the ultimate effects of density on birth, death, and population homeostasis. Perhaps the most important direct connection of these findings to humans is that the results on medfly mortality are consistent with a number of studies on mortality rates in high-stress human populations showing that dire stress over a long period does not influence the *rate* of senescence during that time (Finch 1990). The study also reported that the high stress caused by heightened cage densities did not influence the rate of change in mortality with age. Generally speaking, manipulations of density in experimental systems may eventually be viewed as one of the cleanest and most quantifiable methods available to gerontologists interested in studying the effects of stress on mortality and longevity.

5.5. Summary

There is no question that the large-scale studies on density-effects described in this chapter would not have been conducted if density effects were not considered a possible explanation for the large-scale life table study on slowing of mortality at older ages. The main reason that so few studies on density have been conducted is that the results have been considered primarily in a technical rather than a basic context. However, the medfly studies demonstrated that the results need not be only of technical but also of fundamental value. Specifically the results of the medfly density studies contributed the following to the biodemography literature: (1) rejection of the hypothesis that the slowing of mortality at older ages is an artifact of density reduction; (2) framing the cage density concept as consisting of three types of densities—initial, current, and cumulative; (3) testing the hypothesis that the leveling off is due to demographic heterogeneity; (4) revealing that the sex-specific response to density differs; (5) discovery that the surge in mortality at young ages that was observed in the 1.2 million medfly study was not a statistical artifact (led to experiment by Müller et al. 1997b); and (6) development of new analytical techniques.

6

Dietary Effects

THE MAJORITY of research on the effects of diet on aging and longevity falls into three general categories: (1) *caloric restriction*—experiments (usually only on adults) that are concerned with the modes and mechanisms of longevity extension due to reductions in caloric intake (Masoro 1988; Masoro and Austad 1996; Weindruch 1996); (2) *nutrition*—experiments concerned with the effects on overall health and longevity of different food types such as the influence of different micronutrients, vitamins, proteins, and carbohydrates (Casper 1995; Chippindale et al. 1993); and (3) *starvation*—studies conducted in the context of stress resistance assays that are designed to measure the ability of individuals to survive in the absence of food (Chippindale et al. 1996; Service et al. 1985). In this chapter I describe the results of studies on the medfly that do not fit neatly into any of the three general categories but involve components and concepts drawn from all three.

6.1. Early Mortality Surge in Protein-deprived Females

Life expectancy or average life span of individuals is an important quantitative characteristic for comparing the survival and thus frailty of cohorts subjected to various genetic and/or environmental influences. Genetic factors play an important role in determining life expectancy, as exemplified by the well-known sex differential in life expectancy (Carey, Liedo, Orozco, Tatar, and Vaupel 1995a; Finch 1990; Hazzard 1986; Hazzard 1990; Hazzard and Applebaum-Bowden 1990; Tatar and Carey 1994a). Although life expectancy is an overall measure of longevity, the specific dynamics of mortality and the aging process are reflected in the trajectory of age-specific mortality.

In this section I report the results of an experiment concerning the survival of cohorts of male and female medflies under a full diet (sugar plus yeast) and under protein deprivation (sugar only). The motive for this study was an interest in determining why hazard rates in female medflies began to level off at an earlier age and a lower level than hazard rates in male medflies. These mortality patterns were observed in the 1.2 million medfly cohort studies, all of which were protein deprived. The objective of the study was to see whether this phenomenon was due to protein deprivation and whether protein deprivation has a sex-dependent effect on life expectancy and hazard rates.

The medflies were kept on one of two dietary regimes: (i) a sugar-plus-protein diet in which flies were provided with sucrose and a protein source (yeast hydrolysate); and (ii) a protein-free diet in which flies were provided with a diet consisting only of sucrose (protein-deprived). A total of 416,289 medflies (roughly equal males and females) were maintained in group cages (i.e., 33 cages for protein-free and 33 cages with full diet).

6.1.1. Reversal of Life-expectancy Sex Differential

The first finding from the study by Müller and his coworkers (1997b) was that life expectancy is reduced in both male and female medflies if they are deprived of protein. This reduction is much greater in females (26.7%) than in males (5.9%), and this male-female asymmetry in the amount of reduction leads to a reversal of the female advantage in life expectancy. Whereas under the full diet, female life expectancy exceeds male life expectancy by 1.30 ($\pm$ 0.27 days) $\bar{x} \pm$ SEM) SEM $= s / \sqrt{n}$, under protein deprivation, female life expectancy is shorter than male life expectnacy by 2.24 $\pm$ 0.18 days (table 6.1). This finding was statistically significant ($P < 10^{-3}$).

This appears to be the first study to establish a reversal of the male-female life expectancy differential under dietary manipulation. This reversal demonstrates that this differential is subject to influences exerted by external factors and may be reversed under environmental change. Dietary effects on longevity have been studied by many authors (Chippindale et al. 1993; Graves 1993; Jacome et al. 1995).

6.1.2. Hazard Rates and Early Surge in Mortality

To assess the dynamics underlying the reversal of the sex differential in life expectancy, Müller and coworkers (1997b) obtained hazard rate estimates for the following four groups of medflies (figure 6.1): (i) females, protein deprived; (ii) females, full diet; (iii) males, protein deprived; and (iv) males, full diet. The second major finding in their study is that female mortality is strongly influenced by diet. The female protein-deprived group shows a strong peak of size, 0.17, at day 12 in the hazard rate, and this peak in the hazard rate is absent in the female full-diet group. The peak in hazard rates for protein-deprived females corresponds to a surge in early mortality, which emerges as the cause of the reversal of the sex differential in life expectancies under protein deprivation. After day 25 the hazard rates of the two female groups cross.

In contrast, protein deprivation for male medflies leads to a uniform slight increase in the hazard rate with only minor shape changes. Until about day

TABLE 6.1
Life Expectancies at Eclosion in Days for Male and Female Medflies

Diet	*No. of Cohorts*	*Life Expectancy Males*	*Life Expectancy Females*	*Difference Female-Male*
Sugar plus protein	33	15.2 ± 2.4	16.5 ± 2.5	1.30 ± .81
Sugar-only (protein-deprived)	33	14.3 ± 2.1	12.1 ± 1.8	−2.24 ± .28
Difference sugar-plus-protein Minus sugar only	33	0.84 ± .56	4.4 ± .53	

Note: Reported are samples means ± SEM. Medflies are maintained on either protein-deprived or sugar-plus-protein diets. Total number of flies is 416, 289. The change in the difference of female minus male life expectancies from the sugar-plus-protein diet to the protein-deprived diet demonstrates the reversal of the life expectancy sex differential ($P < 10^{-3}$). Life expectancies for males do not differ significantly between sugar-plus-protein and sugar-only groups ($P = 0.14$), whereas for females they differ significantly ($P < 10^{-3}$). (Müller et al. 1997b).

11, the hazard rate of the female full-diet group stays close to the hazard rates of the two male groups. But it is considerably lower at days 20–30, which helps to explain the higher female life expectancy for the full-diet groups. After day 30, the hazard rate of full-diet females joins and perhaps exceeds that of males. Early on, the female protein-deprived group has by far the highest hazard rate of all groups until day 15, when its hazard rate crosses the hazard rates of the other groups and finally flattens out at older ages, being substantially below that of all other groups. Estimated hazard rates for all 132 cohorts of medflies studied in this experiment are presented in figure 6.2 for each of the four groups. The early surge of mortality with a strongly expressed peak in the hazard rate is found to occur in all 33 female protein-deprived cohorts without exception, providing strong evidence for the consistency of this feature.

Peaks in mortality curves with later timing are found for the cohorts belonging to the other three groups as well. Average timing, peak size, and spread of these mortality peaks are similar for the two male groups, which have late peaks and therefore relatively high peak levels of hazard rate with substantial spread in timing and size. The female groups have much earlier and less variable peaks, with the female protein-deprived group having the earliest peak locations with relatively large sizes. The values for mean peak locations ± SEM were as follows: for the female full-diet group, 16.73 ± 0.31/0.1253 ± 0.0094; for the female protein-deprived group, 11.52 ± 0.31/0.1663 ± 0.0086; for the male full-diet group, 21.91 ± 0.86/0.2092

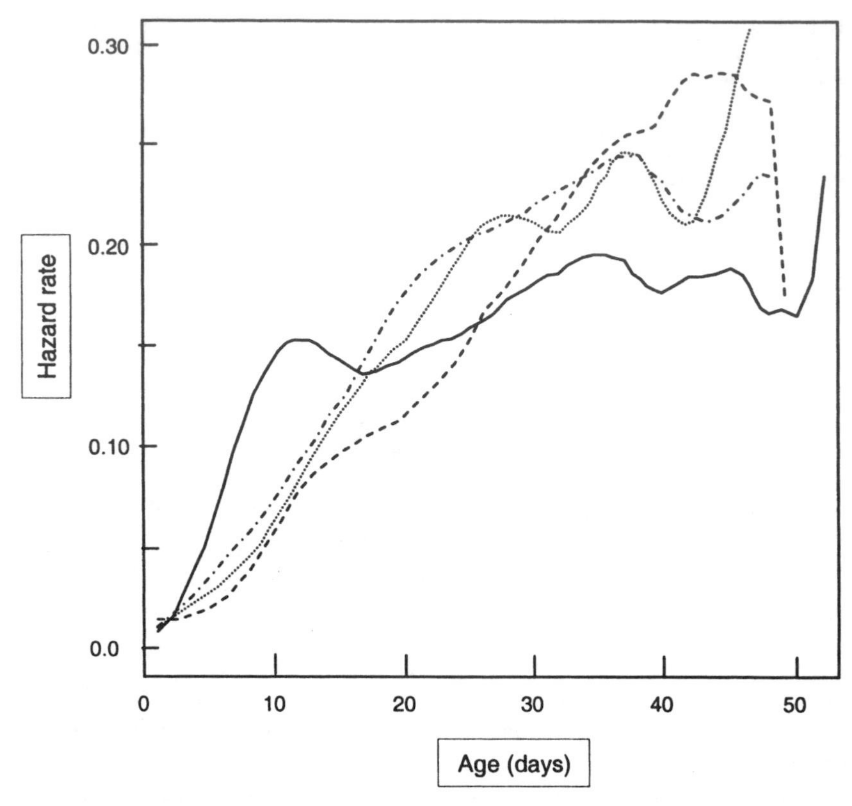

FIGURE 6.1. Estimated hazard rates for 4 groups of medflies: females, protein deprived (solid); females, sugar-plus-protein diet (dashed); males, protein deprived (dash-dot); males, sugar-plus-protein diet (dotted). Each curve is based on an initial number of more than 100,000 individuals (Müller et al. 1997b).

$\pm$ 0.0114; for the male protein-deprived group, 21.18 $\pm$ 0.71/0.2017 $\pm$ 0.0089. The differences between the full-diet and protein-deprived groups were significant for females both for peak location and amplitude ($P < 0.01$) and not significant for males ($P > 0.05$).

6.1.3. Vulnerable Periods

The existence of vulnerable periods during which the hazard rates surge, with associated higher risk and decreased chance of survival, can be inferred. The major example is the early surge in mortality for protein-deprived females. It is remarkable to what extent this surge in mortality is confined to a strictly limited time period, as is seen in figure 6.2A. Other,

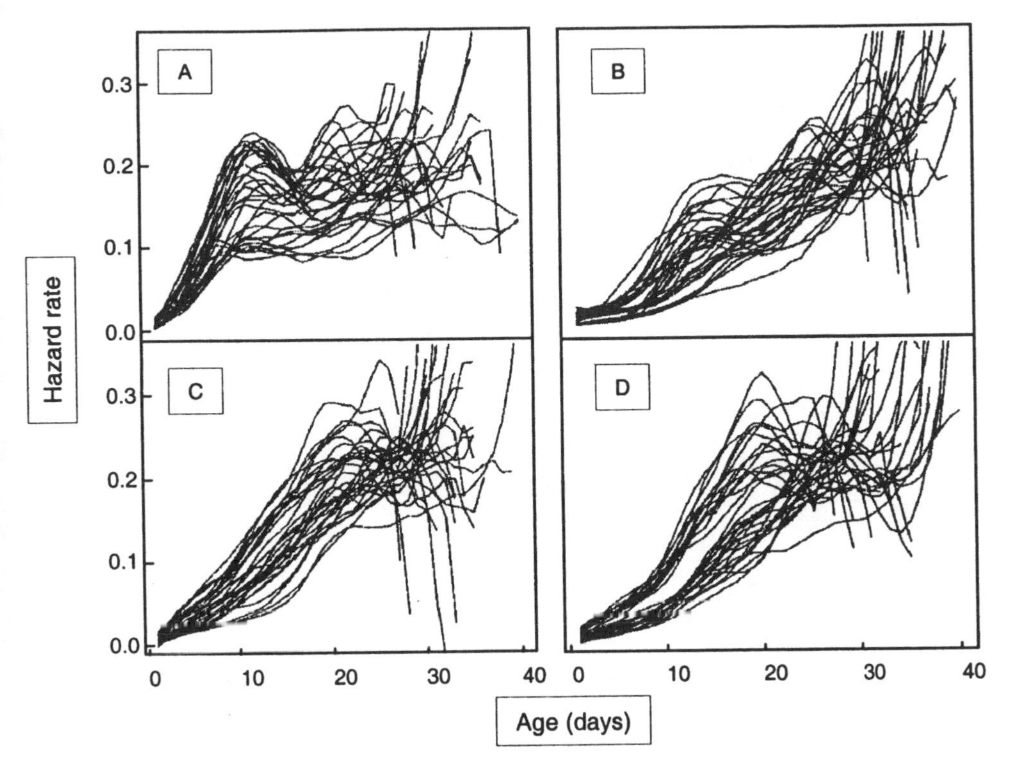

FIGURE 6.2. Estimated hazard rates obtained separately for each of 132 cohorts of medflies. Each cohort contains approximately 3,000 medflies. Each panel consists of 33 estimated hazard functions for the female protein-deprived group (A), the female sugar-plus-protein group (B), the male protein-deprived group (C), and the male sugar-plus-protein group (D) (Müller et al. 1997b).

less expressed vulnerable periods seem to occur for both male groups around day 38 and for the protein-deprived female group around day 35, corresponding to discernible peaks in the hazard rates (figures 6.1 and 6.2).

A more detailed study of early mortality for the different groups can be based on peak-aligned hazard rate estimates (figure 6.3). Focusing on the first 25 days, the pronounced early mortality surge at day 11 for the female protein-deprived group under peak alignment corresponds in size and width almost to a late surge in male mortality observed at day 21. All hazard rate trajectories are close together at day 15. Before this junction, the female protein-deprived group stands out, as these medflies go through a major vulnerable period. Afterward, the female hazard rate trajectories continue to differ, whereas the male trajectories continue to stay close together.

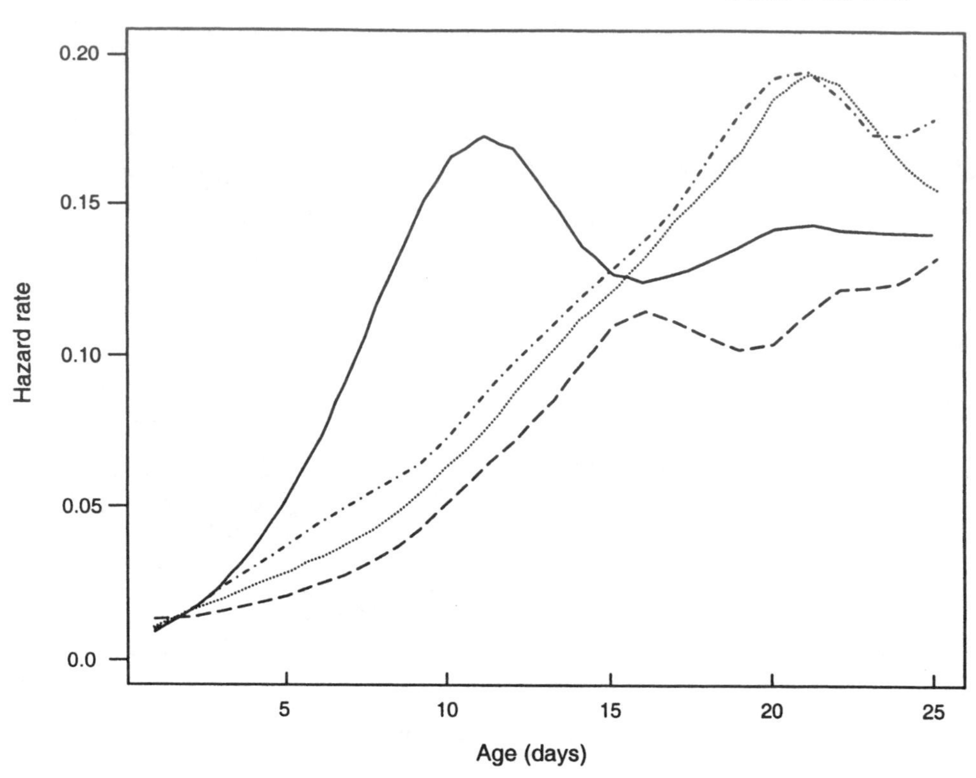

FIGURE 6.3. Peak-aligned estimated hazard rates from 0 to 25 days. Hazard rates are shown separately for the four groups: females, protein deprived (solid); females, sugar-plus-protein diet (dashed); males, protein deprived (dash-dot); males, sugar-plus-protein diet (dotted). Each curve is based on an initial number of more than 100,000 individuals (Müller et al. 1997b).

6.1.4. Implications and Conclusions

These analyses demonstrate that the normally present female advantage in life expectancy is mainly caused by a markedly lower female hazard rate compared with the male hazard rate in the age range 20–30 days. The reversal of the sex differential in life expectancies is caused by the early surge in female mortality under protein deprivation, which is so strong that the later decline in the hazard rate cannot make up for the resulting loss in life expectancy. In contrast, male hazard rate trajectories are considerably less affected by protein deprivation.

The early surge in female mortality under protein deprivation may be due to the weakening effect caused by the transfer of proteins and other essential nutrients required for egg production that would otherwise be used for maintenance and repair. This finding is consistent with the hypothesis that the early surge in mortality of protein-deprived females is a consequence of stress induced by attempting to produce eggs under conditions that are not commensurate with both reproduction and maintenance (Roff 1992). These findings thus point to a vulnerable period with specific nutritional demands for females during reproduction, and a resulting surge in mortality if these specific demands are not met. Before day 15, protein deprivation has a major differential effect only for females, but after day 15, an overriding sex differential predominates, although a lingering diet effect remains for females.

Three lines of evidence support the hypothesis that nutritional demands caused by the activation of the reproductive system for individual females are the underlying reason for the early vulnerable period and corresponding surge in mortality in protein-deprived females:

1. High rates of mating and egg production in female medflies are observed during the first 7 to 10 days (Carey 1982; Carey 1984; Carey, Krainacker, and Vargas 1986; Krainacker et al. 1987). It is likely that the reproductive system of females is "programmed" to begin egg production and maturation shortly after eclosion. If no protein is available, then adults must draw on larval reserves.

2. This early surge is not observed in protein-deprived females when they are subjected to ionizing radiation (Carey and Liedo 1995a). The likely reason for this is that the radiation renders females sterile and, therefore, unable to produce and mature eggs.

3. Male cohorts did not exhibit this pattern irrespective of diet. We note that sucrose-fed females of a related species, the apply maggot fly (*Rhagoletis pomonella*), were able to produce a few eggs (Webster and Stoffolano 1978; Webster et al. 1979). This was likely due to the transfer of protein and vitamins from immature stages to the adult.

Using advanced statistical curve estimation methods (Müller et al. 1997a, 1997b), we found and analyzed an early surge in mortality for protein-deprived female medflies. This surge in early mortality was shown to cause a reversal of the life expectancy sex differential. This points to a vulnerable period in the life cycle of female medflies that is narrowly confined in time and corresponds to the time of reproduction and egg laying.

The results presented in this section provide evidence that changes at the level of the individual account for changes observed in hazard rates for cohorts of individuals. Such changes at the individual level can account for the observed leveling off of hazard rates at older ages in fruit flies (Carey, Liedo, Orozco, and Vaupel 1992; Curtsinger et al. 1992). Other explanations for this leveling off based on density effects (Carey, Liedo, and Vaupel 1995b), demographic selection and heterogeneity (Vaupel and Carey 1993; Vaupel et al. 1979), or epigenetic stratification (Jazwinski 1996) do not have to be invoked.

6.2. Female Sensitivity Underlies Sex Mortality Differential

Differences in the reproductive biology of males and females (Davey 1999) underlies sex-specific nutritional requirements in virtually all species (Gillot 1999; Huebner 1999; Warburg and Yuval 1996). Whereas female fruit flies are required to manufacture a large quantity of eggs with high-protein content (Carlson and Harshman 1999; Carlson et al. 1998), males are required to produce only a small volume (relatively speaking) of sperm consisting of minimal amounts of protein. Because of sex differences in energetic and protein requirements related to reproductive demands, Carey, Liedo, Müller, Wang, Love, Harshman, and Partridge (2001) hypothesized that changes in diet will have a substantial effect on the mortality trajectory of females but will have little or no effect on the mortality trajectory for males. Consequently female sensitivity to dietary conditions will underlie changes in the sex-mortality differentials. In this section I report the results of a test of this hypothesis involving male and female Mediterranean fruit flies subjected to sterilizing doses of cobalt^{60} irradiation (plus intact controls) and maintained on both sugar-only and full (i.e., protein-enriched) diets.

6.2.1. Experimental Framework

Approximately 3,800 medflies (total of both sexes) were placed in 15 $\times$ 60 $\times$ 90 cm aluminum frame cages. Each of 4 cages collectively constituted a replicate for a 2 $\times$ 2 design (2 diets, irradiated and not irradiated). For irradiated flies, two days before emergence pupae were subjected to a sterilizing dose of 14 krad CO^{60} in hypoxia (Williamson et al. 1985). The diets were sugar-only and full diet consisting of a 1:3 ratio of enzymatic yeast hydrolisate and sugar. Although irradiated females and males both experienced increased oxidative damage to their DNA, protein, and lipids in a dose-dependent manner (Balin and Vilenchik 1996), the primary reason for irradiation treatment was to destroy their gamete-producing gonadal germaria and therefore eliminate the nutritional demands for egg production in females. Irradiation would likely have little impact on reducing the reproductive costs in males since their baseline reproductive requirements are minimal (Blay and Yuval 1997).

A total of 35 different cages (replicates) were used for each of the 4 treatments with a grand total of 140 cages containing over 536,000 flies in the study. The statistical methods included multivariate analysis of variance (MANOVA) for life expectancies. The experimental unit was the cage (sample size $n = 140$ cages) and the two outcomes per cage were the lifetimes of the female and male cohort; a MANOVA was used to allow for dependencies between male and female cohorts living in the same cage. Such dependencies may be caused by shared environmental conditions such as fly density. A second class of methods involved the nonparametric estimation of hazard functions form life tables with additional methodological details provided in Müller et al. 1997a (see also Wang et al. 1998).

6.2.2. Sex-specific Life Expectancy

The between-treatment variation for female medfly life expectancies greatly exceeds that for males (table 6.2). For example, the life expectancies for females range from a low of 13.7 days for intact flies maintained on sugar-only diets to 18.4 days for sterilized flies maintained on a full diet—a difference of 4.7 days. In contrast, the life expectancy of males ranged from a low of 14.3 days for sterile males maintained on a full diet to a high of 15.1 days for flies maintained on sugar-only diets—a difference of less than 1 day. Note that longevity is the greatest for females but the lowest for males on the "sterile full-diet" treatment. Table 6.3 contains the MANOVA table (decomposition of sum of squares), which shows that the fertility-by-diet interaction is not significant ($F \leq 0.01$; $p = .996$). Based on a significance level of $P < 0.001$, fertility status and diet are significant factors for mean lifetimes. The results on treatment effects of diet and fertility separated by sex are shown in table 6.4. Since interactions were overall insignificant, they were not considered. Fertility status and diet significantly affect mean lifetimes for females but do not affect mean lifetimes for males apart from very small changes that are not significant and are likely due to chance variation. Specifically, female mean life expectancy is changed by 2.9 days depending on fertility status, whereas fertile flies have significantly lower mean life expectancies. Female mean life expectancy is changed by 1.84 days depending on diet, whereas sugar-only flies have the lowest mean life expectancy.

The sex life-expectancy ratio is reversed from favoring males to favoring females, not only when the diets are switched from sugar-only to protein as previously reported (Müller et al. 1997b), but also in cohorts of irradiated flies regardless of diets. Specifically there is a 10% male advantage in life expectancy for intact flies maintained on sugar-only diets. However, there is a 10% male disadvantage when both sexes are maintained on a full diet and a 10% and a 30% male disadvantage in life expectancy, respectively, when

TABLE 6.2

Expectation of Life in Days (e_0), Standard Deviation (SD), and Number of Cages (n) for Male and Female Medflies Subjected to 4 Different Treatments

Sex	*Fertility Status*[1]	*Diet*	e_0	*SD*	n
Female	Intact	Sugar	13.70	1.73	35
		Full	16.65	2.02	35
	Sterile	Sugar	15.58	2.42	35
		Full	18.44	3.74	35
Male	Intact	Sugar	15.12	2.13	35
		Full	14.76	2.26	35
	Sterile	Sugar	14.68	2.46	35
		Full	14.28	2.91	35

Note: A total of over 536,000 flies or 265,000 individuals of each sex were used in the study (Carey, Liedo, Müller, Wang, Love, Harshman, and Partridge 2000).

[1]sterile and intact refer to irradiated and nonirradiated, respectively.

TABLE 6.3

MANOVA Table for Mean Lifetimes, Subject to Different Diets and Irradiation Treatments

		Wilks's Lambda		*Numerator*	*Demominator*	
Main Effects	df	Λ*	F	df	df	*P-value*
Fertility	1	.4758	74.3614	2	135	<.0001
Diet	1	.6541	35.6931	2	135	<.0001
Interactions	1	.9999	.0045	2	135	.9955
Residuals	136					
TOTAL	139					

Note: Experimental unit is the cage, and the dependent variable is the bivectors consisting of mean life time of the female and male cohort in a cage (n = 140 cages).[a] Fertility refers to the fertility status of flies (intact vs. sterile) depending upon whether they were irradiated or not. While main effects are highly significant, the interactions are not significant (Carey et al. 2000).

[a]The underlying MANOVA model is: $(X_{ijk1}, X_{ijk2}) = (\mu_1, \mu_2) + (\alpha_{i1}, \alpha_{i2}) + (\beta_{j1}, \beta_{j2}) + (\alpha_{i1}\beta_{j1}, \alpha_{i2}\beta_{j2}) + (\alpha_{i1}\beta_{i1}, \alpha_{i2}\beta_{i2}) + (\varepsilon_{ijk1}, \varepsilon_{ijk2})$. These are bivectors where each component corresponds to one sex and $I = 1, 2$ denotes fertility, $j = 1, 2$ denotes diet, $k = 1, \ldots, 35$ denotes cage, $(\varepsilon_{ijk1}, \varepsilon_{ijk2})$ are bivariate normal errors with zero mean, and the interaction terms for diet and fertility are given by the products.

TABLE 6.4
Results on Treatment Effects of Diet and Fertility Separated by Sex

		Female			*Males*		
Effect[a]		*Mean Value Life Expectancy*	*SD*	*P-Value*	*Mean Value Life Expectancy*	*SD*	*P-Value*
Overall Mean		16.09			14.71		
Fertility Effect	fertile	−1.45	0.31	<.001	.19	.29	.54
	sterile	1.45	0.31		−.19	.29	
Diet Effect	sugar	−.92	0.31	<.01	.23	.29	.31
	full	.92	0.31		−.23	.29	

Note: The fitted MANOVA models: $(X_{ijk1}, X_{ijk2}) = (\mu_1, \mu_2) + (\alpha_{i1}, \alpha_{i2}) + (\beta_{j1}, \beta_{j2}) + (\varepsilon_{ijk1}, \varepsilon_{ijk2})$, $i = 1, 2$ (fertility), $j = 1, 2$ (diet), $k = 1, \ldots, 35$ (cage) for given fertility-diet combination (without interactions). The quantities in the model are bivectors, with one component for
tion (without interactions). The quantities in the model are bivectors, with one component for
each sex. The fitted means are the overall mean effect + the fertility effect (sterile or fertile) + the diet effect (sugar or full diet). The estimated effects, their standard deviations (SD) and significance levels are listed above, separately for male and female flies. Differences from the means listed in Table 1 are due to residual effects not included in the MANOVA model (Carey et al. 2000).

[a]Checking for interactions between sex and diet respectively sex and fecundity within the MANOVA model, the null hypotheses of no interactions would be $\alpha_1 = \alpha_2$ and $\beta_1 = \beta_2$ respectively. Since the 99.9% confidence interval for $\alpha_1 = \alpha_2$ is found to be (.9025, 2.3775) and that for $\beta_1 = \beta_2$ is found to be (.4125, 1.8875), we conclude that both interactions are highly significant ($P < .001$). The exact P-values are for the significance of the differences were for the alphas a P-value of 3.0442×10^{-13} and for the betas a P-value of $1.5164 * 10^{-7}$. More details about the MANOVA can be found in (Johnson and Wichem 1998).

flies are irradiated and maintained on both a sugar-only and a full diet. Note from table 6.2 that not only are female life expectancies more variable in response than males, but also their responses to irradiation are in the opposite directions—sterilized males do worse but sterilized females do better.

6.2.3. Sex Differences in Mortality Trajectories

To assess the dynamics underlying the life expectancy differentials, smoothed mortality curves were plotted for all 35 cohorts by sex for each of the 4 treatments (figures 6.4 and 6.5). Three aspects of these figures merit comments. *First,* although slopes of the male mortality schedules in each of the 4

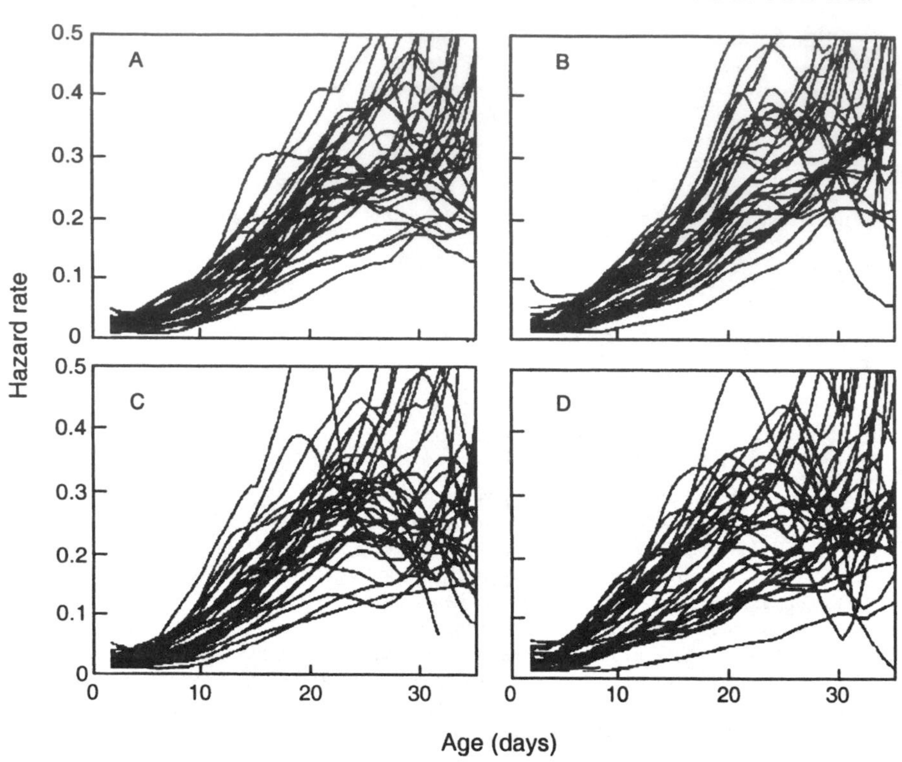

FIGURE 6.4. Smoothed hazard rates for 35 male Mediterranean fruit fly cohorts in each of 4 treatments: A—sugar-fed, intact; B—sugar-fed, irradiated; C—full diet, intact; D—full diet, irradiated. Each curve is based on deaths in approximately 1,900 flies (Carey, Liedo, Müller, Wang, Love, Harshman, and Partridge 2001).

treatments vary widely, the trajectories themselves are remarkably similar—monotonically increasing through 20–30 days and then leveling off. Mortality in some male cohorts decreased at older ages (figure 6.4A–D). *Second*, the opposite is seen in female medfly cohorts inasmuch as their mortality patterns are unique in each of the 4 treatments and bear little resemblance to any of the male mortality patterns (figure 6.5A–D). *Third*, the patterns of female mortality in cohorts of both nonirradiated and irradiated flies are similar if they are maintained on sugar-only diets (figure 6.5A,B). However these patterns are substantially different in the nonirradiated and irradiated groups maintained on a full diet (figure 6.5D). For example, mortality in the nonirradiated cohorts maintained on a full diet is initially low but then increases. However, mortality in cohorts of irradiated females maintained on a full diet is low at both young and old ages.

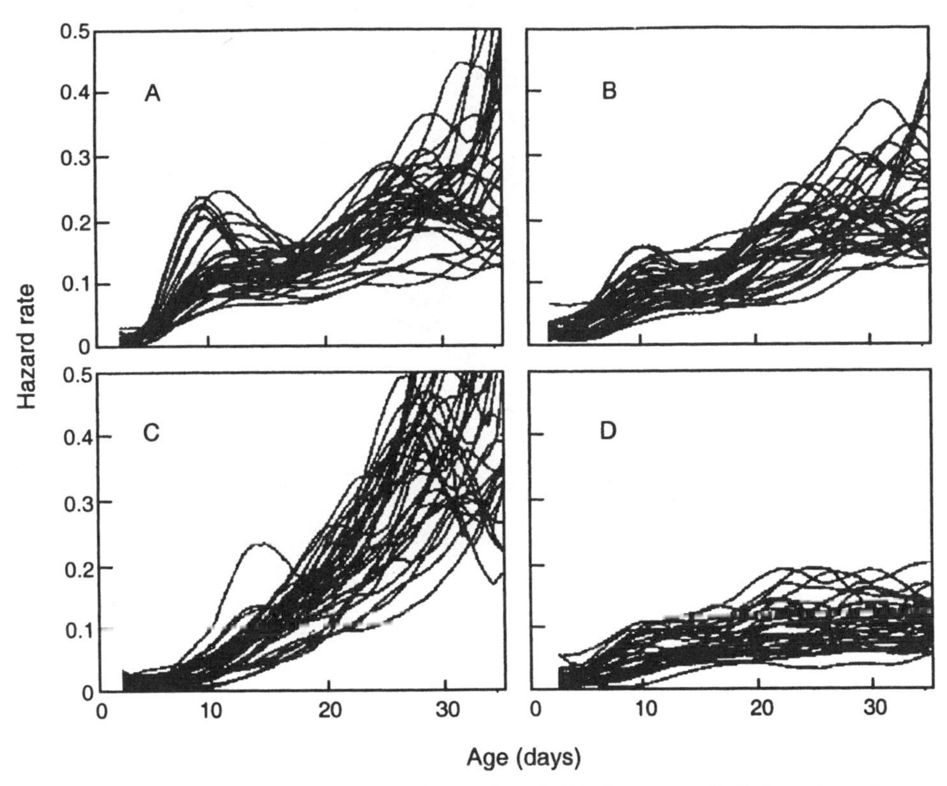

FIGURE 6.5. Smoothed hazard rates for 35 female Mediterranean fruit fly cohorts in each of 4 treatments: A–D—same as figure 6.4 (Carey et al. 2001).

6.2.4. Mean Sex-mortality Trajectories

The mean curves of the 35 cohorts in each of the treatments provide a collective summary of the broad sex-mortality patterns (figure 6.6). Although the mortality patterns for both intact and irradiated female flies maintained on sugar-only diets are similar, their overall levels differ—mortality in cohorts of irradiated females is lower than in cohorts of intact females. After day 20, cohorts of both sterilized males and females maintained on full diets experience the lowest mortality of any treatment. Not only are the qualitative patterns (slopes) of male mortality similar among the 4 treatments, their levels are also nearly identical through day 20. This similarity in male mortality is remarkable considering the high doses of radiation and the large differences in the nutritional quality of the diets.

Similar mortality patterns exist for both intact and sterilized female cohorts that are maintained on sugar-only diets. There is a surge in mortality at

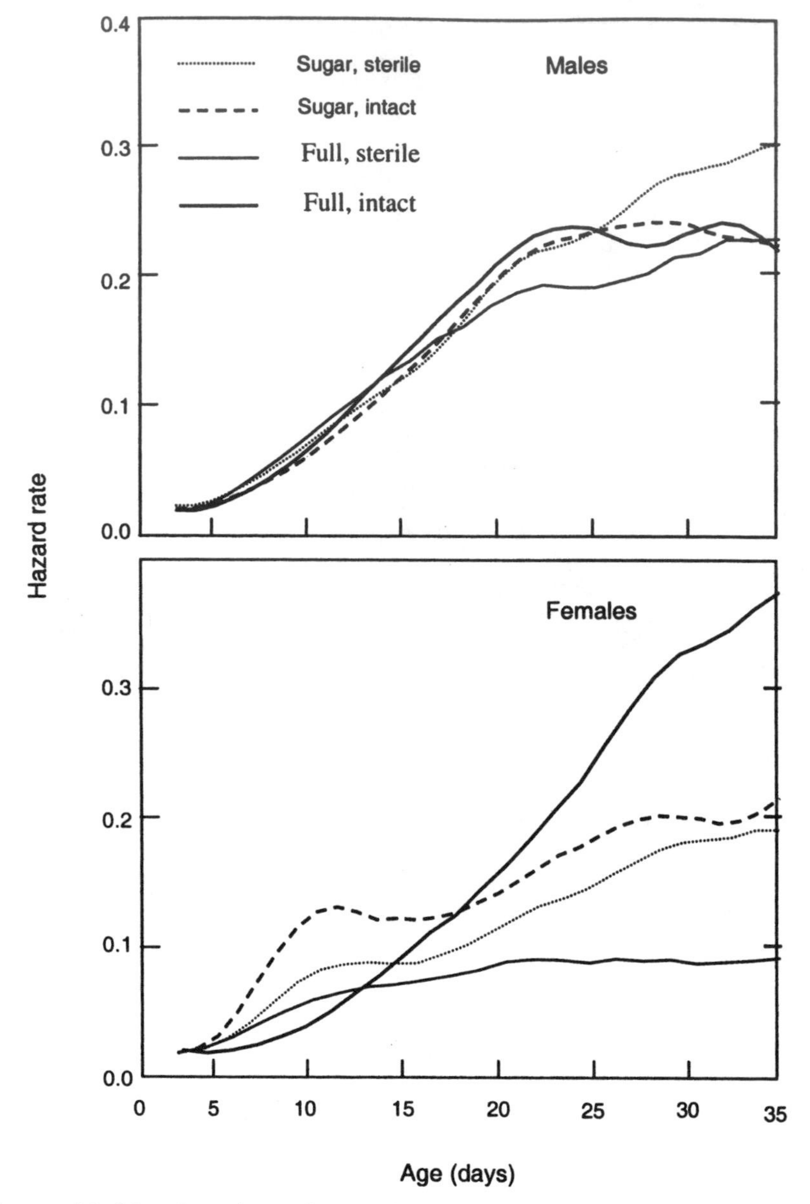

FIGURE 6.6. Mean hazard rates for the 35 male (top) and female (bottom) Mediterranean fruit fly cohorts shown in figures 6.4 and 6.5. Each curve is based on deaths in a total of over 66,000 individuals of each sex. Note that the solid lines denote flies on full diets and broken lines flies on sugar-only diets (Carey et al. 2001).

young ages followed by a shoulder around 10 days and then a gradual increase in mortality thereafter. An interesting feature of these parallel patterns is that not only does the mortality in sugar-fed, intact females exhibit a "shoulder" as observed previously but sugar-fed sterile females did as well. This raises interesting questions about the physiological mechanism underlying this surge in mortality at young ages.

6.2.5. *Relative Cost of Reproduction in Females*

Differences in the levels and patterns of the female mortality schedules shown in figure 6.6b provide insights into the relative cost of reproduction (Hsin and Kenyon 1999; Reznick 1985). For example, the difference between curves in the bottom graph of figure 6.6 shows the cost of

1. actually maturing eggs—mortality differences between sugar-fed, intact (S-I) and full diet, sterile (F-S). Thus S-I females must draw on larval reserves to mature eggs as well as to use for maintenance, whereas F-S females have external sources of protein for maintenance and have no egg production demands.

2. attempting to mature eggs under sterility—mortality differences between sugar fed, sterile (S-S) and full diet, sterile (F-S). Neither S-S nor F-S females in this case have protein demands associated with egg production since both are sterile. However, S-S females must draw on larval protein reserves only for maintenance but F-S females have external source of protein for maintenance. The smaller difference between these mortality curves (relative to differences in former comparison) suggests that maintenance costs are substantially less than those for egg production, as would be expected.

3. maturing eggs with full diet— mortality differences between full diet, intact (F-I) and full diet, sterile (F-S). The crossover at day 13 between the F-I and F-S mortality curves suggests that fertile flies are initially protected, while sterile flies are not. This is reflected in the sharper rise in mortality under full diet and speaks for a protective effect caused by the presence of eggs (Hsin and Kenyon 1999). While the MANOVA did not show a significant interaction pattern between diet and fertility status, figure 6.6 shows that the mortality trajectories of female flies are much more affected when the flies are on full diet rather than when on restricted diet. This interaction demonstrates a close association between reproduction (which is affected by diet) and longevity. It is not captured in mean life time analysis but clearly in the dynamics of mortality and shows the importance of analyzing the entire mortality trajectory.

6.2.6. *Implications*

The experimental study by Carey and co-workers (Carey, Liedo, Müller, Wang, Love, Harshman, and Partridge 2001) produced three main findings:

(1) qualitative differences exist in the sex-mortality response of medflies subjected to dietary manipulations and irradiation; (2) the female mortality response is linked to increased vulnerability due to the nutritional demands of reproduction; and (3) greater female than male vulnerability to perturbation underlies changes in the sex-mortality differential. Whereas considerable variation exists in the mortality trajectories of female cohorts over a wide range of experimental conditions including density (Carey, Liedo, and Vaupel 1995b), diet, mating status (Carey and Liedo, unpublished data), ionizing radiation (Carey and Liedo 1995a), and periodic starvation (Carey et al. 1999), the mortality patterns for male cohorts are similar between treatments (see also Vaupel et al. 1998).

The expression of a mortality shoulder in *both* irradiated and intact female cohorts maintained on sugar suggests that this vulnerable period is not due solely to weakening of protein-deprived females attempting to produce eggs from their larval-derived proteins (Müller et al. 1997b). The prevalence of the shoulder (or peak) can be seen in figure 6.5A,B for protein-deprived females: it is much more pronounced in fertile female cohorts but also occurs for most sterile female cohorts. It also appears occasionally in the mortality curves for full-diet female medflies. Since irradiated females cannot produce eggs, the shoulder in irradiated female cohorts maintained on sugar must be due to increased vulnerability resulting from protein demands in other reproductive contexts such as ovarian requirements unrelated to the manufacture of eggs (e.g., manufacture of accessory gland products [Gillot 1999]). The absence of a mortality shoulder in any male cohort suggests that the shoulder in female cohorts is also not related to the lack of a basic metabolic protein requirement that is independent of sex; otherwise the shoulder would have been expressed in protein-deprived male cohorts.

The results of this study have four implications. *First*, an implicit assumption underlying sex-mortality differentials is that the mortality of *both* sexes is environment-specific. However, our findings demonstrate that male medfly mortality is independent of at least one type of environmental change (i.e., dietary manipulation). Therefore changes in the sign and magnitude of male-female life expectancy differentials are linked to the mortality response to dietary change in *females*. *Second*, the large differences in the sex-specific responses cast doubt on the transferability of the findings from life table studies on one sex to the other. The longevity response of one sex may be substantially different from the response of the other and therefore the outcome of a longevity selection study on females may not apply to males (Christensen and Vaupel 1996). *Third*, greater female sensitivity to changes in environmental conditions will create sex-mortality crossovers; that is, when age-specific death rates favor one sex at younger ages but the other sex at older ages. Understanding mortality crossovers between two cohorts is important because crossovers often point toward fundamental differences in the underlying biology between two cohorts (Olshansky 1995)—the protective effect of eggs at young ages in intact females maintained on a full diet versus females subject to all of the other treatments. *Fourth*, the conventional explanations for differences in male-female mortality, including the behavioral (high-risk/high-stakes male strategies [Zuk 1990]) and chromosomal (homogametic sex advantage [Greenwood and Adams 1987; Partridge and Hurst 1998]) hypotheses may be misleading since the outcome is context-specific. Indeed, the results in the current paper reinforce earlier findings that it is impossible to classify one sex as longer lived than the other without considering the environment in which they are maintained or the treatments to which each sex is subjected (Carey and Liedo 1995a).

6.2.7. Conclusions

This subsection reports the results of the study by Carey and coworkers (2001) that used large-scale experimental methods and advanced statistical curve estimation techniques to record and analyze fundamental differences in the sex-specific responses of medflies to dietary manipulations and cobalt[60] irradiation treatments that were linked with differences in dietary demands of reproduction. The results revealed that mortality is independent of treatment in male medfly cohorts but dependent on treatment in female cohorts and indicated that female sensitivity to dietary change underlies sex-mortality dynamics. Physiological changes at the level of the individual may account for both similarities and differences in the sex-mortality response that are linked to sex-specific reproductive processes.

6.3. Mortality Oscillations Induced by Periodic Starvation

Organisms have evolved a wide range of physiological adaptations to increase their survival when food is scarce, including the ability to enter arrested states (Audesirk and Audesirk 1996), draw on fat reserves (Hoyenga and Hoyenga 1981), reduce metabolic rates (Chapman 1969), and postpone aging (Holliday 1997). Despite the large amount of research that has been conducted on physiological and actuarial responses of organisms to nutritional stress (Chippindale et al. 1993; Masoro and Austad 1996; Weindruch 1996), virtually no data exist on the mortality trajectories of animals subjected to periodic starvation (Giesel 1976; White 1978).

The purpose of this section is to report the results of a series of experiments on large cohorts of the Mediterranean fruit fly in which individuals of both sexes were maintained on 1-of-2 different diets and subjected to 1-of-3 cyclical starvation patterns. The results provided answers to the questions: Is the sex-specific longevity advantage reversed when cohorts are subjected to cyclical (presence-absence) nutritional regimes? Does the sex-specific response to starvation depend upon diet? Does starvation debilitate flies? Answers to these and related questions are important to aging research because they shed new light on the relationship between longevity and conventional studies on nutritional stress (e.g., dietary restriction, starvation), provide a link to the literature on the ecology and natural history of aging (Begon 1976; Bouletreau 1978; Courtice and Drew 1984; Drew et al. 1983; Hendrichs et al. 1993), and build on the results of previous medfly large-scale life table studies (Carey 1997; Vaupel et al. 1998).

6.3.1. Experimental Framework

An average of approximately 4,000 medflies (both sexes) were placed in each of 13 mesh-covered, 15 × 60 × 90 cm aluminum cages for each of the six treatments and two controls (i.e., 104 cages containing approximately 400,000 medflies). Treatments consisted of 3 patterns of 1-day food deprivation cycles—2-, 3- and 4-day cycles. Adult diets consisted of either sugar only or a 3-to-1 mixture of sugar and yeast hydrolysate, the later referred to as a "full diet." Each day dead flies were removed, counted, and their sex was determined. Smoothing techniques for estimating hazard functions are presented in (Müller et al. 1997a; Wang et al. 1994). Further statistical methods such as analysis of variance (ANOVA) were implemented with S+ (Anon. 1997). In particular, interaction plots were used to demonstrate graphically the effect of interactions between the factors (predictors), including gender, diet (sugar only or full diet), and treatment (one of three starvation schemes). The statistical units for the ANOVA were the $n = 208$ cohorts of approximately 2,000 male and female medflies within a cage. The average lifetime of each such cohort was a dependent observation in the ANOVA.

6.3.2. Statistical Summary

The results of an ANOVA that treats life expectancies of the 208 cohorts as dependent variables and gender, diet, and starvation scheme as factors are shown in tables 6.5 and 6.6. Table 6.5 contains the mean life expectancies and standard deviations for the cells formed by combinations of the various factor levels, and table 6.6 contains the usual ANOVA table (decomposition of sum of squares). Note that the 3-way-interaction was not significant ($F = 0.07$; $P = .975$). Therefore, the 3-way interaction was included in the error sum of squares. Based on a significance level of 5%, gender and starvation schemes are significant factors for mean life times, and while diet is not a

TABLE 6.5
Average of the Expectations of Life at Eclosion ($\bar{e}_0$) in Days and Standard Deviations (SD) for 13 Cohorts of Medflies Subjected to 1-of-3 Different Cycles of Food Availability and 1-of-2 Different Diets (i.e., Total of 104 Cohorts for Each Sex)

	Male		*Female*	
Diet/Treatment	$\bar{e}_0$	*SD*	$\bar{e}_0$	*SD*
Sugar-only				
2-day cycle	10.0	3.79	9.3	2.44
3-day cycle	10.8	3.03	10.5	2.09
4-day cycle	11.3	2.86	11.0	2.23
Control	13.1	3.48	9.9	2.20
Full Diet				
2-day cycle	7.4	2.56	10.0	3.10
3-day cycle	8.2	2.17	11.8	3.09
4-day cycle	8.6	1.99	12.0	3.07
Control	13.8	3.55	15.0	3.55

Note: Total number of male medflies used in the sugar only and full-diet treatments was 105,641 and 103,964, respectively, and total number of female medflies used in the sugar-only and full-diet treatments was 99,707 and 91,372, respectively (Carey, Liedo, Müller, Wang, and Chiou 1999).

TABLE 6.6
ANOVA Table for Mean Lifetimes by Sex in the Medfly Subject to Different Diets and Starvation Schemes

Effects	*df*	*Sum of Squares*	*Mean Squares*	*F-Value*	*P-Value*
Main Effects					
Gender	1	32.60	32.60	3.98	0.047*
Diet	1	0.86	0.86	0.10	0.746
Starvation Scheme	3	394.11	131.37	16.03	<0.001*
Two-way Interactions					
Gender:Diet	1	188.57	188.57	23.00	<0.001*
Gender: Starvation scheme	3	56.68	18.89	2.30	0.078
Diet: Starvation scheme	3	136.93	45.64	5.57	0.001*
Residuals	195	1598.50	8.20		

Note: N = 208 cohorts (Carey et al. 1999).
*significant at the 5% level.

significant factor in itself, the interactions between gender and diet and between diet and starvation schemes are significant. This means that the effects of diet and of starvation scheme on mean lifetimes are both modulated by diet. These various effects are now discussed in greater detail.

6.3.3. Treatment Effects on Longevity

Regardless of starvation treatment, life expectancies of males maintained on the sugar-only diets were always greater than life expectancy for females (table 6.5). However, the reverse was true for flies maintained on the full diets; the average female given access to full diets lived longer than the average males under these conditions. The significance of this finding is contained in the gender:diet and gender effect interaction in the ANOVA table (table 6.6). These results for the control flies are consistent with earlier findings.

The sex differentials in life expectancy for flies that were periodically starved *decreased* if they were maintained on a sugar-only diet but *increased* if they were maintained on a full diet (table 6.5). For example, the life expectancy for control males maintained on the sugar-only was 13.1 days, whereas the life expectancy for control females maintained on this same diet was 9.9 days ($P < 0.001$). When flies were subjected to the 3- and 4-day sugar-only treatments (i.e., either every 3rd or 4th day food was removed), the male advantage in the gender gap was reduced to only 0.3 days. In contrast, the life expectancy of control females in the full-diet treatments exceeded male life expectancy by 1.2 days. With the exception of females subjected to the 3- and 4-day sugar-only cycles, life expectancies of control (*ad libitum*) flies exceeded that of treatment flies. Life expectancies of both males and females increased with decreasing frequency of no-food days because mortality was always higher for both sexes on the starvation days.

Patterns of sex-specific life expectancy over all treatments (plus controls) depended on diet, as is shown in (figure 6.7). They show that (1) the life expectancy of about a third of male cohorts maintained on sugar-only diets exceeded the life expectancy of females (top); whereas (2) the life expectancy of females exceeded that for males in 100% of the full-diet treatment cohorts (bottom). In other words, the female advantage was uniformly greater in cohorts maintained on a full diet but variable in cohorts maintained on sugar-only diets.

6.3.4. Interactions

The interactive effect on life expectancy of gender and starvation pattern (averaged over both diets) is shown in figure 6.8a ($P = 0.078$). Life expec-

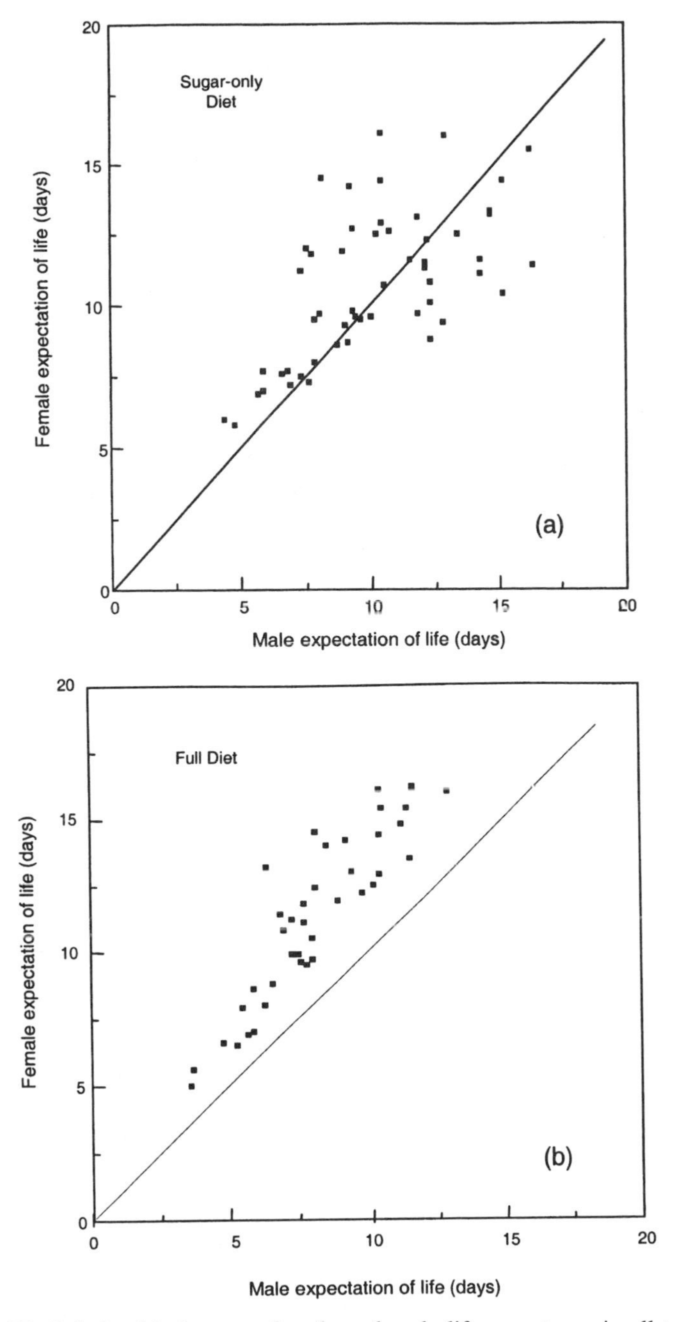

FIGURE 6.7. Relationship between female and male life expectancy in all treatment cohorts for (top) sugar-only treatments and (bottom) full-diet treatments (Carey, Liedo, Müller, Wang, and Chiou 1999).

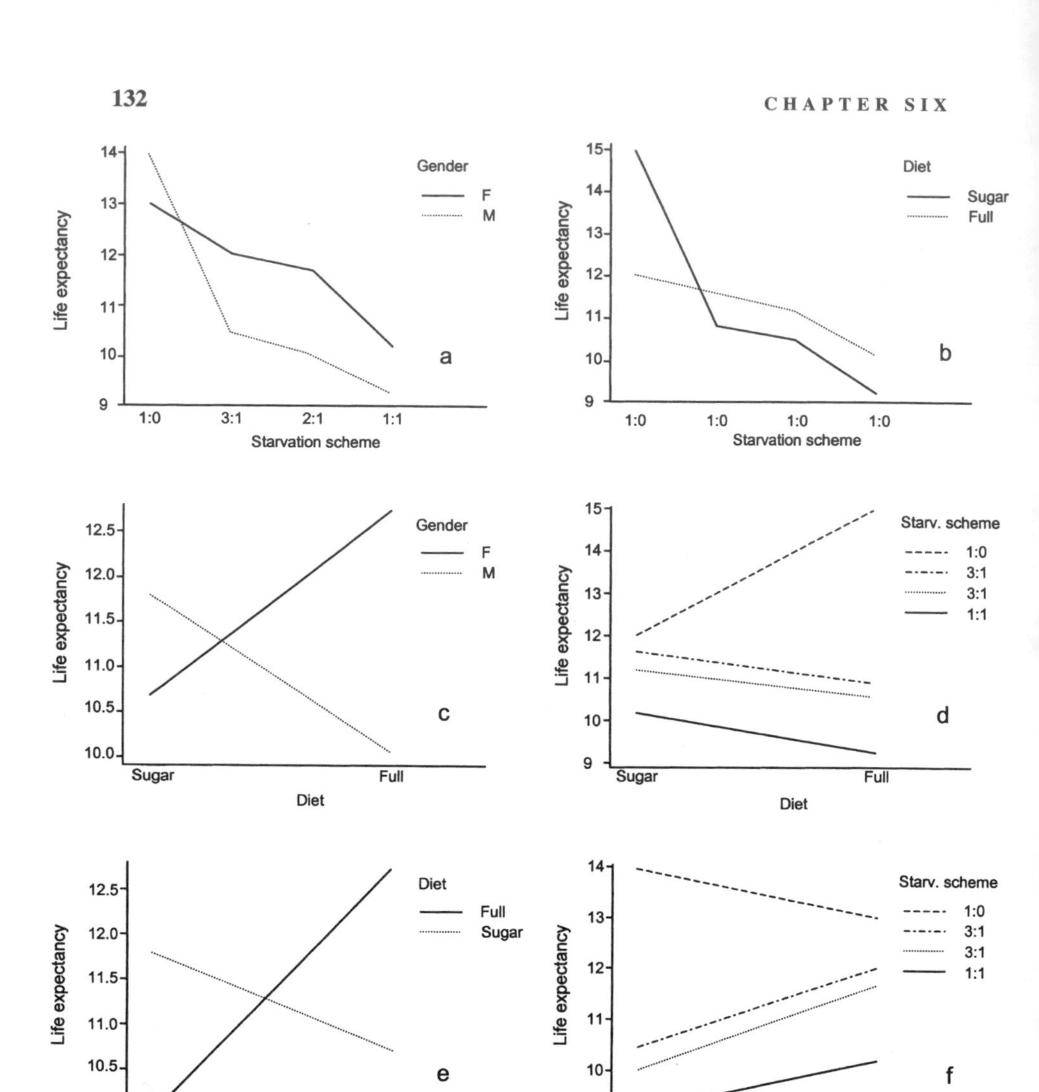

FIGURE 6.8. Interaction plots for mean lifetime response in medfly diet experiments: a—gender-by-starvation scheme; b—diet-by-starvation scheme; c—gender-by-treatment; d—starvation scheme-by-diet; e—diet-by-gender; f—starvation scheme-by-gender (Carey et al. 1999).

tancy favors males in *ad libitum* environments but favors females when cohorts are periodically starved. Life expectancy in female medflies decreases with increasing frequency of starvation (dose) from the highest in the *ad libitum* controls to the lowest for alternate day (1:1) starvation. These results suggest that, relative to female rates, the death rates of male medflies are much more sensitive to perturbations in food availability. In other words, male survival is more dependent upon the *constancy* of food availability than on its *quality*. That diet and starvation patterns profoundly interact can be seen from figure 6.8b ($P < 0.001$). This reveals that the range of life expectancies for flies maintained on sugar-only diets was only 3 days (i.e., life expectancies ranged from 9 to 12 days) whereas the range of life expectancies for flies maintained on a full diet was 6 days (i.e., life expectancies ranged from 9 to 15 days) or double the life expectancy differential of cohorts maintained on sugar-only diets (both sexes combined), regardless of starvation regime.

Pooling data over all starvation schemes (figure 6.8c) reveals that life expectancy favors females when they are maintained on a full diet but favors males when they are maintained on a sugar-only diet ($P < 0.001$ for the gender:diet interaction). This is in accordance with previous findings discussed earlier that protein-deprived females suffer from a surge in mortality during a vulnerable period occurring at the early reproductive stage. Averaging over both sexes when flies are fed *ad libitum*, mean life expectancy increases by almost 2.5 days when a sugar-only diet is compared with a full diet as shown in figure 6.8d ($P < 0.001$). However, life expectancy decreases when making this comparison for any of the starvation diets. Therefore the overall effect of diet is not significant. The reason for this pattern is the substantial decline in longevity for males when on an average scheme under full diet (figure 6.8e) as compared to sugar diet. Exactly the opposite is observed for females. Differences in longevity are much more expressed under full diet than under sugar-only diet inasmuch as starvation effects are more pronounced under full diet than they are under sugar-only diet (see figure 6.8a). Thus it is not surprising that averaging over the different starvation schemes reveals that females do better than males on full diet and males better than females on a protein-deprived diet (figure 6.8e). This type of effect has been described before as a reversal of the sex differential in life expectancy in Müller et al. (1997b). Males have a life expectancy that is about 1 day greater than that of females under *ad libitum* diets if averaged over the two types of diet. However, female longevity is greater than male longevity under all three starvation schemes (figure 6.8f) when averaging over diet. This greater longevity also shows that the variability of the female response between these various feeding schemes is noticeably less pronounced than that for males.

This effect suggests the existence of a second reversal phenomenon for the sex differential in life expectancy, brought on by starvation rather than by diet. The reversal is such that a male advantage in life expectancy under *ad libitum* feeding turns into a female advantage under periodic starvation. While the reversal of life expectancy due to diet could be pinned down to a vulnerable period with heightened mortality in early reproduction for females under protein deprivation (Müller et al. 1997b), the reversal due to periodic starvation observed here is due to a heightened vulnerability of males during the starvation periods. This vulnerability is particularly strong under full diet.

6.3.5. Within-cycle Distributions of Deaths

Daily mortality for both sexes is concentrated more on days when food is absent than on days when which food is present (table 6.7). The percentage of deaths on non-food days is higher than predicted by chance alone (i.e., randomly distributed). For example, the predicted percentage of all deaths to occur on the non-food days for the 4-, 3- and 2-day cycles is 25, 33, and 50%, respectively if there is no immediate effect of food deprivation on mortality. However, the observed percentages departed from this prediction for both sexes and all treatments in two respects. *First*, the immediate effect of starvation on male mortality is substantially greater than for females. The departure of the observed death rate from the predicted one is conditional on diet and treatment ranging from 10 to 17.5% for males but only 4 to 10% for females. *Second*, the immediate effect on mortality on the same day that the full-diet flies were starved is much greater than the immediate effect on mortality on the same day that the sugar-only flies were starved. For example, the departure in observed daily mortality (both sexes) from predicted daily mortaility ranged from 4 to 11% when flies were provided sugar-only diet, but average daily death rates ranged from 5 to 17.5% in full-diet treatments. One explanation for this unexpectedly large increase in mortality is that access to a full diet sets in motion physiological processes in males that depend on a continuous supply of protein but that these processes cannot be stopped immediately. The "physiological momentum" of the process has a weakening effect on individuals and their death rates increase if the protein source is removed. Another explanation is "mating momentum." Other researchers have demonstrated that males maintained on a full diet mate more frequently than males maintained on a sugar-only diet (Blay and Yuval 1997). Perhaps this increased mating activity is carried over to days when food is not present. Therefore the increased metabolic demand of the flies due to continued mating activity exceeds their "normal" capacity to withstand the short period of starvation.

The relative distributions of deaths by cycle-day for the 4- and 3-day cycles are presented in table 6.8. The prediction of a uniform distribution of deaths for all post-starvation days is based on the assumption that there is no residual mortality post-starvation. However, the results shown in table 6.8 reveal the existence of a carryover effect for 1-day post-starvation. For example, 38.3% of all deaths occurred on the day of food deprivation and

TABLE 6.7
Number of Deaths by Sex on Days When Food Was Present (Food-Days) and Not Present (Non-Food Days) and the Percent of All Deaths on Non-Food Days

	Sugar-Only Cycle			*Full-Diet Cycle*		
	4-Day	*3-Day*	*2-Day*	*4-Day*	*3-Day*	*2-Day*
Male Deaths						
Non-food days	9,553	10,996	16,812	11,475	13,169	15,058
Food-days	16,938	15,011	10,783	15,502	13,588	10,983
Totals	26,491	26,077	27,595	26,977	26,757	26,041
Percent (%) deaths on non-food days	36.1	42.3	60.9	42.5	49.2	57.8
Observed						
Predicted	25.0	33.0	50.0	25.0	33.0	50.0
Difference	11.1	9.3	10.9	17.5	16.2	7.8
Female Deaths						
Non-food days	7,673	9,053	14,949	8,385	10,000	11,768
Food-days	16,807	15,207	10,930	15,514	13,144	9,677
Totals	24,480	24,260	25,879	23,899	23,144	21,455
Percent (%) on non-food days	31.3	37.3	57.8	35.1	43.2	54.9
Observed						
Predicted	25.0	33.0	50.0	25.0	33.0	50.0
Difference	6.3	4.3	7.8	10.1	10.2	4.9

Note: The percent of predicted deaths is based on the assumption that absence of food on non-food days had no effect on mortality. For all entries in the table, the differences observed versus predicted are highly significant ($P < 10^{-2}$) (Carey et al. 1999).

27.6% on the following day in the 4-day cycle. These percentages can be contrasted with the 17.3 and 16.8% of all deaths that occurred 2 and 3 days post-deprivation. The similarity of these percentages suggests that the carryover (residual) effect of food deprivation only lasted one day. The daily distribution of deaths in the 3-day cycle suggests a similar trend—the residual effect of starvation diminished with time.

6.3.6. Mortality Patterns and Oscillations

The smoothed and unsmoothed age-specific mortality for both sexes, all treatments, and the controls are presented in figure 6.9. It is remarkable that a strong oscillation is introduced in the raw mortality under starvation schemes. This is caused by much higher mortality on starvation days as compared to non-starvation days. Males experience a catastrophic increase in

TABLE 6.8
Percent of Total Deaths That Occurred in Experimental Cohorts at Different Stages of the Food-Deprivation Cycle

Days Post-Deprivation	Male		Female		
	Sugar	*Full*	*Sugar*	*Full*	*Predicted*
4-day cycle					
0	38.3	45.0	32.8	37.1	25
1	25.6	30.3	23.9	30.7	25
2	18.3	13.1	21.1	16.8	25
3	17.8	11.6	22.2	15.4	25
Total	100.0	100.0	100.0	100.0	100.0
3-day cycle					
0	43.9	51.0	38.4	44.9	33.3
1	33.0	33.1	31.5	34.5	33.3
2	23.1	15.9	30.1	20.6	33.3
Total	100.0	100.0	100.0	100.0	100.0

Note: Day 0 of the cycles denotes the day of food deprivation (Carey et al. 1999).

mortality (> 40%) at young ages in all full-diet treatments but in particular on starvation days in the 3- and 4-day cycles. This period of high vulnerability was not observed in females of any age or in males maintained either on sugar-only diet or for males at older ages maintained on the full-diet.

The impact of food removal on the amplitude of mortality either remained approximately the same with age or increased slightly and, with the exception of males maintained on the full diet, was similar between the sexes. These effects can be seen from Table 6.9, for log mean amplitudes. An ANOVA reported in table 6.10 shows significant effects of gender and starvation schemes and a marginally significant effect of diet on the hazard rate oscillations. Interestingly, all interactions are significant. Thus the hazard rate oscillations are influenced by the same effects as the mean life times. Particularly noteworthy is the strong effect of the starvation scheme on the oscillation, which is not unexpected, and the highly significant effect of the gender:diet interaction, which was also noted for mean lifetimes. In order to investigate the mechanism of the effect of periodic starvation on mean lifetime, an oscillation amplitude was defined by averaging the ups and downs of the original life table values over the first 20 days for each cohort. The relation between mean lifetime and log mean amplitudes (figure 6.10) is seen to be approximately linear. The regression equation is mean lifetime = 1.7488 − 4.5678 (long mean amplitude), with $r^2 = 0.62$ and a highly significant slope parameter ($P < 0.001$). It follows that the increase in mortality oscillations caused by food deprivation is associated with decreased

TABLE 6.9
Means and Standard Deviations (SD) for Log Amplitudes of Hazard Rate Oscillations by Sex for Intact or Irradiated Medflies Maintained on 1-of-2 Diets

	Male		*Female*	
Treatment	*Mean*	*SD*	*Mean*	*SD*
Sugar-only Diet				
2-day cycle	−2.1325	0.6870	−2.1643	0.5995
3-day cycle	−1.9982	0.6475	−2.1595	0.5359
4-day cycle	−2.1239	0.3667	−2.2000	0.4744
Control	−2.4279	0.7850	−2.0553	0.6476
Full Diet				
2-day cycle	−1.7788	0.4182	−2.0920	0.5078
3-day cycle	−1.5209	0.5404	−2.2249	0.3364
4-day cycle	−1.5652	0.3664	−2.0250	0.3823
Control	−2.4141	0.6319	−2.5446	0.4902

Source: Carey et al. 1999.

mean lifetime. Mortality oscillations are a local effect in the hazard rate that causes a global effect in terms of reduced longevity. A similar phenomenon of a local oscillation in the hazard rate with an impact on longevity has been observed previously in Müller et al. (1997b). Reduced longevity associated with larger oscillations is caused by the mortality peaks, which are more pronounced in large oscillations.

TABLE 6.10
ANOVA Table for Mean Log Amplitudes of the Hazard Rate Oscillations by Sex in the Medfly Subject to Different Diets and Starvation Schemes

Effects	*Df*	*Sum of Sq*	*Mean Sq*	*F-Value*	*P-Value*
Main Effects					
Gender	1	1.83814	1.838141	6.3521	0.0125
Diet	1	0.97624	0.976242	3.3736	0.0678
Starvation Scheme	3	5.24778	1.749260	6.0449	0.0006
Two-Way Interactions					
Gender:Diet	1	2.37854	2.378535	8.2195	0.0046
Gender:Starv.Scheme	3	2.10661	0.702204	2.4266	0.0668
Diet:Starv.Scheme	3	2.64965	0.883216	3.0521	0.0297
Residuals	195	56.42835	0.289376		

Note: N = 208 cohorts (Carey et al. 1999).

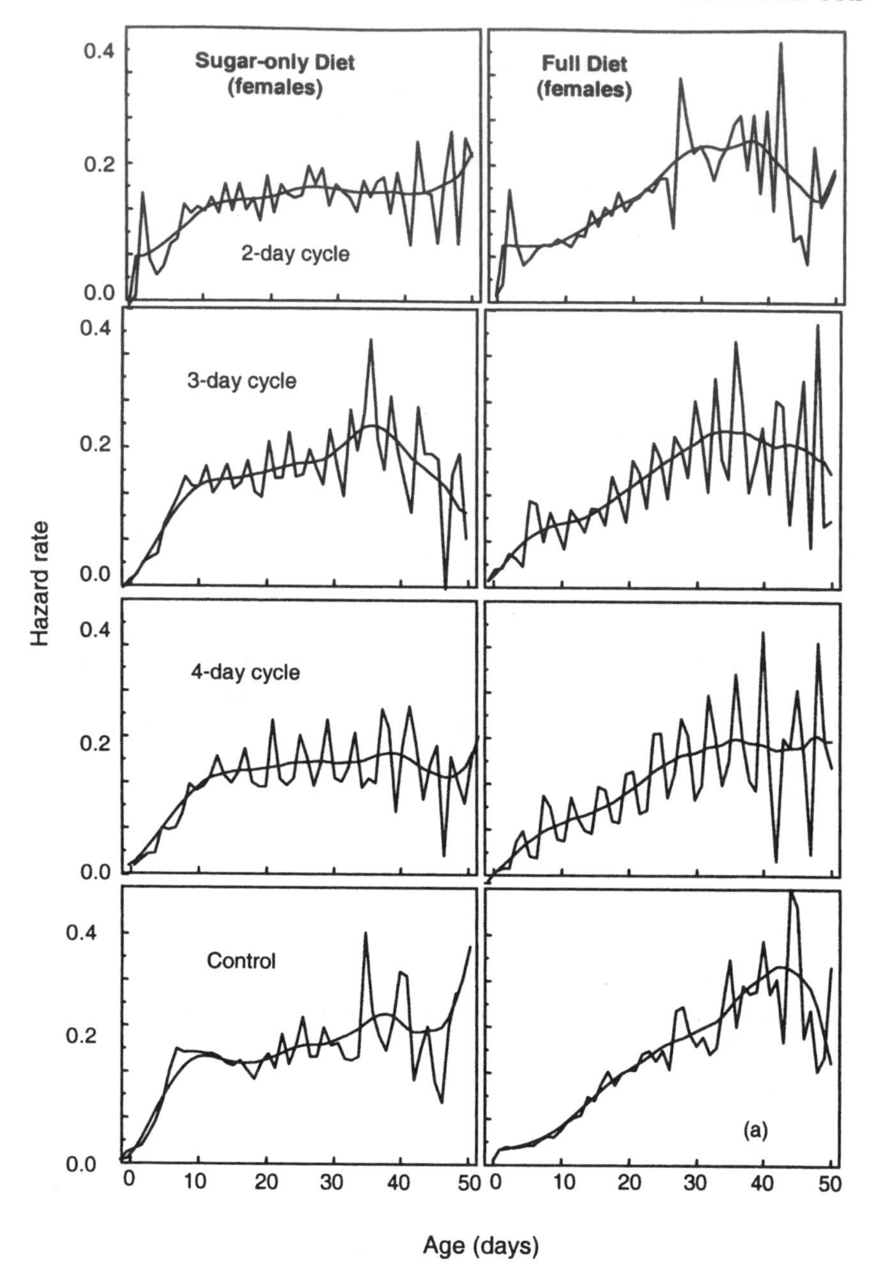
Sugar-only Diet
(females)
Full Diet
(females)
2-day cycle
3-day cycle
4-day cycle
Control
(a)
Hazard rate
0.0
0.2
0.4
0
10
20
30
40
50
Age (days)

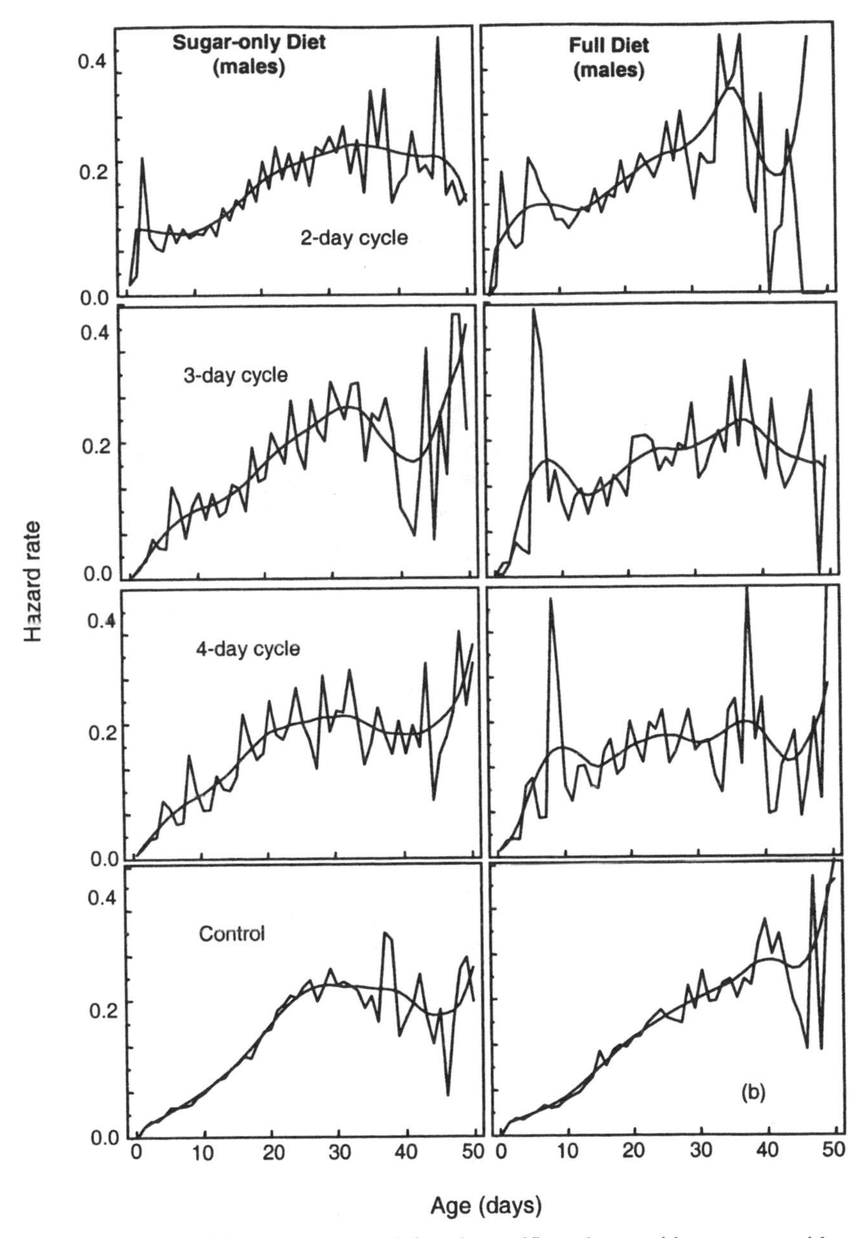

FIGURE 6.9. Hazard rates response of female medfly cohorts with access to either sugar-only (left) or full (right) diets in 1-of-3 different cyclical patterns or control (graphs appearing in first two columns); and of male cohorts under the same conditions (graphs appearing in second two columns) (Carey et al. 1999).

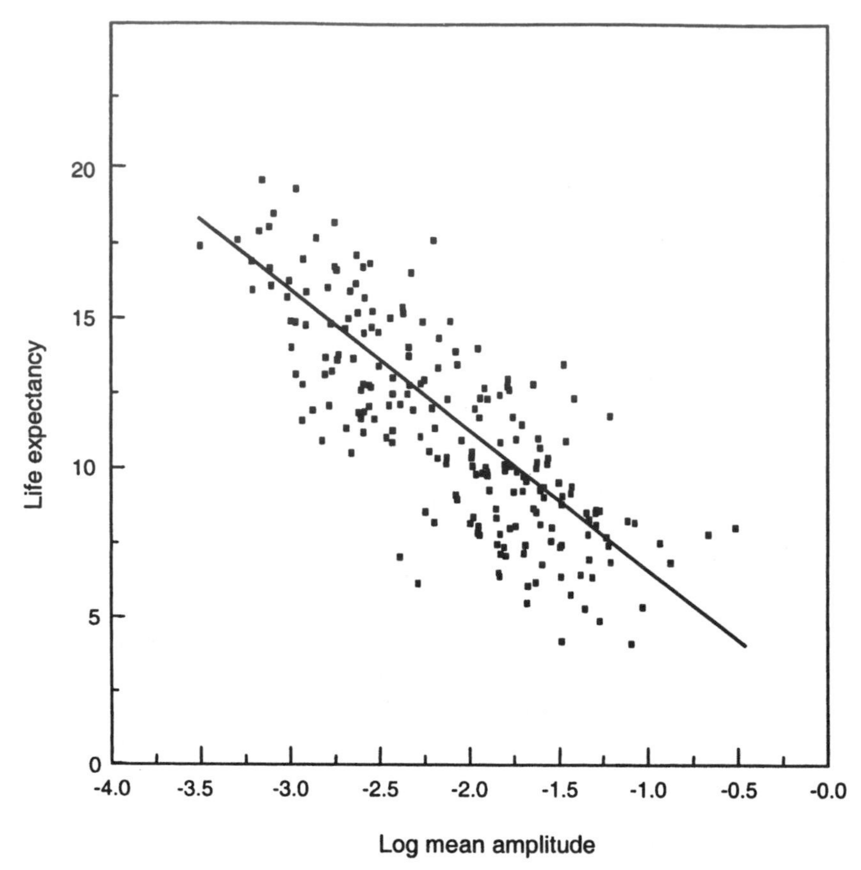

FIGURE 6.10. Log mean amplitudes measuring the oscillations in life table data shown for all cohorts ($n = 208$). The relationship is approximately linear (regression line: mean lifetime = 1.75–4.57 mean log amplitude; $p < 0.01$, $r^2 = 0.62$) (Carey et al. 1999).

6.3.7. Implications and Conclusions

The study described in this section appears to be the first large-scale investigation on any species to document the impact of periodic starvation on the long-term age trajectory of mortality, the research by Kopec (1928) on intermittent starvation in *Drosophila melanogaster* not withstanding. Five new results emerged from this study that appear not to have a precedent in the biogerontology literature:

1. female medflies live longer than male medflies when they are subjected to periodic starvation. This advantage appears to be because female medflies are better able to (a) withstand both the immediate and the short-term (debilitation) effects of starvation (Vaupel et al. 1988) and (b) suspend physiological processes that are deleterious when dietary protein is absent;

2. whereas males have a life expectancy advantage over females if they are allowed constant access to a sugar-only diet (Müller et al. 1997b), this life expectancy advantage disappears when cohorts are subjected to periodic starvation. Thus periodic starvation constitutes a second actuarial mechanism for reversing and amplifying sex longevity differences (Müller et al. 1997b);

3. male medflies maintained on a full diet experience a catastrophic increase in mortality if food is removed for a single day at young ages. This mortality spike was not observed for females maintained on either diet or for males maintained on a sugar-only diet. Nor has it been observed in any other medfly life table study for either sex. This phenomenon points to an inexplicable source of extreme vulnerability in males that has heretofore never been reported;

4. mortality oscillations caused by food deprivation accounts for the decreased life expectancies in both males and females—life expectancy is inversely related to the amplitude of mortality oscillations; and

5. the large perturbations in mortality at younger ages caused by periodic starvation have little or no effect on the amplitude of mortality at older ages. This finding is important because it reveals that demographic selection—the changing composition of frailty with age due to winnowing—does not account for the slowing of mortality at older ages (Vaupel and Carey 1993) as was observed even in cohorts that are subjected to periodic starvation.

The general finding that female medflies are more resistant than males to starvation conditions is consistent with the results of several studies on *Drosophila*, including the work by both Pearl and Parker (1924b) and Service et al. (1985), who reported that starved *D. melanogaster* females lived substantially longer than starved males, and Chippendale et al. (1993), who reported small differences in starvation survival times between male and female *D. melanogaster* of both long-lived (O-strain) and shorter-lived (B-strain) lines. Inasmuch as the results from several previous studies on different worldwide strains and species of *Drosophila* suggested that starvation resistance is an evolved trait (Herrewege and David 1997; Hoffman and Parson 1989), it is likely that the sex-specific differences which we observed in medflies, and which others have observed in *Drosophila*, is also due to natural selection. Greater resistance of females than of males to starvation is also reported in the vertebrate and human biology literature (Hoyenga and Hoyenga 1981; McCurdy 1994; Widdowson 1976). The overall results reveal the sensitivity of sex-mortality differentials to environmental perturbations, such as short-term starvation, and reinforces earlier findings on sex-longevity differences in the medfly that it was not possible to classify one sex as longer lived by a single criterion (Carey and Liedo 1995a).

In conclusion, the heretofore unexplored experimental concept that periodic starvation induces mortality oscillations has shed new light on several aspects of actuarial aging, including the effect of short-term starvation on life expectancy, on sex-mortality differentials, and on residual mortality. The current findings also complement the recent discovery that the life history of medfly females is characterized by two physiological modes when fed either a sugar-only (waiting mode) or a full (reproductive mode) diet (Carey, Liedo, Müller, Wang, and Vaupel 1998c). The complexity of the mortality response of medflies to food type and availability underscores the need for additional studies designed to determine how diet affects longevity and aging. In particular, future research should consider experiments designed to build on, extend, and integrate the results derived from research on the effects of dietary restriction on aging (Yu 1994), where the experimental treatments involve the manipulation of diet quantity rather than the manipulation of food type and availability.

6.4. Summary

Although the results of the medfly studies described in this chapter are interesting in the specific context of dietary manipulation, including the mortality surge in young females, the oscillations in both males and females of all ages, and the differential plasticity between the sexes, the collective results from these dietary studies is that the concept of a fixed, species-specific mortality trajectory is unfounded. The findings show conclusively that both the level and the trajectory of mortality can be substantially modified by changing an environmental variable such as diet. This broad finding is important because it identifies a specific factor that can be easily studied and modified in all animals. Understanding the effect of diet on health and longevity is also important because it is a significant factor in qualitative and quantitative aspects of reproduction—itself a key pacemaker in aging. Thus diet manipulations can also be used to explore the linkage between longevity and reproduction—the topic of the next chapter.

7

Linkages between Reproduction and Longevity

EVOLUTIONARY biologists have long known that reproduction and longevity are linked, as documented in the substantial literature on what is termed "cost of reproduction"—an increment in reproduction at some age may result in a decrement in expected reproduction and an increase in mortality at later ages (Reznick 1985). Numerous experimental studies have demonstrated that reproduction can reduce survival or future fertility, or both, in a wide variety of taxa (G. Bell and Koufopanou 1986; Partridge 1986; Partridge 1987; Partridge and Andrews 1985; Partridge et al. 1987; Williams 1957). It is generally agreed that some sort of experimental manipulation is necessary to demonstrate reproductive costs (Partridge and Harvey 1985; Reznick 1985; Stearns 1992). Both environmental and genetic manipulations have been used, including virginity, X-irradiation, poor nutrition, lack of behavioral exposure to males, and reduced mating frequency have all been shown to extend the life span of *Drosopohila* females (Fowler and Partridge 1989). Costs of reproduction are important because they constrain the possible combinations of survival and fertility that can occur at different ages, and shape the phenotypic and evolutionary responses of the life history to environmental variables such as nutrition. They occur because different costly activities such as growth, reproduction, and somatic repair compete for resources, so that they cannot all be maximized simultaneously. There has been some work on the interaction of the cost of reproduction with nutrition. Chapman and Partridge (1996) showed that the cost of mating is more severe in *Drosophila melanogaster* females that are maintained on a high level of nutrition. This could be because they mate more often, or could indicate a higher mortality cost of each mating, or both.

7.1. Dual Modes of Aging

7.1.1. Hypothesis

As a strategy for prolonging survival while maintaining reproductive potential, most organisms suspend reproduction in periods of food scarcity by entering a different physiological mode (Audesirk and Audesirk 1996). The genes that regulate the transition to and survival in waiting mode in nematode worms (dauer states) and yeast (stationary phase) are closely linked to longevity (Finch and Tanzi 1997; Vaupel et al. 1998). My colleagues and I hypothesized that fruit flies may also be able to enter a waiting mode and that they can live to and reproduce at extreme ages when subject to a lack of dietary protein. To test this, an initial pool of 2,500 male and female Mediterranean fruit flies from the Moscamed rearing facility were maintained in single-pair cages on sugar and water. At 30, 60, and 90 days subgroups of 100 pairs were provided with a full diet *ad libitum*. Pairs were housed in $6.5 \times 6.5 \times 12$ cm clear plastic containers. Dead males were replaced with virgin males of the same age. Their reproduction and survival were monitored until the last female died. Lifetime reproduction and survival were also monitored from eclosion for two 100-pair control cohorts—one maintained on sugar-only (Control A) and the other on a full diet (Control B).

7.1.2. Life Expectancies

In all three treatments, the remaining life expectancy of flies increased on the days when they were given a full diet after having been maintained on a sugar-only diet (table 7.1). Access to a full diet increased the remaining life expectancy of a 30-day-old sugar-fed fly 1.3-fold, that of a 60-day-old fly 2.3-fold, and that of a 90-day-old sugar-fed fly over 12-fold. In addition, the life expectancy of the full-diet control flies at eclosion (age zero) was similar to the remaining life expectancy of the treatment flies at the ages when they were first given a full diet. In most cases, the remaining life expectancies of these cohorts were also similar 30 and 60 days after they were first given a full diet, even though their absolute ages differed substantially. Finally, remaining life expectancy declined rapidly after medflies were switched to a full diet. Remaining life expectancies in cohorts on full diets ranged from 8 to 16 days after 1 month on full diet and from 1 to 10 days after 2 months on a full diet. In contrast, the remaining life expectancy of the sugar-only cohort was 22 days after 1 month and 11 days after 2 months. When all possible comparisons are simultaneously considered—the life expectancy for protein groups $t = 0$, 30, 60, and 90 day—and when the 95% confidence intervals using Tukey's method for the differences in the mean post-protein is applied, lifetime differences between any two means are at most 9.21 days. For example, for the differences between protein groups $t = 0$ and $t = 30$, $t = 60$, and $t = 90$ we obtain the 95% confidence intervals 3.22 ± 4.72, 4.49 ± 4.72, and $-.70 + 4.72$, respectively (all units in days). This general response of a short-term gain in survival but a long-term reduction in life expectancy for flies switched from sugar to a full diet was true for all three treatment cohorts.

TABLE 7.1
Longevity and Reproductive Data for Medfly Cohorts Given Access to a Full Diet at Different Ages

	Lifetime Sugar-Only	*Age Given Full Diet (t)*			
		0	*30*	*60*	*90*
Maximum Age (days)	92	61	93	123	173
Remaining Life Expectancy[a]					
e_0	40.3	**32.8**	—	—	—
e_{30}	22.4	*8.0*	**29.6**	—	—
e_{60}	10.9	1.0	*8.8*	**25.1**	—
e_{90}	2.0	0	3.0	*8.3*	**25.8**
e_{120}	0	0	0	3.0	*15.8*
e_{150}	0	0	0	0	10.4
Maximum Lifetime Eggs	130	1,686	1,603	1,127	1,042
Reproductive Rates (eggs/ female)[b,c,d]					
Gross	27.1	794.1	715.3	540.2	394.1
Net ($x = t$)	23.8	657.8	456.3	160.1	105.7
Net ($x = 0$)	23.8	657.8	305.7	25.6	1.1
% zero-days	94.7%	46.5	46.5	63.8	77.1
Lifetime Egg Production (% of cohort)[e]					
0 eggs	43%	3	9	33	53
1–500 eggs	57	40	46	57	42
> 500 eggs	0	57	45	10	5

Note: Data for the Control A flies are given in the "sugar-only" column and for the Control B flies under the t = 0 column where t refers to the age at which flies were switched from a sugar-only to a full diet. Life expectancies for the ages when a cohort was first given access to a full diet are highlighted in bold and for life expectancies at 30 and 60 days later are italicized and underlined, respectively (Carey, Liedo, Müller, Wang, and Vaupel 1998c).

[a]the number of days remaining to the average individual age x, denoted e_x.

[b]gross reproductive rate is the number of eggs laid by a hypothetical female that lives to the last day of possible life; this measure characterizes reproduction in a cohort in the absence of mortality.

[c]$x = t$ denotes the number of eggs laid by the average female alive at age t. This net rate is the average number of eggs laid by a female that survived to the age at which her diet was switched from sugar to full.

[d]$x = 0$ denotes the number of eggs laid by the average female maintained on sugar-only diet to age t and maintained on full diet thereafter. This net rate gives the average number of eggs laid by a female that was subject to mortality rates on a sugar-only diet from eclosion and then switched to a full diet at time, t.

[e]only 3 females produced over 100 eggs when maintained on sugar-only diet.

7.1.3. *Reproduction*

Female medflies maintained on a sugar-only diet and then switched to a full diet were capable of producing eggs (figure 7.1), but lifetime egg production decreased as the age at the time of the switch increased (table 7.1). Many sugar-only females were infertile. Most suppressed reproduction at ages younger than 30 days and all suppressed reproduction at older ages (figure 7.2). Egg production for females switched to a full diet on day 30 was similar to that of the flies maintained on a full diet from eclosion except for the absence of a 4- to 6-day post-reproductive period in the latter group. One of the most remarkable discoveries was that 4- to 5-month old medflies were capable of producing moderate numbers of eggs if they had been maintained on a sugar-only diet for the first three months.

7.1.4. *Mortality Trajectories*

Age-specific central death rates, defined as $m_x = 2d_x/(n_{x-1} + n_x)$ were smoothed to obtain hazard function estimates, using case weights n_x. Here d_x is the number of observed deaths and n_x the number at risk at the xth day. The 95% pointwise confidence bands were calculated on the variance of these estimates. The lower bound of the 95% confidence band for the hazard function for the Control A (sugar-only) is above the upper bound of the 95% confidence band for the hazard function for the $t = 30$ cohort from day 33 through 53 (with a few days of minor overlapping). Similarly, the lower bound of the confidence band for Control A is consistently above the upper bound of the confidence band for the $t = 60$ cohort from day 63 through 80.

Age-specific death rates for the cohorts that were switched to a full diet at $t = 30$ and $t = 60$ days and for the Control B cohort ($t = 0$) revealed striking similarities in their trajectories after they were given access to a full diet, even though their chronological ages differed by up to 2 months (figure 7.3). The cohorts switched to a full diet on days 30 and 60 shifted to the same mortality trajectory as that of the cohort fed a full diet from emergence (figure 7.3 inset). Mortality in the cohort that was switched to a full diet on day 90 was similar to that experienced by the cohort maintained on a sugar-only diet from emergence. The switch to a full diet on day 90 appears to have set the mortality clock back 90 days. From this point on, we observe the slowly rising trajectory of the sugar-only cohort rather than the more rapidly rising trajectories of the other cohorts switched to full diets. The flies that survived to day 90 on a sugar-only diet may have had special genetic or induced physiological properties that enable exceptional longevity. These

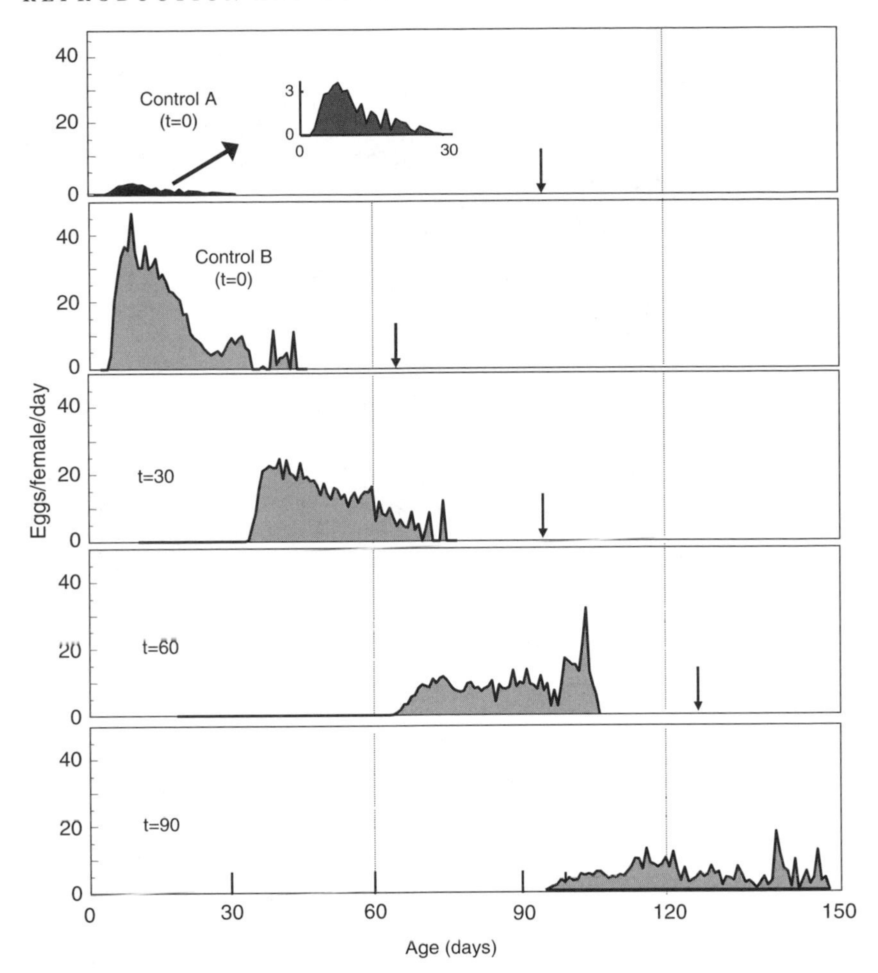

FIGURE 7.1. Age-specific schedule of reproduction for female medflies maintained on a sugar-only diet (Control A—inset; note scale), on a full diet (sugar + protein hydrolysate) throughout their lives (top panel; $t = 0$), or on a sugar diet from eclosion and then switched to a full diet on days 30, 60, and 90. The vertical arrows specify the oldest age attained by the cohort (Carey, Liedo, Müller, Wang, and Vaupel 1998c).

findings suggest that the life history of female medflies includes two distinct modes: (1) a waiting mode with low mortality, in which few or even no eggs are produced; and (2) a reproductive mode with prolific egg laying and low initial mortality followed by an acceleration in mortality and a reduction in egg laying.

The observed patterns of mortality and reproduction in the medfly may be associated with reproductive costs: an increment in reproduction at some age

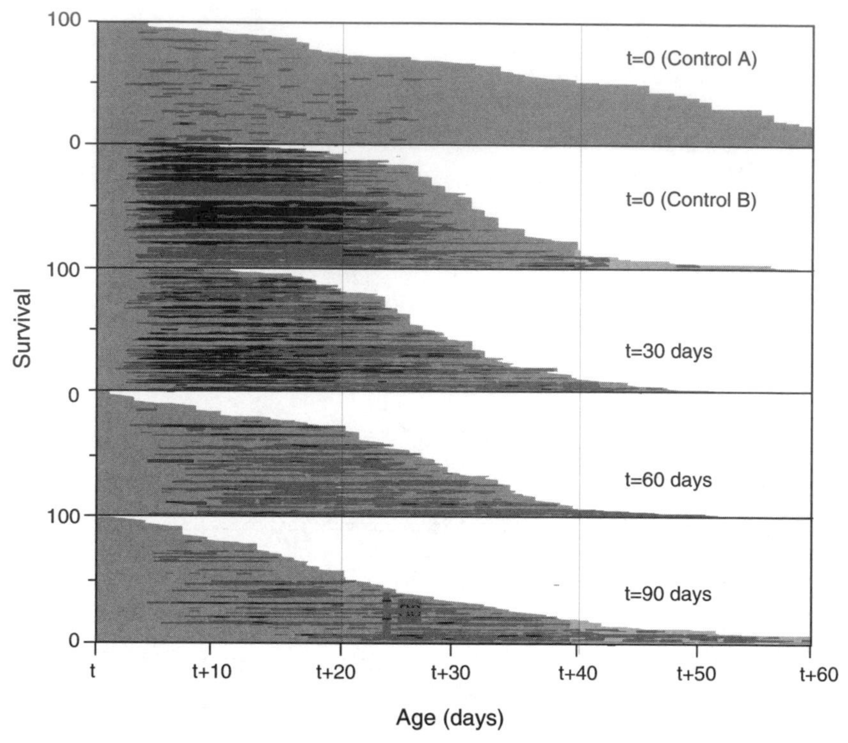

FIGURE 7.2. Event history diagrams for survival and reproduction in the 5 cohorts of 100 females. Each horizontal line represents the life course of an individual fly, the length of which is proportional to its life span. Segments within each lifeline are color coded to depict egg-laying levels at each age—light gray = 0 eggs, medium gray = 1–40 eggs, black = > 40 eggs. Controls A (top panel) and B (second-from-top panel) refer to flies maintained throughout their lives on sugar-only and sugar-plus-protein diet, respectively. The symbol *t* denotes the time at which sugar-fed cohorts were fed a full-protein diet (Carey et al. 1998c).

may result in a decrement in expected reproduction and an increase in mortality at later ages. Mortality may increase because reproduction diverts resources from somatic repair and maintenance, especially if the diet contains no protein (Müller et al. 1997b). The observed patterns may also be attributable to reproductive determinism (i.e., fixed endowment) if the total supply of reproductive units (eggs) is limiting (Bell and Koufopanou 1986). The mammalian analog for reproductive determinism is the ovarian exhaustion of follicles as the pacemaker of reproductive senescence (Finch 1990).

The finding that suppression of reproductive activity prolongs survival in the medfly is consistent with the results of other studies. Mortality decreases in male *Drosophila* when they are denied access to mates (Partridge 1986;

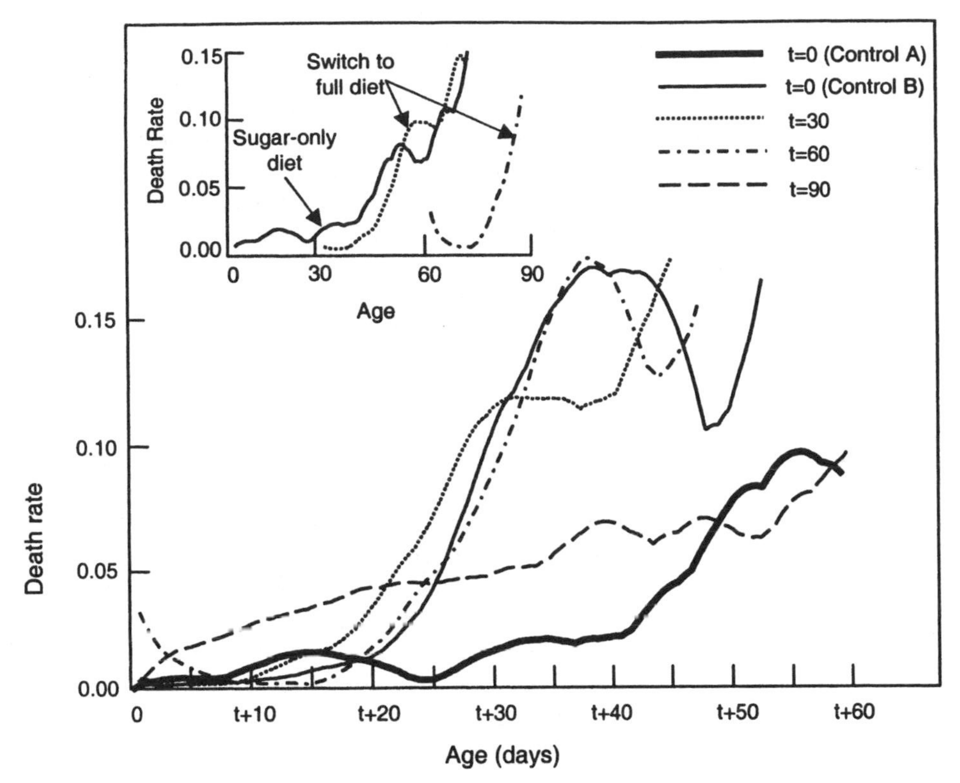

FIGURE 7.3. Smoothed age-specific probabilities of death for the 5 study cohorts starting at the time (*t*) when they were first switched from sugar to a full diet (except for the Control A flies, which were fed sugar-only throughout their lives). Inset in upper-left corner shows the shift in the mortality trajectory that occurred for the cohort which received sugar only to day 30 and was then given a full diet (solid line) (Carey et al. 1998c).

Partridge and Andrews 1985) and in female medflies when they are deprived of either mates or oviposition hosts (Aigaki and Ohba 1984a; Carey, Krainacker, and Vargas 1986; Partridge et al. 1987) or when they are sterilized (Carey and Liedo 1995a; Roitberg 1989; Tatar and Carey 1995). Female rodents maintained on a restricted diet at young ages but subsequently fed *ad libitum* are, like medflies, capable of producing offspring at advanced ages (Ball et al. 1947; Holehan and Merry 1985; Masoro 1988; Merry and Holehan 1979; Visscher et al. 1952). The effects of reduced food intake on reproductive aging may be mediated by retardation of the rate at which ovarial follicles are depleted (DePaola 1994).

7.1.5. Implications

These findings have several important implications. First, the assumption that there are two modes of aging provides an ecological and evolutionary context for aging research; the modes link the reproductive fate of individuals to the availability of food, mates, and hosts (Holliday 1989; Masoro and Austad 1996). Second, the two modes enable flies to increase their longevity—individuals that experience both modes survive longer than those that remain in one mode exclusively. Third, when resource availability is uncertain, older individuals may contribute more offspring than very young ones (Carey and Gruenfelder 1997; Wachter 1997). For medflies, protein is usually scarce in the wild (Bouletreau 1978; Hendrichs and Hendrichs 1990; Hendrichs et al. 1993), and thus it is likely that many female flies are quite old before they find protein and can reproduce. Induced demographic schedules of the kind observed in this medfly experiment do not fit the simplistic formulation of the Lotka equation (1928), and thus new equations will have to be developed to incorporate the observed plasticity of fertility and survival (Tuljapurkar 1989). Fourth, the effects on longevity of dietary restriction may be mediated by gonadal activity or through the rate of ovarial depletion. The causal mechanism underlying the dietary restriction response that has been observed in a wide range of species (Weindruch 1996) may be linked with physiological adaptations to nutritional stress in yeast (stationary phase) and nematodes (dauer stage) and to host and mate deprivation in insects.

7.2. Reproductive Clock

The purpose of the study by Müller and coworkers (2001) described in this section is twofold: to establish the patterns of reproductive activity as measured in terms of egg-laying for a cohort of female Mediterranean fruit flies (medflies) and then to correlate those patterns with longevity. As it turns out, the egg-laying trajectories on the individual level follow simple exhaustion or decay dynamics. Müller and coworkers (2001) demonstrate that the best predictor for subsequent mortality is the rate of decline of egg laying, i.e., the rate at which the egg supply is exhausted, rather than intensity of reproduction.

The cost of reproduction concept has been established by many researchers (Abrams and Ludwig 1995; Chapman et al. 1998; Kirkwood and Rose 1991; Partridge and Farquhar 1981; Westendorp and Kirkwood 1998; Williams 1957). In this section Müller and his colleagues (2001) argue that individual egg-laying data for medflies point to the critical role played by the remaining reproductive potential that quantifies the degree of egg depletion for an individual fly. As the reproductive potential declines, subsequent mortality increases. This adds an important dimension to the concept of cost of reproduction for medflies. The classical cost of reproduction concept envisions a damage incurred by reproduction that leads to shortened life span. In light of the findings by Müller and his group, reproduction itself leads to a decline in reproductive potential, which then is associated with increased subsequent mortality. In particular, they find that the rate of decline of reproduction and not intensity of reproduction proves to be the best predictor of subsequent mortality. Thus an unqualified concept of cost of reproduction incurred by competition for limited available resources between reproduction and maintenance proves too simplistic.

Individual egg-laying counts for 1000 mated female medflies were recorded daily at the mass-rearing facility in Metapa, Mexico. In addition, time of death was recorded for each fly. From these flies, 531 egg-laying subjects were selected who lived beyond day 26. The flies were held in individual cups and were fed a full diet of protein and glucose *ad libitum*. For the statistical analysis of the data, the Cox proportional hazards model, nonlinear least squares, bootstrap tests, and smoothed hazard function estimates were used. Details for the two bootstrap tests, which were developed to test for an association between reproductive clock and life span, are given in Appendix A in Müller et al. (2001).

7.2.1. Constant Rate of Egg-laying Decline

Individual egg-laying trajectories sharply rise after egg-laying begins (5–17 days after emergence), reach a peak, and then slowly decline. The rate of decline varies among individuals but one of the findings is that this rate is approximately constant for each individual.

The age-trajectory of reproductive decline for each fly is accordingly modeled by the exponential function

$$f(x) = \beta_0 \exp(-\beta_1(x - \Theta)), \qquad (7.1)$$

where $f(x)$ is the fecundity (measured by daily egg count) of the fly at age x days and Θ is the age at peak egg laying (sample mean 11.09 $\pm$ 3.55(SD)).

The two parameters β_0 (mean 57.25 $\pm$ 17.70), the peak height of the trajectory, and β_1, the rate of decline (mean 0.090 + 0.093) varied considerably from fly to fly (figure 7.4). A modest but significant negative correlation between β_0 and $\beta_1 (r = -.15, P < .05)$ indicated that fecundity tends to decline more slowly for flies with higher peak fecundity. The protracted decline in egg-laying after the initial sharp rise is reasonably well predicted by the exponential model (figure 7.4).

A consequence of this simple egg-laying dynamics is that for any age x,

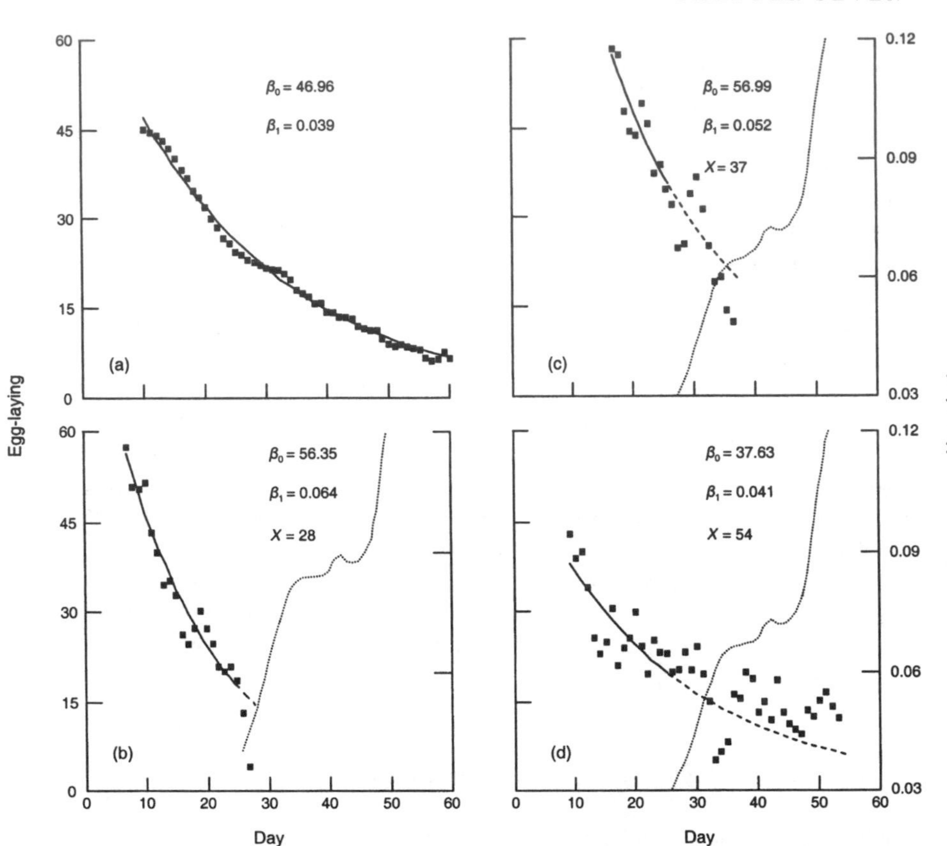

FIGURE 7.4. Trajectories of fecundity and mortality. The trajectories of fecundity are fitted to data from the peak to day 25 (solid line) by nonlinear least squares and predicted thereafter (dashed line). Age at death is indicated by X. (a) Trajectory fitted to average counts of daily eggs for the sample of 531 medflies. Daily egg counts, reproductive trajectory, and predicted smoothed (Müller et al. 1997a) individual hazard function (dotted line) for a fly with life span at (b) the 10% quantile, (c) the 50% quantile, and (d) the 90% quantile (Müller et al. 2001).

the fraction of remaining eggs can be predicted relative to the total number of eggs, by

$$\pi(x) = \frac{\int_x^\infty f(s)ds}{\int_\theta^\infty f(s)ds} = \exp(-\beta_1(x - \theta)). \tag{7.2}$$

This function, with values declining from 1 to 0 as the fly ages, provides a simple measure of reproductive exhaustion at age x in terms of remaining (relative) reproductive potential. It could be loosely described as an individual's reproductive clock, which advances at a speed determined by the rate of decline β_1.

7.2.2. Mortality and Exhaustion of Reproductive Potential

In the experiment the likelihood that a fly would die increased as the fly's reproductive potential was exhausted and the reproductive clock advanced. The finding of an association between mortality and exhaustion of reproductive potential is the main result. It leads to a new perspective on the relation between reproduction and longevity. The association between remaining reproductive potential and longevity was established in three different ways. The first is a bootstrap test using randomly resampled lifetimes and quantifying the number of eggs that would have been produced under a random exchange of lifetimes. The idea is that if there is no association between reproductive potential and life span, a random exchange of lifetimes would not alter the total number of eggs produced by all flies in the sample.

The second analysis provides a similar quantification of total number of eggs laid if flies in randomly formed pairs exchanged lifetimes. Again, if there is no association the number of eggs would remain unaffected by such an exchange. The third analysis provides for a direct prediction of subsequent mortality of an individual fly, based on a hazard regression model and using reproductive potential as predictor variable.

Obviously, fecundity and mortality are strongly correlated with age: As fecundity decreases with older age, mortality increases. Hence, to prevent confounding effects from this association, we fit the trajectories by using only data prior to day 25, whereas longevity is measured as remaining lifetime after day 25. Thus, the fitted trajectories of fecundity are predicted after age 25, based on the above model. This guarantees that the fitted trajectories are not influenced by a fly's life span, and allows bona fide predictions of subsequent mortality.

7.2.3. Confirming the Association between Reproductive Potential and Longevity

Suppose that instead of dying at the actually observed age at death x, each fly is assigned a new life span of x^*, chosen at random from the sample of 531 observed life spans. If longevity and reproductive clock, or equivalently, remaining reproductive potential at death, are linked, then this random reassignment should tend to increase the remaining reproductive potential at death on average (measured by the average value of μ).

The null hypothesis that no such change occurs corresponds to no link between reproductive potential and lifetime. Utilizing the bootstrap method (Efron and Tibshirani 1993; Manly 1997), we devise a bootstrap test for this null hypothesis (see Appendix A in Müller et al. 2001). This bootstrap test provides strong evidence against the null hypothesis of no link ($P = 0.0004$) and in favor of the alternative that the occurrence of death becomes ever more likely as the reproductive potential of a fly is exhausted.

7.2.4. Graphical Confirmation of the Association

The result on the association between reproductive clock and lifetime is illustrated by an event history diagram figure 7.5. The event history diagram (Carey et al. 1998b) is based on fitting exponential trajectories to egg laying over the entire life span. This diagram demonstrates graphically that there will be a loss in actual eggs when lifetimes are randomly rearranged. The event history diagram demonstrates the close relationship between reproductive potential and life span for all 531 flies. For example, almost half of the flies (262/531) died with fewer than 20% of their eggs left (gray zones indicating an advanced state of the reproductive clock) and 91% (482/531) died with fewer than 50% of their eggs left (dark gray zones). Thus early death is less likely in the presence of a large remaining reproductive potential, and random reassignment of lifetimes will tend to increase remaining reproductive potential at death.

As flies in the upper half of the graph with relatively short life spans are likely to be assigned increased life spans in a random reshuffling of life spans, not much in terms of additional egg laying will be achieved for these flies due to the near exhaustion of their egg-laying potential. The same reshuffling likely includes the random assignment of shorter life spans to the relatively long-lived flies in the lower half of the graph. Their deaths will then occur while they are still in the intermediate zones with sizeable remaining egg-laying potential. Their lifetime output in eggs will therefore decline sharply. These losses in terms of eggs not laid relative to egg-laying potential will dominate the at best meager gains that the flies in the upper half might achieve. The net result is therefore a decrease in the conversion of egg-laying potential into actual eggs, on the average.

7.2.5. Exchanging Lifetimes between Flies in Randomly Selected Pairs

In another thought experiment, assume that flies are randomly grouped into pairs and that for each pair life spans are exchanged but reproductive

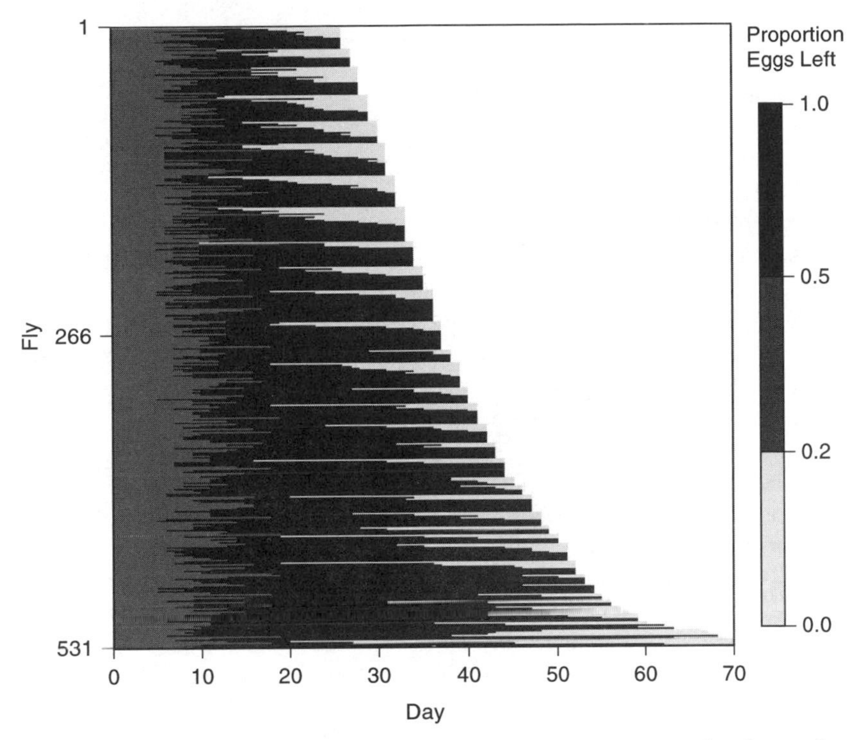

FIGURE 7.5. Lifetime and proportion of eggs left. All 531 flies are ordered according to life span, which defines the length of the colored horizontal bar for each fly. Taken together, the bars provide the empirical survival function. The colors within each bar indicate the remaining relative reproductive potential. The initial phase (light gray) marks the time elapsed between eclosion and peak reproduction. Flies with the same lifetime are ordered according to remaining reproductive potential (Müller et al. 2001).

clocks and fecundity trajectories are not. Testing the null hypothesis of no change in the average value of π at death against the alternative of an increase in this value, we find that an exchange of life spans is detrimental to the flies' total egg-laying output ($P = 0.00002$). For example, for the pair of flies whose egg-laying data are shown in figure 7.4b,c, their average remaining egg-laying potential at death (corresponding to eggs not laid) would increase by 4.8% under life span exchange. Their predicted combined output in terms of eggs would decline by about 130 eggs, a loss of 9.6%, as the longer-lived fly of figure 7.4b would lose more in egg output than the shorter-lived fly of figure 7.4c would gain as a consequence of the hypothetical exchange of lifetimes.

7.2.6. Predicted Subsequent Egg Laying

A Cox proportional hazards model for mortality after day 25 was fit with the function π as a time-varying predictor, obtaining a highly significant ($P < .001$) relative risk function of $\exp(-1.37\ \pi(x))$ (Andersen et al. 1993; Cox 1972). Flies with only 5% of their egg-laying potential left are 3.42 times as likely to die as flies with 95% of their potential remaining, and 1.85 times as likely to die as flies with 50% left. Using the predicted mortality for days 26 to 30 to form low, medium, and high risk groups of 177 flies each, we found observed death rates of 14/177, 26/177, and 61/177, respectively ($P < .005$). This demonstrates highly significant predictions of subsequent mortality from early reproductive patterns, solely based on the rate of exhaustion of reproductive potential.

In accordance with other medfly experiments (Müller et al. 1997b), hazard rates rise rapidly, reach a shoulder, and then rise further (figure 7.4b–d). It is noteworthy that the hazard regression model with π as predictor of mortality was better than alternative models, according to the Akaike information criterion (see Müller et al. 2001), and also in terms of prediction error. Therefore, it appears that the remaining egg-laying potential function π is indeed providing the link with longevity. It is thus the dynamic features of egg-laying and not the absolute number of eggs laid that matters.

7.2.7. Discussion and Implications

A substantial body of theoretical and experimental research on the costs of reproduction (Partridge and Farquhar 1981) and on reproductive determinism (Bell 1984; Bell and Koufopanou 1986; Carey, Krainacker, and Vargas 1986; Minchella and Loverde 1981; J. M. Smith 1958) has shown that reproduction can decrease survival and that exhaustion of reproductive capacity plays a role in aging. Our findings point to a fundamental link between reproductive dynamics and survival. Recently a "delayed wave of death" following reproduction was reported (Sgro and Partridge 1999) and while this finding reinforces the concept of a reproductive clock, our findings do not support the idea that there is a direct cost of reproduction. The link between mortality and reproduction is carried by the dynamics of reproduction and not by the absolute magnitude of reproduction, as measured in number of eggs produced. For example, a high reproduction rate with slowly declining reproductive potential is associated with longer life span according to our findings. In contrast, the classical cost of reproduction hypothesis would associate high reproduction rates with shortened life spans. In particular, our analysis provides a detailed description of the nature of the linkage between the dynamics of the reproductive trajectory and subsequent mortality. We establish the primacy of the rate of reproductive decline over absolute levels of reproduction regarding this link. It is quite amazing that based solely on knowledge of early reproductive patterns, our approach allows a reasonable prediction of the increase in subsequent death rates, at the level of the individual. A possible interpretation of this finding is that the rate of reproductive decline is a good indicator of the speed of aging of the organism. In this sense, the reproductive clock is synchronized with an individual's biological age as contrasted to the chronological age. Reproductive decline thus serves as an indicator not only of reproductive exhaustion and gonadal aging, but also of senescence. Individual medflies experience age-specific fecundity and mortality trajectories that are linked in such a way that flies generally are enabled to lay most of their potential eggs before death. Flies that exhaust their egg-laying potential fast tend to die early, while flies that experience slowly declining egg-laying trajectories live longer. These flies have a more slowly advancing reproductive clock and their increased longevity coupled with higher levels of daily egg-laying leads to an abundance of eggs as compared to flies with a faster advancing reproductive clock who tend to live shorter with rapidly declining daily egg-laying yields. One is tempted to classify individuals into groups displaying various degrees of "vitality" or "frailty," which expresses itself in both longevity and level of reproductive activity.

The pace of an individual's reproductive exhaustion and a fly's survival chances may be jointly determined by pleiotropic genetic factors. It is also possible that both egg laying and mortality patterns are affected by micro-environmental conditions early in life, conveying physiological strength or frailty (Giesel 1976; Wagner and Altenberg 1996). Such phenotypic adaptability is plausible as this link enables a fly to produce more offspring on average and because experiments have shown that depending on whether conditions are favorable to reproduction, medflies can switch between radically different egg-laying and mortality trajectories (Carey et al. 1998c).

7.3. Food Pulses

Protein availability in nature is a fundamental determinant of individual fitness and population growth because amino acids (the basic ingredients of proteins) are required by individuals for both physiological maintenance and offspring production. Although the accessibility of protein-rich diets for most species in nature is highly variable, the vast majority of life history studies have focused on the birth and death response of individuals that have had constant access to *ad libitum* diets. Very few studies have been conducted in which access to dietary protein by individuals is pulsed. For example, it is not known whether the life expectancy of female fruit flies subjected to periodic pulses of protein-rich food increases relative to cohorts with unlimited access to protein-rich food, whether flies rapidly use the dietary protein to synthesize and lay new eggs but at the expense of little or no subsequent egg laying, or whether their days of reduced egg laying due to the absence of protein are recovered later when protein again becomes available. The reason why answers to these questions are not known is that historically most investigations concerned with the relationship between food and birth and death rates have focused on either caloric restriction (Austad 1989; Kirk 2001; Masoro 1988; Masoro and Austad 1996; Sohal and Weindruch 1996; Weindruch 1996), reproductive physiology (Engelmann 1970), biodemography (Carey 1998; Carey, Liedo, Müller, Wang, and Chiou 1998a, 1998b; Sgro and Partridge 1999), cost of reproduction (G. Bell 1988; Bell and Koufopanou 1986; Partridge 1987; Reznick 1985; Roitberg 1989), life history trade-offs (Jacome et al. 1995; Tatar and Carey 1995), and the effects of "egg load" on foraging behavior (Prokopy et al. 1994).

Understanding the impact of variability in access to high quality (high protein-content) food on birth and death rates is important for several reasons. *First*, conditions in which food availability fluctuates are more consistent with the conditions to which most animals are exposed in nature. Therefore the results of experiments involving variable food levels will provide new insights into the demographic response of animals subjected to more realistic conditions in nature (Begon 1976; Bouletreau 1978; Courtice and Drew 1984; Drew et al. 1983; Hendrichs et al. 1993; Webster and Stoffolano 1978). In particular, the results may reveal if and to what extent animals subjected to shortages of high quality food can extend their lives. *Second*, food regimes that vary with time will necessarily generate patterns of egg production that vary with age. Therefore analysis of both the local and lifetime dynamics of egg production and mortality in females denied access to protein-rich food sources part of the time will deepen understanding of the reproductive physiology (Engelmann 1970) and mortality dynamics of insects in general and fruit flies in particular. *Third*, earlier studies demonstrated that both food (Carey, Liedo, Müller, Wang, and Vaupel 1998c) and host (Carey, Krainacker, and Vargas 1986) deprivation in the Mediterranean fruit fly (*Ceratitis capitata*) will extend longevity. However, these and similar studies on *Drosophila* (Chippindale et al. 1996; Chippindale et al. 1993) were conducted using constant and not variable conditions. It is currently not known whether longevity is increased further with variable food availability as a result of a caloric restriction effect or whether it will be decreased due to increased stress and/or partial starvation. *Fourth*, one of the central questions in the host foraging ecology of adult insects involves trade-offs between "egg load" and foraging time (Prokopy et al. 1994; Rivero-Lynch and Godfray 1997; Roitberg 1989; Rosenheim 1996, 1999). Knowledge of the relationship of both egg production and survival to access to protein sources will shed light on this important area of population dynamics. *Fifth*, studies on the effects of diet on aging have a long history and have included the topics of healthy living (Casper 1995; Willett 1994) and caloric restriction (Masoro and Austad 1996; McAdam and Millar 1999; Sohal and Weindruch 1996; Weindruch 1996). Thus the results of investigations concerned with the effects on vital rates of variable food availability will complement the literature on the effects of diet on aging, shed light on the underlying mechanisms aging, and provide new insights into the dynamics of the cost of reproduction.

The current study builds on previous investigations on the mortality (Carey and Liedo 1999b) and reproductive (Carey, Liedo, Müller, Wang, and Chiou 1998a; Carey, Liedo, Müller, Wang, and Vaupel 1998c) response of the medfly subject to changing dietary conditions. In this section, I present the results of a study on the Mediterranean fruit fly designed to determine the impact of periodically alternating food quality (food pulses) on its reproductive dynamics and longevity. Three general questions are addressed: (1) Is the life expectancy increased in females that have periodic access to a full (protein-rich) diet? (2) What are the general dynamic properties of egg laying in females subjected to regular pulses of protein-rich food? and (3) To what extend does the inability to produce eggs at young ages due to the lack of dietary protein preempt and/or enhance egg laying at older ages?

7.3.1. Experimental Methods

A total of 100 medfly females were subjected to 1-of-9 different dietary regimes classified in 1-of-3 ways:

1. *Cyclical*. This group consisted of 4 treatments in which food was available to flies on every 2nd, 4th, 6th, 11th, or 21st day. We coded these as 1:1, 1:3, 1:5, 1:10, and 1:20, respectively (i.e., the second number indicates the number of days flies were provided sugar-only diet after the one day of full diet).
2. *Cyclical-Lag*. This group consisted of 2 treatments in which flies were given one day of protein at eclosion followed by 30 days of sugar-only diet and then followed by either a 1:3 or a 1:5 pattern of food availability.
3. *Controls*. Flies were monitored in three sets of controls, including (i) *ad libitum* sugar-only (Control A); (ii) *ad libitum* full diets (Control B); and (iii) full diet first day after eclosion flies, and sugar-only diet thereafter (Control C).

Pair of flies were housed in 6.5 × 6.5 × 12 cm clear plastic containers. The full adult diet consisted of yeast hydrolysate and pure sucrose (1:3 ratio by volume), whereas the sugar-only diet consisted of pure sucrose only. The yeast hydrolysate (ICN Biomedicals Inc.) contained 60% protein along with vitamins and minerals (Vargas et al. 1997).

7.3.2. Descriptive Statistics

Summary statistics for the reproductive and longevity responses of female medflies subjected to the 7 treatments and 3 control conditions are presented in tables 7.2 and 7.3. The results merit several comments. *First*, it is clear that dietary pattern has a profound influence on longevity. Life expectancy at eclosion among the treatments differed by 35% or nearly 13 days, ranging from 37.0 days in the 1:1 cyclical treatment to nearly 50 days in the 1:30–1:3 lag treatment. Interestingly, neither the shortest nor the greatest life expectancy was observed in the controls. This suggests that, with respect to survival, constancy of either protein availability or sugar-only alone is better under some circumstances but worse under others. *Second*, proportionally reproduction was affected to much greater degree by dietary pattern than was life expectancy. For example, net reproduction ranged from slightly over 50 eggs/female in the sugar-only controls (Control A) to slightly less than 900 eggs/female in the *ad libitum*, full diet control (Control B). Clearly, reproduction is far more sensitive to fluctuations in the availability of protein-rich food than is life expectancy. *Third*, flies in the two treatments with the lowest life expectancy (< 40 days for full diet Control B and the 1:1 treatment) experienced the highest lifetime reproduction. Gross reproduction was only 30% higher than net reproduction for the *ad lib* control but 70% higher for the 1:1 treatment. This indicates that many more flies died in the 1:1 treatment that were still capable of producing a substantial number of eggs than those in the *ad lib* controls. In other words, rate of reproductive senescence was more closely linked to somatic senescence in the *ad lib* treatment than the cyclical treatment. *Fourth*, life expectancy was greatest for treatments in which females were subjected to long periods of sugar-only diet and lowest for treatments in which females had access to short periods of sugar-only diet. For example, the longest-lived flies were those subjected to one day of protein followed by 30 days of sugar-only diet and then followed by a 1:3 cycle (i.e., the 1:30–1:3 treatment). Generally speaking, the demographic summary measures reveal that reproduction increases with increasing access to a full diet, life expectancy increases with decreasing access to a full diet, and the longevity trade-offs with reproduction are complex due to the disparity between the rate of aging of the reproductive system and the rate of aging in the fly as a whole.

7.3.3. Cohort Survival

Patterns of cohort survival (l_x) differed across treatments, as seen in figure 7.6. The 4-day difference in life expectancy for females in Control A vs.

TABLE 7.2
Life Expectancy, e_0 (Days), and Lifetime Reproduction (Eggs/Female) for Medfly Cohorts Subject to Different Food Cycles

	Life Expectancy		*Reproduction*		
Treatment	e_0	*SD*	*Mean*	*95% CI (L)*	*95% CI (U)*
Controls					
Control A (sugar)	44.1	26.83	65.0	53.5	76.5
Control B (full)	39.7	13.94	914.4	817.2	1011.6
Control C (1:0)	46.8	30.64	762.0	664.7	859.3
Cyclical Treatments					
1:1	37.0	17.56	532.0	453.7	610.4
1:3	43.6	34.57	508.5	444.9	572.0
1:5	47.5	25.28	317.4	272.9	361.8
1:10	49.6	28.73	182.5	154.5	210.5
1:20	47.9	33.22	332.0	275.7	388.2
Lag Treatments					
1:30–1:3	49.9	31.44	220.9	181.9	259.8
1:30–1:10	47.6	31.00	136.7	120.4	152.9

Note: SD denotes standard deviation, and CI (L) and CI (U) denote lower and upper confidence intervals, respectively. Codes indicate ratio of days in which flies were given access to a full diet (sugar + yeast hydrolysate) relative to days with sugar-only (Carey, Liedo, Harshman, Liu, Müller, Partridge, and Wang 2002a).

Control B (table 7.2) was due to age differences in the patterns of survival—the survival rate of young flies maintained on *ad libitum* full diet was high initially but then rapidly declined after one month (figure 7.6a). In contrast, survival of flies given access to a sugar-only diet was nearly linear through the last fly's death. Differences existed between survival patterns of flies subjected to the different cycle treatments. For example, female survival among treatments where flies were given access to a full diet only on either the 2nd, 4th, or 6th days (treatments 1:1, 1:3, and 1:5, respectively) was similar through day 20 (figure 7.6b) but diverged thereafter. Survival after day 20 was inversely related to the frequency at which flies had access to a full diet—flies with more frequent access to a full diet died more quickly (treatment 1:1), whereas flies with less frequent access to a full diet died off more slowly (treatments 1:3 and 1:5). Differences in female medfly survival for treatments with long full-diet intervals were relatively small, as shown in figure 7.6c for treatments in which flies had access to a full diet either every 11th day (treatment 1:10) or every 21st day (treatment 1:20). Similarly differences in survival among flies subjected to the lag treatments shown in

TABLE 7.3
Mean and Standard Error for First Egg Peak and Its Location (Age)

	Number Eggs		*Age*	
Treatment	*Mean*	*SE*	*Mean*	*SE*
Controls				
Control A (sugar)	33.8	2.110	6.0	0.785
Control B (full)	41.2	1.369	10.33	0.891
Control C (1:0)	39.8	1.588	11.09	1.136
Cyclical Treatments				
1:1	45.4	2.189	8.63	0.908
1:3	49.7	2.415	6.89	0.609
1:5	49.3	2.813	6.46	0.888
1:10	44.1	3.107	5.97	0.842
1:20	50.8	3.000	8.31	0.848
Lag Treatments				
1:30–1:3	47.3	3.137	7.80	1.246
1:30–1:10	36.6	2.331	6.35	0.883

Source: Carey et al. 2002a.

figure 7.6d were also relatively minor. In general, it appears that survival patterns of medfly subjected to different cycles of protein availability was mediated primarily by qualitative (switch on or off) rather than quantitative (egg production levels) reproductive efforts.

7.3.4. Age Patterns of Reproduction

The age-specific egg production in the cohort of flies in the *ad lib* full-diet treatment (figure 7.7a) was nearly indistinguishable from the daily pattern of egg production in the cohort of flies that had access to full diet only on alternate days, treatment 1:1 (figure 7.7b). In other words, despite medfly females having access to a full diet half as many days in the 1:1 treatment relative to the *ad libitum* control, females were still capable of producing eggs at virtually the same rate as flies with access to a full diet 100% of the time. This result shows that medflies are capable of compensating for the absence of food for 24-hour periods with very little loss of lifetime productivity. One reason for the nearly identical lifetime egg production between the *ad libitum* control and the 1:1 treatment was that flies in the 1:1 treatment laid slightly fewer eggs at younger ages but compensated for this slight decrease by laying more eggs at older ages.

Distinct peaks and troughs in egg production begin to appear in the over-

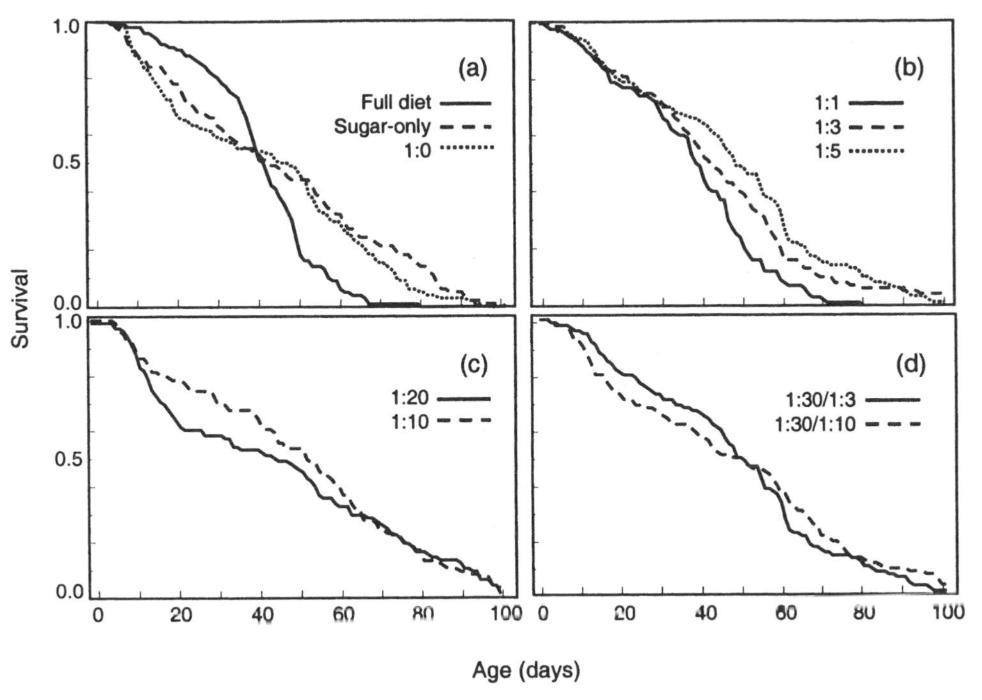

FIGURE 7.6. Survival (l_x) schedules for flies in each of the 7 treatments and 3 controls: a—three control cohorts including *ad libitum* full diet, sugar-only diet, and first day full diet followed by sugar-only diet for remainder of cohort life (1:0); b—three short-cycle treatments including full diet every 2nd, 4th, and 6th day for treatments 1:1, 1:3, and 1:5, respectively; c—two long-cycle treatments including full diet every 11th and 21st day for treatments 1:10 and 1:20, respectively; d—two lag-cycle treatments including full diet on first day followed by 29 days of sugar-only diet followed, in turn by either full diet every 4th or every 11th day (Carey, Liedo, Harshman, Liu, Müller, Partridge, and Wang 2002a).

all reproductive schedule when flies are denied access to full diet for 3 or more days, with both the peaks and the troughs directly related to the length of the period. This is evident in figure 7.7c–f, which shows the reproductive rates for flies subjected to the 1:3, 1:5, 1:10, and 1:20 treatments, respectively. Note that the peaks and troughs vary by only around 5 eggs/day in the shortest cycle treatment (3 days without full diet) but by up to 15 eggs/day at young ages in the longest cycle treatment (20 days without full diet). The cycle highs and lows in egg production generally decrease with age. Also, cohorts of flies that were denied access to a full diet at younger ages were capable of producing eggs at much older ages than were flies fed *ad lib* (figure 7.6a) or fed a full diet on alternative days (figure 7.7b).

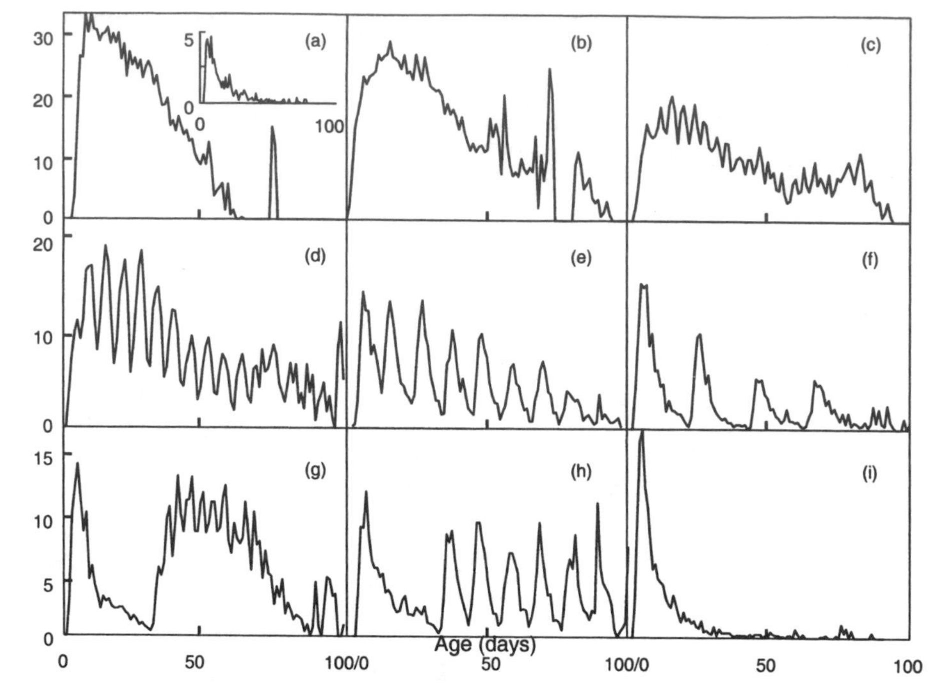

FIGURE 7.7. Average number of eggs/female/day in the medfly study consisting of 7 treatments and 3 control cohorts: a—*ad libitum* full diet (inset—sugar-only diet); b—full diet every other day (1:1); c—full diet every 4th day (1:3); d—full diet every 6th day (1:5); e—full diet every 11th day (1:10); f—full diet every 21st day (1:20); g—full diet on first day followed by sugar-only diet for 29 days, followed by full diet every 4th day; h—same as g except full diet every 11th day starting on day 30; i—full diet on first day followed by sugar-only diet (Carey et al. 2002a).

The reproductive schedules of females subjected to the lag cycles shown in figure 7.7g–h reinforce the observation in other treatments that flies denied access to a full diet at younger ages are capable of producing substantial numbers of eggs at older ages relative to *ad lib* flies. For example, the egg production rates of the female cohorts at ages beyond 50 days, shown in figure 7.7g–h, are substantially higher than egg production rates of the females subjected to similar treatment cycles but that were not denied access to a full diet at young ages (figure 7.7c). The egg-production rate did not fall to zero until females were denied access to a full diet for at least 20 days (figure 7.7f). This suggests that medfly females have adopted a reproductive strategy in which they retain the ability to produce at least a few eggs over long periods when they do not have access to dietary protein.

The data on daily reproduction for individual medfly females was used to construct event history diagrams, shown in figure 7.8a–i. These diagrams are useful because they provide insights into the within-cohort variation in age patterns of egg laying for the flies for each of the treatments. The charts reveal that the reproduction of flies denied access to a full diet part of the time was lower than that for flies with unlimited access to a full diet; the former flies experienced both a reduction in the average daily egg output, and particularly in the number of high egg-laying days, and an increase in the frequency of zero-egg-laying days. For example, females with the highest lifetime egg production were maintained on *ad lib* full diet (Control B). These females produced at least some eggs over 72% of the time and laid in excess of 30 eggs/day over 33% of the time. In contrast, the females with substantially lower lifetime egg production were those maintained on the 1:5, 1:10, and 1:20 treatments and that produced one or more eggs/day 58, 48, and 33% of the time, respectively, and produced over 30 eggs/day 6.8, 3.4, and 2.5% of the time, respectively. It thus appears that the reproductive strategy for medflies under conditions of food scarcity is to produce a small number of eggs over a sustained period rather than to produce a large number of eggs over an abbreviated period, *ceteris paribus*.

7.3.5. Within-cycle Reproductive Patterns

Composite within-cycle schedules of egg production in medfly females for selected treatments were constructed by averaging the per capita fecundity for each day of the cycle over all ages. These composite schedules thus represent the daily pattern of egg production of a hypothetical female who lived to the last day of possible life. These schedules are the within-cycle equivalent of the gross reproductive rate in classical demography—the number of eggs a hypothetical female would lay in her lifetime if she lived to the last day of possible life. Three of these hypothetical schedules are shown in figure 7.9 (for the treatments 1:5, 1:10, and 1:20) and reveal several important patterns regarding the reproductive dynamics of medflies in general and the relationship between protein availability and egg production in particular.

First, in all three treatments the maximum number of eggs were laid an average of exactly 4 days after females were given access to a full diet. The same time lag between when females had access to a new source of dietary protein and their maximal egg production suggests that the rate of production of new eggs is regulated solely by when *new* protein becomes available.

Second, in contrast to the timing of maximal egg production relative to the availability of dietary protein, the egg production *level* is dependent upon the cycle length. This is evident in the differences in the peak heights between treatments. For example, peak egg production in the 1:5 treatment was greatest (at 13.1 eggs/female) after 4 days, but in the other two treatments, peak

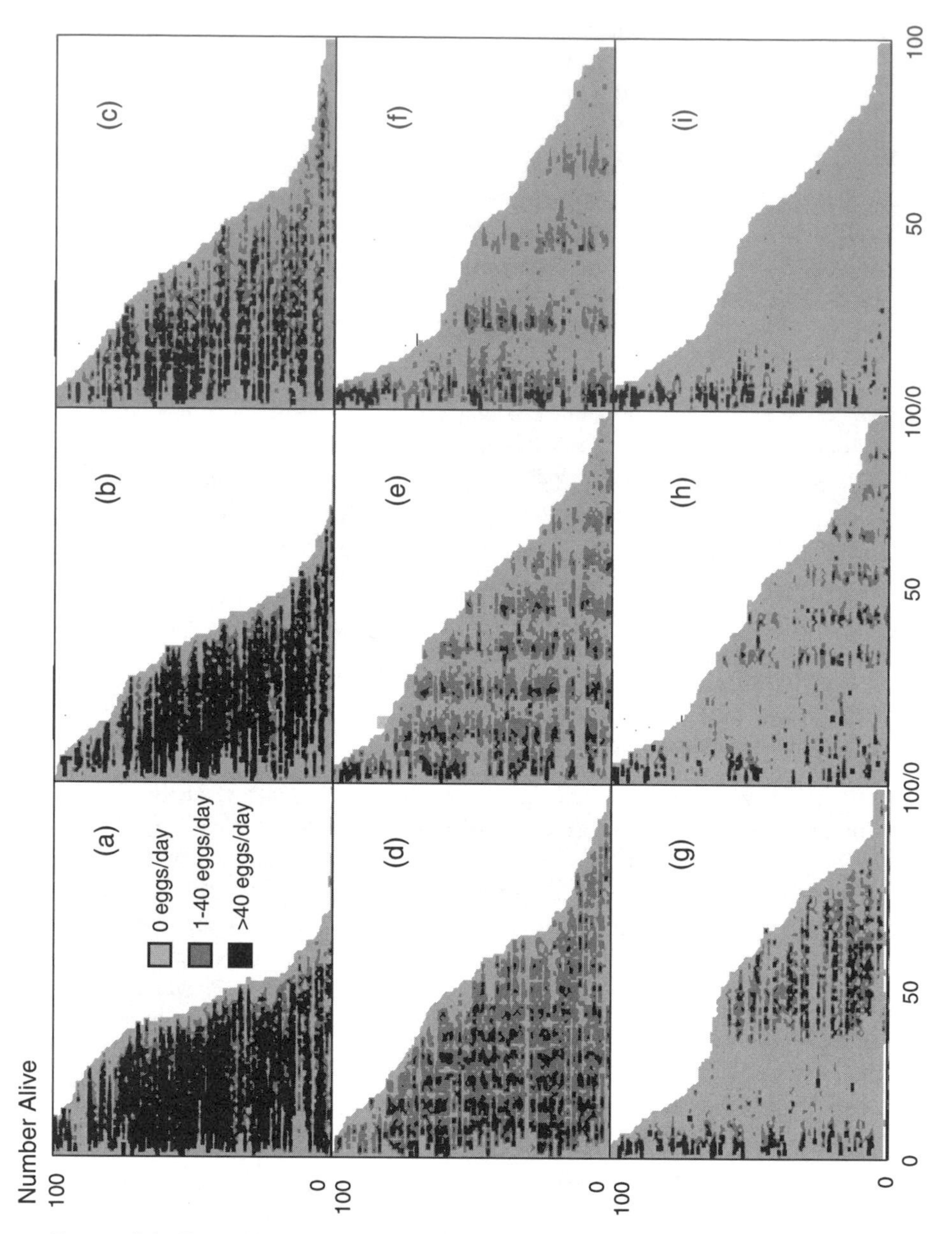

FIGURE 7.8. Event history graphs of individual female reproduction in each of the 10 treatments. Labels a–i represent same dietary treatments as specified in 7.7a–i. Each individual female within a treatment is represented by a horizontal "line" proportional to her life span (Carey, Liedo, Müller, Wang, and Vaupel 1998c). Each day of a female's life is coded according to whether she laid 0 (light gray), 1–40 eggs (medium gray), or greater than 40 eggs (black) (Carey et al. 2002a).

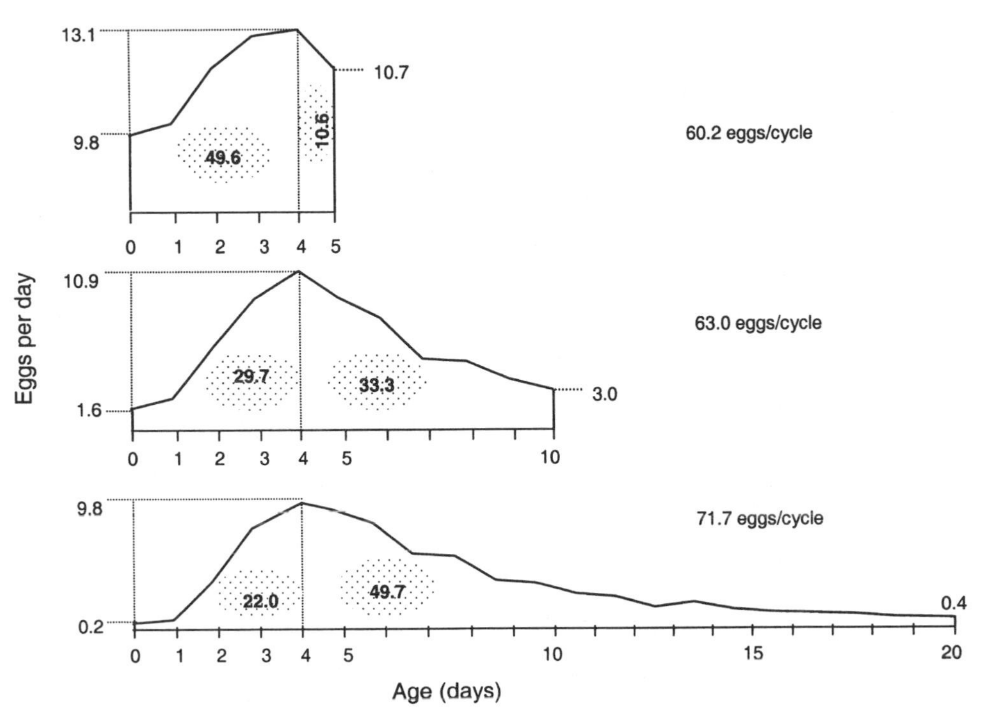

FIGURE 7.9. Patterns of egg production for composite egg-laying cycles averaged over the life course of all flies. These composites correspond to a within-cycle, gross reproductive rate—the average number of eggs laid per day (within cycle) for a female who lives to the last day of possible life. Note that the number of eggs at the start of the cycle and its peak are given for each treatment on the left of the schematic and the number of eggs at the end of the cycle are given on the right of the schematic. The average total number of eggs both before and after the peak are contained within each graph and the total eggs per cycle is given to the right of each schematic (Carey et al. 2002a).

egg production was less than 11 eggs/female. This suggests that a female's current blood protein level determines whether she has enough reserves immediately to synthesize proteinaceious and lipid yolk for egg maturation (Engelmann 1970).

Third, the *rate of increase* in egg production depends on the level of egg production on the day when females are given access to new protein. For example, egg production in the 1:5 treatment increased by only 3.4 eggs per 4 days from food-day to peak, whereas egg production in the 1:10 and 1:20

treatments increased by 9.3 and 9.6 eggs per 4 days, respectively. Also note that egg production began increasing immediately after the flies had access to new protein.

Fourth, the number of eggs per "food-cycle" differs by only around 20% between the 6-day (1:5) cycle and the 21-day (1:20) cycle. The difference suggests that both short-term and lifetime fecundity is more food-limited rather than time-limited, *ceteris paribus*. That is, females subjected to the 6-day dietary cycle produced 60 eggs per cycle but females subjected to a cycle over 3-fold greater produced only 12 eggs more.

7.3.6. Modeling Relationships and Graphical Analysis

The dynamic relationships were examined between feeding regimen, egg production, peak egg laying, age at peak egg laying, and longevity using several statistical regression models, including a linear model fitted by least squares with lifetime number of eggs as response, a Poisson regression model fitted by maximum likelihood with lifetime (age-at-death) as response, and a variety of nonparametric regression fits to describe the relationships between pairs of variables.

MODEL 1

This model relates the total number of eggs (EggNo) as response variable with the predictors lifetime (age-at-death), no. of eggs at peak egg production (peaklevel), age at peak egg laying (peakage), and eight treatments. The eight treatments considered in Model 1 are coded by seven treatment indicator variables, assuming the values 0 or 1, as follows, where the designations for the treatments are the same as those used in table 7.1:1 tr(1:1) = 1 for the 1:1 cyclical diet (full diet on alternate days); tr(1:3) = 1 for the 1:3 cyclical diet; tr(1:5) = 1 for the 1:5 cyclical diet; tr(1:10) = 1 for the 1:10 cyclical diet; tr(1:20) = 1 for the 1:20 cyclical diet; tr(C) = 1 for Control C (one day protein, then all other days sugar); and tr(A) = 1 for Control A (sugar only). Since the egg-laying patterns for the treatments with switching diets—the lag treatments of table 7.1—were quite different, the two lag treatments were not included in this model. If the fly is under a specific treatment, the respective treatment indicator is = 1 while all other treatment indicators are = 0. The case where all treatment indicators are zero corresponds to the baseline treatment, which is Control B (*ad lib* full diet).

The fitted model involves a log transformation of the response and various transformations of the predictors. As a main criterion for model selection and variable transformations, the Akaike information criterion (AIC) was used. For variable transformations, we investigated Box-Cox transformations and we also checked for interactions between the predictors, which were not included in the final model. The residuals of the model were reasonably normal. The equation of the fitted model is given by

$$\begin{aligned} \text{EggNo} = \{&\exp(1.96 + .0076^* \times \text{lifetime} + .1537^* \\ &\times \log(\text{peaklevel}) - 1363^* \times \text{sqrt}(\text{peakage}) + .0061 \times \text{tr}(1{:}1) \\ &- .1277^* \times \text{tr}(1{:}3) - .2060^* \times \text{tr}(1{:}5) - .4188^* \\ &\times \text{tr}(1{:}10) - .6418^* \times \text{tr}(1{:}20) - .6884^* \\ &\times \text{tr}(C) - .8673^* \times \text{tr}(A))\}^{8/3}. \end{aligned} \tag{7.3}$$

(The asterix [*] indicates that the corresponding predictor has a significant effect on the response, $P < .05$.)

This model reveals three important relationships. *First*, with the exception of the 1:1 cyclical treatment (full diet on alternate days), which is almost indistinguishable from the baseline *ad lib* diet in terms of the effect on lifetime number of eggs, all of the treatments lead to significantly different amounts of total eggs as compared to the baseline control B treatment (*ad libitum* full diet). This is seen as the corresponding treatment indicator variables, which are predictors in the model and have significant regression coefficients. *Second*, again ignoring the 1:1 cyclical treatment, there exists a monotone dose-response relationship in that the number of eggs laid increases in a monotone fashion with the frequency of protein days. In other words, the more frequent the availability of full diet is, the greater is the number of total eggs produced. While increased protein in the diet is also related to shortened longevity with an associated shortening of the egg-laying period, this finding implies that this shortening of the egg-laying period is more than compensated for by an increase in the intensity of egg laying. *Third*, the significance of the coefficients for lifetime (age-at-death) and for both the peak height and the location (age) of the egg production peak indicate that early and high peaks of egg production and a long female lifetime are positively related to lifetime egg production. Highly productive egg-layers can therefore be identified early in life, in agreement with the findings in (Müller et al. 2001).

MODEL 2

Lifetime (age-at-death) as response was related to treatment and peak characteristics (it is important to note that all of these variables are available at an early age of the fly) as predictors. We found that a Poisson regression model, which is a special case of the generalized linear model (McCullagh and Nelder 1986) using the log link function, was adequate for modeling this regression relationship. The predictors are the seven treatment indicators as defined in Model 1, plus two indicators for the two lag treatments, tr(1:30 − 1:3) = 1 if the fly received the lag treatment 1:30 − 1:3, and tr(1:30 −

1:10) = 1 if the fly received the lag treatment 1:30 − 1:10 (table 7.2). As in Model 1, the baseline treatment when all indicators are zero is Control B (*ad lib* full diet). The predictors for the peak characteristics are again peaklevel and peakage, as described in Model 1. The equation of the fitted model is

$$\text{Lifetime} = \exp(3.6651 - 0.0764 \times \text{tr}(1{:}1) + 0.0534 \times \text{tr}(1{:}3) + 0.1681 \times \text{tr}(1{:}5) + 0.2041 \times \text{tr}(1{:}10) + 0.1537 \times \text{tr}(1{:}20) + 0.1403 \times \text{tr}(C) - 0.01058 \times \text{tr}(A) + 0.2116 \times \text{tr}(1{:}30 - 1{:}3) + 0.1145 \times \text{tr}(1{:}30 - 1{:}10) - 0.00192 \times \text{peaklevel} + 0.01837 \times \text{peakage}) \quad (7.4)$$

All of the coefficients in this model turn out to be significant, $P < .05$. The main inferences that we can draw are the following: *First*, longevity is reduced when the peak location of egg laying is earlier and the peak is higher. This is consistent with the general concept of reproductive trade-offs with mortality: early egg laying tends to reduce later egg laying and to reduce longevity. *Second*, less frequent protein days tend to enhance longevity, with two notable exceptions: full diet every other day, the 1:1 cyclical diet, and the sugar-only diet (Control A) are associated with a decrease in longevity. The maximum increase in longevity is obtained for the cyclical 1:10 treatment.

Graphical displays of the egg-laying data (figure 7.10) plotted against female lifetime (top), peak location (middle), and peak size (bottom) help to visualize the modeling results as well as reveal other important relationships. The significance of the relationships was established in models 1 and 2. In these graphs we use nonparametric regression fits to assess the shape of the relationship, which in most cases is nonlinear. The nonparametric regression estimates allow nearly arbitrary shapes for the regression function and are therefore very useful for modeling nonlinear trends that cannot be easily parameterized (McCullagh and Nelder 1986).

Several aspects of these graphs merit comment. *First*, lifetime egg production is positively correlated with female lifetime but only up to a certain age, at which point no correlation exists. The initial correlation is the result of limits on daily egg production constraining lifetime production—females must live several weeks before they have time to lay a substantial complement of eggs. The lack of correlation of lifetime egg production and longevity at high female lifetimes indicates that lifetime egg production may ultimately be limited by an endowment concerning the potential eggs a female is capable of producing in her life time. *Second*, the relationships between egg production and both the location (age) of the initial egg peak and its size are shown in figure 7.10b and c respectively. These graphs reveal that the size of the egg peak is positively associated with lifetime egg production, whereas the location is negatively associated with lifetime egg production; in other words, the older the age at which the initial peak occurs, the lower the number of lifetime eggs produced.

7.3.7. Discussion and Implications

The studies on the response of the medfly to food pulses were designed to both complement and build on the results of previous medfly demographic research, including baseline studies on birth and death rates (Carey 1984; Carey, Liedo, Müller, Wang, and Chiou 1998a; Carey, Liedo, Müller, Wang, and Vaupel 1998c; Carey, Yang, and Foote 1988; Vargas and Carey 1989; Vargas et al. 1997), experimental investigations in which conditions of food or host availability are manipulated (Carey, Krainacker, and Vargas 1986; Carey et al. 1998c; Krainacker et al. 1987), and modeling cost of reproduction (Müller et al. 2001). None of these previous studies provided insights into the life history response of medfly females to cyclical environmental perturbations at the level of the individual, that is, comparative studies in which daily reproduction of individual females was monitored for a wide variety of treatments over their entire lifetime.

The life history response of individual females to these different patterns of environmental manipulations shed important new light on the reproductive dynamics of medflies in particular and insects in general. First, the current findings shed important light on the dynamics of insect reproduction including the strategic trade-offs between reproduction and survival in the medfly—how females retain amino acids in their system as "common currency." Specifically the within-cycle patterns of reproduction suggest that the reproductive strategy of medfly females with respect to protein allocation involves three levels: (1) a fraction of amino acids are always held in reserve in the fat bodies and hemolymph rather than used for the immediate production of yolk protein; (2) even though flies may not have access to new protein sources, a fraction of this reserve is allocated daily to vitellogenesis; (3) new protein food stimulates flies to allocate a greater fraction of the amino acids reserve for egg production than would be allocated if no new protein food was available.

Second, the results reveal that the concept of cost of reproduction (increment of reproduction exchanged for a decrement in survival) is not a straightforward trade-off between reproduction and survival (Reznick 1985). Although medflies appear to be endowed with a fixed number of "reproductive units" (i.e., potential eggs) (Bell and Koufopanou 1986), the trade-off involves at least two components that preempt either all or a fraction of the potential reproduction at older ages and, in turn, reduce lifetime reproduction. These are (1) risk of premature death, which will prevent all reproduction at older ages; and (2) loss of potential eggs due to an apparent age-

Egg-laying data plotted against female lifetime:

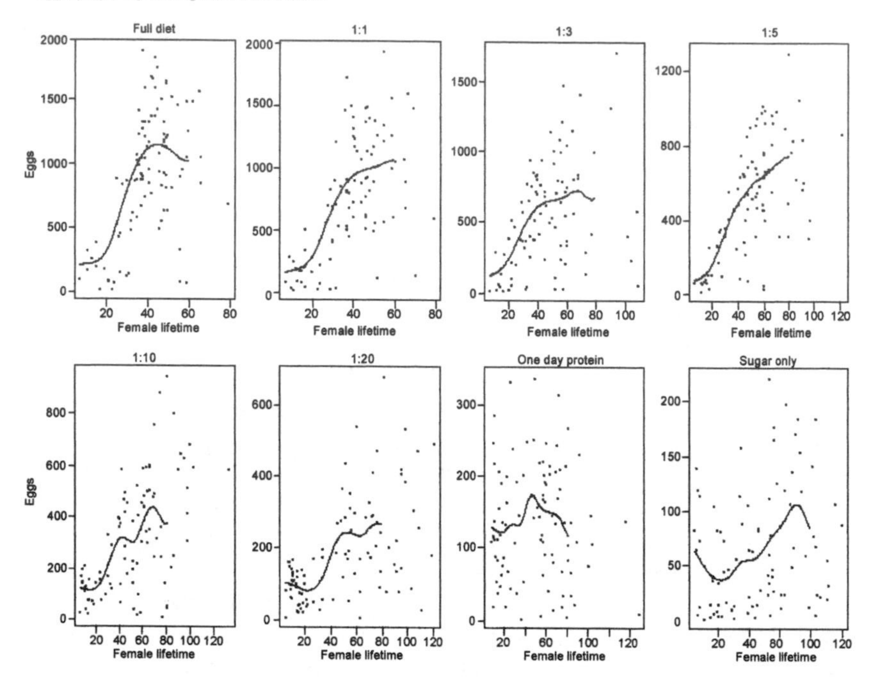

Egg-laying data plotted against peak location:

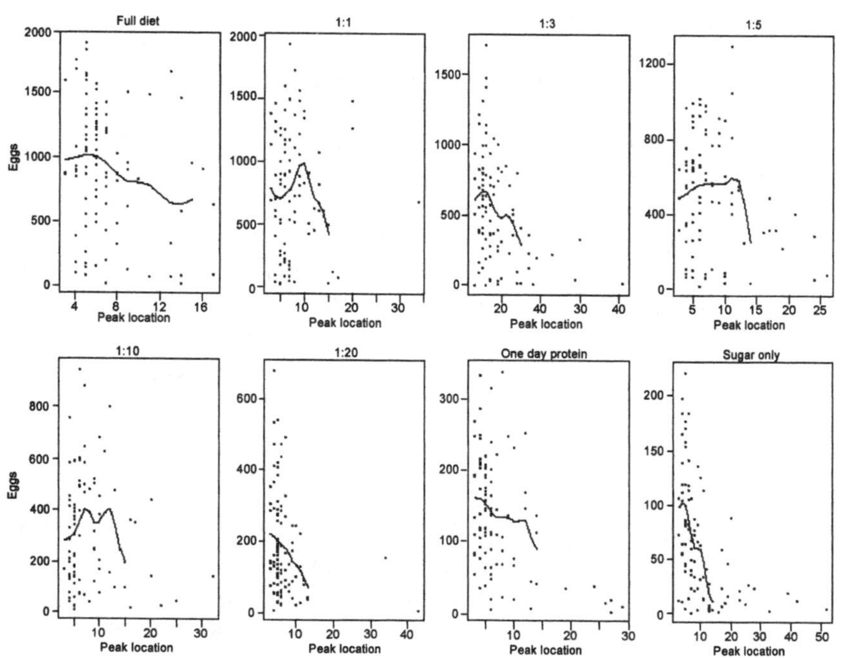

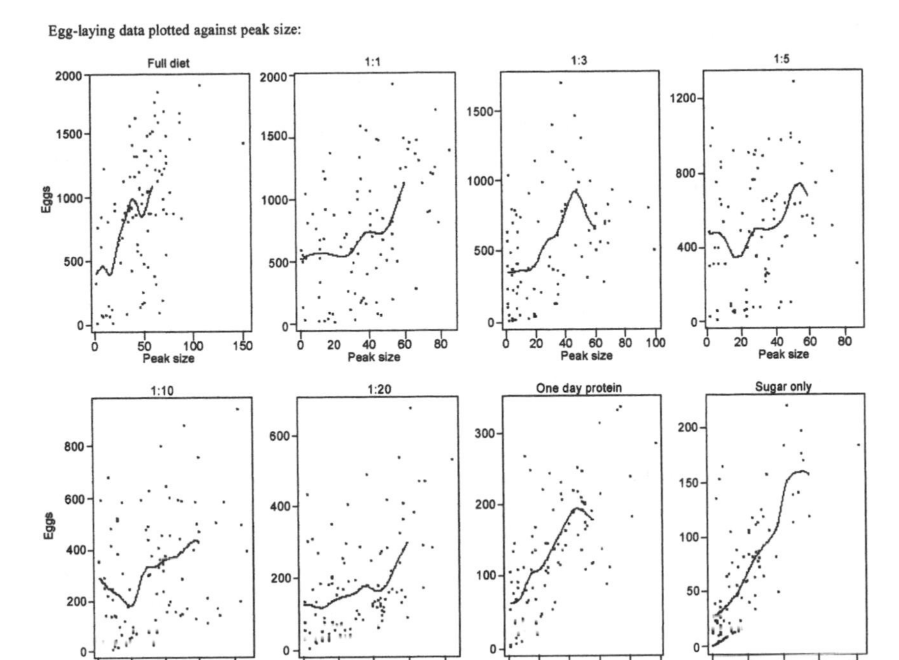

FIGURE 7.10. Relationship of lifetime egg production (response) and the predictors of female longevity (opposite, top), age of initial peak of egg production (opposite, bottom), and the height of the initial egg-laying peak (above). For each of these relations, a scatterplot is shown, indicating the observations made on each individual fly. The graph of a nonparametric regression fit is superimposed, indicating that these relationships are highly nonlinear (Carey et al. 2002a).

related degradation of oogonia (potential eggs). In other words, the level of reproduction at older ages in flies that laid few eggs at younger ages is higher than reproduction in flies that realized their full reproductive potential at younger ages. However, as a female ages, a fraction of her potential reproduction is completely lost due to aging alone, even if she manages to live to an extreme age. This same response to food manipulations was also observed when medfly females were deprived of ovipositional hosts (Carey et al. 1986).

Third, although the food restriction methods used in the current study with medflies (i.e., high-quality food either available or not available on a given day) differ from the food restriction methods used in mammalian caloric restriction studies (i.e., daily access to reduced amounts), the general response was similar to rodent studies (Masoro 1988; Masoro and Austad 1996)—extended life expectancies and reduced reproduction relative to *ad libitum* controls. Therefore we believe that the hypothesis introduced by Holliday (1989) regarding the evolutionary response of animals to reduced food availability is relevant to our medfly studies. Holliday noted that from all the experimental studies that have been carried out in which animals are maintained on a near-starvation diet, breeding is prevented during this period. However, breeding will be resumed when food supply becomes plentiful and females will retain the capacity to breed at significantly later ages than animals with a high-calorie diet throughout their life. These patterns led Holliday to postulate that Darwinian fitness will be increased in mammals and (by extension) other groups of organisms if animals cease breeding during periods of food deprivation and invest saved resources in maintenance. This would increase the probability of producing viable offspring during an extended life span.

Fourth, a great deal of interest in reproductive dynamics exists because egg load influences host foraging by parasitoids in natural populations (Casas et al. 2000), determines stability properties of host-parasitoid models (Rosenheim 1996, 1999), and affects the host-finding behavior in a wide range of herbivorous insects (Prokopy et al. 1994). The results of the current studies suggest that egg load—the number of mature eggs present in a female at a given time—does not necessarily reflect the lifetime reproductive strategy. Rather egg load is likely determined by a combination of both feeding (i.e., access to dietary protein) and egg-laying history.

With a few notable exceptions (Boggs 1986; Boggs and Ross 1993; Chapman et al. 1994), surprisingly little research has been previously conducted on the reproductive response of insects to variable food environments. This study provides new insights into how insects respond to variable nutritional environment, including their cyclical reproductive response to pulses of protein-rich food, their increased longevity in response to reduced reproductive effort, and their lifetime patterns of egg production relative to longevity and level and age of peak egg production.

7.4. Summary

The three main studies described in this section contribute to understanding the linkage between reproduction and longevity in the following respects. *First*, the results of the dual-modes study (section 7.1) revealed the trade-off at the *cohort level* between early and late reproduction and early and late mortality. Cohorts that were provided access to a full diet immediately entered the reproductive mode where the mortality was initially low but then increased rapidly as reproduction continued. *Second*, the results of the reproductive clock study (section 7.2) revealed the trade-off at the *individual level*: the rate of change in reproductive rate with age was the primary determinant of mortality at older ages. In other words, the dynamic of egg laying at young ages mattered more than the intensity (total number). *Third*, the results of the study on food pulses (section 7.3) revealed the general constraints of food manipulation on both reproduction and longevity. Although virtually any method of food manipulation that reduces reproduction will increase longevity, the extent of increase has limits. It was not possible to extend longevity in medflies subjected to variable rates of access to full diets substantially beyond what was achieved by simple manipulations of constant access after certain ages. The collective importance of all of these studies is that all three revealed the complexity of the cost of reproduction concept. The examination of the dynamics of birth and death rates through both the mortality schedules and, in the case of the reproductive clock study, the slope of the individual reproductive schedule, provides greater empirical and theoretical depth to the cost of reproduction concept.

8

General Biodemographic Principles

SCIENCE is the systematic enterprise of gathering knowledge about the world and organizing and condensing that knowledge into testable laws and theories. The first step in this process is the identification of general patterns used to formulate principles and working hypotheses, which, as Cajal (1999) notes, "are the first murmurings of reason in the darkness of the unknown; are the sounding instruments lowered into the mysterious abyss; are the audacious bridges connecting the familiar shore with the unexplored continent" (p. 117). Inasmuch as scientific principles and objective data are bound together in close etiological and epistemological relationships, the store of hard data is of limited use without principles. It follows that without biodemographic principles, the value of the data on mortality, survival, and longevity gathered on the medflies and other model species is never fully realized; accumulation of data is not the same as acquisition of knowledge.

I thus have two broad objectives in this chapter. *First*, to identify or to derive general biodemographic principles using the results of the medfly studies (e.g., mortality deceleration, cost of reproduction) and concepts taken from the general literature on aging (e.g., biology of senescence, comparative demography of life span). *Second*, to situate human longevity in the context of biodemography by summarizing the results of a recent paper on predictions of human life span from primate patterns (Judge and Carey 2000) and by briefly reviewing the literature on proximate determinants of longevity (e.g., healthful living).

This chapter will thus serve as a transition from the presentation and discussion of results from the medfly experimental demography studies contained in chapters 3–7 to the development of general demographic concepts that are more directly relevant to human longevity and life span. This shift will set the conceptual stage for development of a theory of life span extension in humans, which I present in the chapter that follows, and provides information and concepts that I draw upon in the final chapter.

8.1. Why Biological Data Is Important for Deriving General Principles

Conventional demography, defined by Pressat (1985) as "the study of populations and the processes that shape them," is an observational science dependent for its data on observation and recording of events occurring in the external world rather than on experiments under controlled conditions. One of the overriding constraints of any of the observational social sciences such as demography and sociology was referred to by Hauser and Duncan (1959) as "the problem of historicism"—the question of the extent to which generalizations drawn from human data localized in time and space can lead to general principles rather than simply to descriptions of situations unique to a particular time and location. This constraint preempts the use of any human data alone as a source for the derivation of the most basic principles.

8.2. Principles of Senescence

Timiras (1994) notes that despite some minor interpretative differences the terms aging and senescence are often used interchangeably—aging refers to the process of growing old regardless of chronologic age, whereas senescence is a process restricted to the state of old age characteristic of the late years of an organism's life span. Senescence in this context is defined as "the deteriorative process characterized by increased vulnerability, functional impairment, and probability of death with advancing age" (Timiras 1994). In this section I describe two principles of senescence that are fundamental to biodemography because they provide the biological, evolutionary, and conceptual foundation for its constituent disciplines—whereas demography is concerned with the determinants of probabilities of death, biology is concerned with the determinants of vulnerability.

8.2.1. All Sexual Organisms Senesce

G. Bell (1988) established the deep connection between the two invariants of life—birth and death—by demonstrating that protozoan lineages senesce as the result of an accumulated load of mutations. This senescence can be arrested by recombination of micronuclear DNA with that of another protozoa through conjugation. Conjugation (sex) results in new DNA and in the apoptotic-like destruction of old operational DNA in the macronucleus. Thus, rejuvenation in the replicative DNA and senescence of operational DNA is promoted by sexual reproduction. When this is extended to multicellular organisms, sex and somatic senescence are inextricably linked (W. Clark 1996). In multicellular, sexually reproducing organisms, the function of somatic cells (i.e., all cells constituting the individual besides the germ cells) is to promote the survival and function of the replicative DNA—the germ cells (W. Clark 1996). Prior to bacteria, the *somatic* DNA was the *germ line* DNA; prior to multicellular animals, the *somatic cell* was the *germ cell*. Like the macronuclei in the paramecia, the somatic cells senesce and die as a function of their mitotic task of ensuring the survival and development of the germ cells. The advent of sex in reproduction allowed exogenous repair of replicative DNA (Bell 1988), while in multicellular organisms the replication errors of somatic growth and maintenance are segregated from that DNA passed on to daughter cells and are discarded at the end of each generation. Senescence is built into the life history of all sexually reproducing organisms. Thus, the death rate can be altered by modifying senescence, but death itself can never be eliminated. This evolutionary argument concerning senescence is one of the fundamental canons in the emergence of all sexually-reproducing organisms.

8.2.2. Natural Selection Shapes Senescence Rate

Whereas the previous principle provides a theory of how and why senescence evolved, it provides no framework for the evolution of a particular rate of senescence. Hypotheses regarding the onset and rate of senescence were developed independently by Medawar (1952) and Williams (1957), both of whom argued that senescence was essentially the result of deleterious genes not being selected out of a population because the force of selection decreases with age. Many deleterious genes that arise by mutation are not selected out because they are only expressed at relatively late ages; over time the population acquires a load of late expressing deleterious mutations (Medawar 1952). Genes favorable to reproduction when young may be deleterious at older ages. This is referred to as *negative* or *antagonistic pleiotropy* (Williams 1957). The broad concept that somatic cells are produced only as a means for generating more germ cells has been characterized as the *disposable soma theory* (Kirkwood 1977). This theory explains aging as an organismal trade-off between allocation of metabolic resources for somatic maintenance versus producing progeny, both of which continue the germ line.

8.3. Principles of Mortality

The single most important function of the life table is age-specific mortality—the fraction of individuals alive at age x that die prior to age $x + 1$. There are at least three reasons why this function is more important than, for example, cohort survival or life expectancy (Carey 1999):

1. Death is an event constituting a change of stage from living to dead whereas survival is a continuation of the current state. Life table parameters are based on probabilities of measurable events rather than "nonevents" like survival. This is important because death can be disaggregated by cause whereas survival cannot.

2. Age-specific mortality is algebraically independent of events at all other ages and thus changes in age patterns can often be traced to underlying physiological and/or behavioral changes at the level of the individual. With the exception of period survival, this is not true for the other life table parameters.

3. Several different mathematical models of mortality (e.g., Gompertz 1825) have been developed that provide simple and concise means for expressing the life table properties of cohorts with a few parameters (Keyfitz 1982).

In the following section I describe mortality concepts that I believe are both general and relevant to understanding mortality in humans.

8.3.1. Biological Organisms Die Whereas Mechanical Systems Fail

Although the term "death" is often used in metaphorical contexts, as in describing divorce as the death of a marriage and an automobile accident as the death of a car, the term refers literally only to the cessation of life. This distinction between literal and metaphorical is important in the context of mortality or death rate. Whereas marriages and cars can fail and therefore possess a failure rate, only living organisms can die and therefore possess a true death rate.

I believe that this distinction is important in biodemography because of the argument that complexity may explain similarities in the age-specific patterns of death (or failure) in mechanical and biological systems (Abernethy 1979; Vaupel 1997; Vaupel et al. 1998). For example, the Weibull mortality model (1951) is used extensively in both biodemographic and industrial situations to model processes ranging from fatigue life of ball bearings and electron tube failure to human mortality and disease rates (Lee 1992). Gerontologists have applied analytical tools and concepts derived from reliability theory to the "wear-and-tear" theory of human aging (Finch 1990; Gavrilov and Gavrilova 1991).

One of the key differences between a biological organism and a mechanical system is the concept of *development*. Inasmuch as all living organisms grow through stages with multiple interdependencies among life history stages, prolonging one stage changes the proportionality of stages within the overall life cycle and is apt to effect unanticipated changes in other life-cycle stages (Barker 1994). For this reason, organisms are more than complex "machines" and therefore their mortality patterns are manifestations of far more than the complexity of their component parts. This is important when comparing longevity extension in living organisms to longevity extension in machines by eliminating "cause" or by improving the quality of its parts, construction, and maintenance. Whereas a machine is built *de novo* as a nonreproducing "adult" to perform the same task(s) throughout its life, the complexities of stage progression and multiple inter-linkages must be considered when attempting to compare longevity extension in living systems with the extension of "warranty period" in machines and equipment.

8.3.2. Mortality Decelerates at Advanced Ages

Slowing or deceleration of mortality at older ages has been observed in virtually every large-scale life table study on insects. These include *Drosophila* (Clark and Guadalupe 1995; Curtsinger et al. 1992; Fukui et al. 1993), houseflies (Rockstein and Lieberman 1959), medflies (Carey, Liedo, Orozco, and Vaupel 1992), and bruchiid beetles (Tatar and Carey 1994a, 1994b). Similar patterns are shown in human populations (Barrett 1985; Horiuchi and Wilmoth 1998; Kannisto 1988; Klemera and Doubal 1992; Riggs and Millecchia 1992; Strehler and Mildvan 1960; Thatcher 1992). There are three reasons why this general principle is important (Carey et al. 1992; Demetrius 2001; Vaupel et al. 1998; Wachter 1999): (1) it provides a conceptual and empirical point of departure from the Gompertz model of ever increasing, age-specific mortality; (2) it forces demographers and gerontologists to rethink the idea that senescence can be operationally defined and measured by the increase in mortality rates with age; and (3) it suggests that there is no definite life span limit (see subsequent section).

8.3.3. Mortality Is Sex-specific

The prevailing wisdom in gerontology is that the female advantage in life expectancy is a universal law of nature. Carey et al. (1995a) tested whether a female longevity advantage exists for *C. capitata* and discovered that the answer was not straightforward—males exhibited a higher life expectancy at eclosion, but females were 4 times more likely than males to be the last to die. They concluded that there were at least three reasons why it is impossible to state unequivocally that either males or females are "longer-lived" (Carey and Liedo 1995b). *First*, longevity can be characterized in different ways (e.g., life expectancy at eclosion [day 0], life expectancy at day 30, age when 90% of the original cohort is dead [life endurancy], maximal life span, etc.); also, one measure of longevity often favored one sex, whereas another measure favored the other sex. *Second*, there is considerable variation among cohorts for a given longevity measure. For example, neither male nor female longevity was greater in all of the cages regardless of the measure used. And *finally*, relative longevity for the two sexes was conditional on the environment in which they were maintained or the treatment to which they were subjected. Expectation of life for males and females was similar if flies were maintained in solitary confinement but favored males if the flies were maintained in grouped cages. The overall conclusion was that sex-specific mortality responses and, in turn, male–female life expectancy differences cannot be predicted a priori; and that a female-longevity advantage is not universal across species.

8.3.4. Mortality Trajectories Are Facultative

The term "facultative" is used in biology to describe life history traits that have alternative conditions that often vary with environmental conditions. For example, clutch size in some birds, diapause in insects, and diet selection in many animals are all considered facultative. I suggest that the term also applies to mortality patterns in the medfly and most other species because there exists no unique pattern—the specific trajectories frequently depend on the environmental conditions. One of the most compelling findings emerging from the collection of life table studies on the medfly, and one that was not evident even after the first large-scale study was completed (Carey, Liedo, Orozco, and Vaupel 1992), is that the female mortality patterns are extraordinarily plastic. The reason why this elasticity was not evident from the first series of studies is that none involved manipulations that altered the physiology and/or behavior of the flies. It is now apparent that manipulations that affect components of a fly's life history, such as irradiation, diet, or mating, have a profound effect on the trajectory of mortality in females and less of an effect on male trajectories (Carey, Liedo, Müller, Wang, Love, Harshman, and Partridge 2000).

In light of these findings on the medfly, I believe that Olshansky and Carnes (1997) are incorrect in their assertion that species possess a characteristic "signature" and thus an "irreducible" mortality component defined by Tuljapurkar and Boe (1998) as "a component of mortality that will never be reduced by human intervention." The existence of such a component would effectively set a limit to life span. Rather, I believe that the experimental evidence in virtually all species for which large-scale data sets on mortality have been examined (including humans) suggests that mortality trajectories are facultative; that there is no component of a species' mortality trajectory that is intractable.

8.3.5. Selection Shapes Mortality Trajectories

The concept of subgroups endowed with different levels of frailty is known as *demographic heterogeneity*, and the winnowing process as the cohort ages is referred to as *demographic selection* (Carey 1997; Vaupel et al. 1979). As populations age, they become more selected because groups with higher death rates die out in greater numbers than those with lower death rates, thereby transforming the population into one consisting mostly of individuals with low death rates (Carey and Liedo 1995a; Keyfitz 1985; Rogers 1995; Vaupel et al. 1979). The actuarial consequence of cohorts consisting of subsets, each of which possesses a different level of frailty, is that the mortality trajectory of the whole may depart substantially from Gompertz rates even though each of the subgroups displays Gompertz mortality rates. Vaupel and Carey (1993) fitted observed *C. capitata* mortality patterns with mixtures of increasing Gompertz curves and demonstrated that twelve subgroups were sufficient to capture the observed pattern of medfly mortality using a range of frailty values and initial proportions of subgroups. Demographic selection winnows the frail and leaves the robust and, thus, shapes the mortality trajectory as cohorts age.

8.3.6. Mortality Rates Are Undetectable below **1/n**

A major concern of any study designed to estimate mortality rates at older ages in a cohort is sample size. The number of individuals at risk of dying becomes progressively less with age due to attrition. Consequently the sample size needed to measure mortality at older ages may be insufficient. The problem of insufficient sample size also may apply to the measurement of mortality at young ages, even when the number of individuals at risk is at or near the initial number n. This occurs whenever the "actual" mortality rate is less that $1/n$ and has been referred to as the "left-hand boundary problem" (Promislow et al. 1999). For example, a mortality rate of $\mu = 0.001$ (an average of 1 out of 1,000 individuals die in the interval) cannot be detected with a sample size of $n = 100$ inasmuch as a single death yields an estimate of $\mu = 1/100 = 0.01$. Thus the combination of both age-specific cohort size *and* age-specific mortality determine whether mortality at a given age can be detected. In conclusion, this problem with "demographic sampling error" can bias estimates of rates of aging, age at onset of senescence, costs of reproduction, and demographic tests of evolutionary models of aging (Promislow et al. 1999).

8.3.7. Mortality Variance Increases with Cohort Age

Chiang (1984) noted that the life table is often presented as a subject peculiar to public health, demography, and actuarial science, and thus its development has received little attention in the field of statistics. One reason for this is that human life tables, at least at the younger ages, are constructed from thousands and sometimes tens of thousands or millions of deaths. Statistical properties such as the standard deviation and variance of mortality at each age are small and thus inconsequential. However, in biology this is often not the case—frequently life tables are based on substantially fewer deaths, either by design or because some animals are long lived, rare, and expensive to maintain (or to follow in the wild), and thus life table information is scarce. Such statistical problems are becoming increasingly important in human populations, particularly at the older ages where the numbers at risk of dying are often exceedingly small.

The number of individuals in a cohort that are subject to the risk of dying in all studies of *real* cohorts decreases with age due to attrition. There will be few individuals left alive and thus subject to the risk of dying at extreme ages in cohorts that began with small initial numbers. Consequently the confidence intervals of mortality will become greater with age. The equation for the 95% confidence interval for age-specific mortality is given as

$$CI_{95\%} \equiv \hat{q}_x \pm 1.96 S_{\hat{q}_x} \tag{8.1}$$

where $S_{\hat{q}_x}$ denotes the standard deviation of the death rate at age x. The formula for $S_{\hat{q}_x}$ is

$$S_{\hat{q}_x} = \hat{q}_x \sqrt{\frac{1}{D_x}(1 - \hat{q}_x)} \tag{8.2}$$

where $\hat{q}_x$ is the age-specific mortality rate at age x and D_x is the number of deaths at age x. Diverging intervals is a characteristic of all mortality schedules that are based on the absolute number dying at each age (rather than subsamples). This is because the number at risk decreases with age due to attrition and therefore the variance increases. For example, the mortality variance when $\hat{q}_x = 10^{-4}$ and the number at risk of dying is one million individuals is nearly the same as when $\hat{q}_x = 10^{-1}$ but only 100 individuals are at risk of death.

8.4. Principles of Longevity

Longevity refers to the period between birth and death of an individual. It is operationalized in several different ways, including *expectation of life at birth*—the average number of years (days, weeks, etc.) that a newborn will live; *median life span*—the age at which half of an initial cohort is dead (or alive); *life endurancy*—the age at which 90% of the original cohort is dead; and *record life span*—the age at which the longest-lived, observed individual died. As a life history trait longevity covaries with other traits such as body size, brain size, ability to fly, possession of armor, subterranean habits, and sociality (S. Austad 1997; S. N. Austad and Fischer 1991; Sacher 1978, 1980a, 1980b, 1982).

8.4.1. Longevity Is Adaptive

In evolutionary biology, an "adaptation" is a characteristic of organisms whose properties are the result of selection in a particular functional context (West-Eberhard 1992). Just like different bird beaks are adaptations for exploiting different niches and must be balanced with the other traits such as body size and flight propensity, the longevity of an animal is also an adaptation that must be balanced with other traits, particularly with reproduction. As Hirshfield and Flaws (1999) note, the variations in the relationship between reproduction and longevity can make sense only when placed within the context of such factors as demographics, duration of the infantile period, number of young, and the species' ecological niche—the organism's overall life history strategy. Indeed, the longevity potential of a species is not an arbitrary or random outcome of evolutionary forces but rather an adaptive one that must fit into the broader life history of the species (Charnov 1993; Harvey et al. 1989; Read and Harvey 1989; Stearns 1992).

A graphical summary of the longevity data for each of four broad vertebrate groups is presented in figure 8.1 (Carey and Judge 2000a). Longevity is positively correlated with body size between orders (e.g., the smaller rodents are shorter lived than the larger primates) though not necessarily within orders (e.g., longevity not correlated with body size in the seals and walruses [pinnipeds] or in the small bats). Longevity is also positively correlated with certain unique traits (Austad 1997), including flight ability (birds and bats), possession of armor (turtles, armadillos) and subterranean lifestyle (moles, mole rats). Analysis of the database shown in figure 8.1 reveals that life spans differ by a factor of over 50 in mammals, herps, and fish and by over 15-fold in birds. It also provides important biological and evolutionary context for human longevity—primates are long-lived mammals, great apes (gorillas, chimpanzees) are long-lived primates, and humans are long-lived great apes. Indeed, the analysis reveals that human longevity exceeds nearly all other species both relatively and absolutely. This finding is important because it suggests that extended longevity should be considered along with features such as large brain, bipedalism, and language as a key trait of our species.

8.4.2. Maximal Age Is Influenced by Sample Size

The value of the parameter "maximal age" is dubious because the magnitude of the largest characteristic life span grows with an increase in the number of

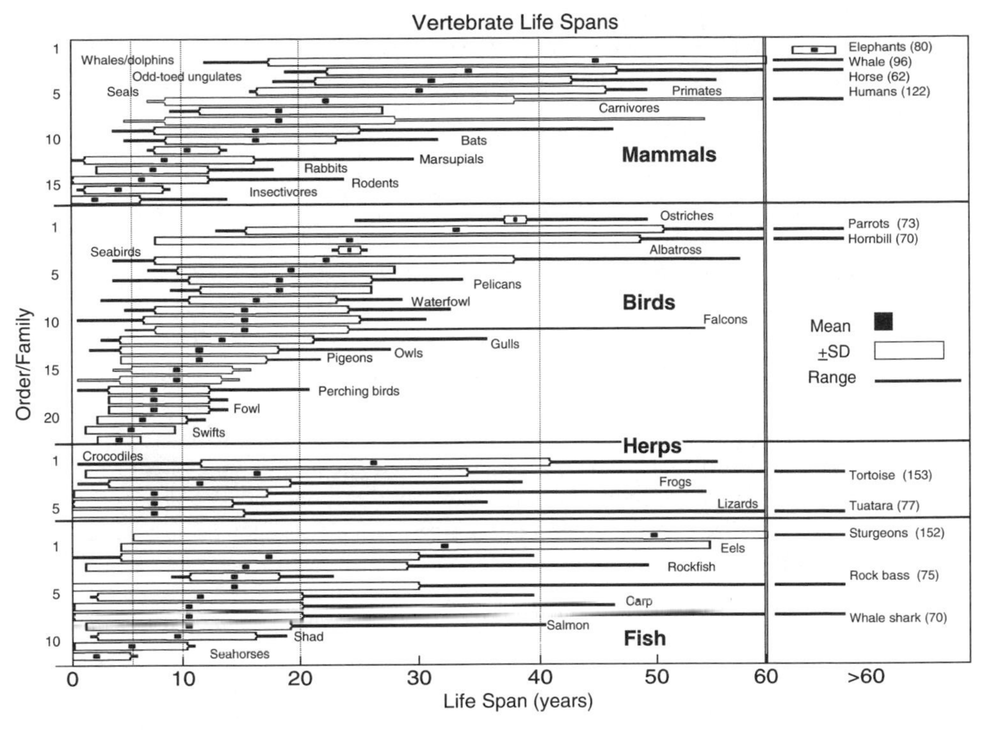

FIGURE 8.1. Summary chart of all longevity data contained in *Longevity Records: Life Spans of Mammals, Birds, Reptiles, Amphibians and Fish* by Carey and Judge (2000a). Subgroups such as Orders (mammals) within each section were ranked by mean longevity within the group. Statistics—including mean, standard deviation, and ranges—were determined using the record life span for each species regardless of sex or whether this age was determined from individuals in the wild or from laboratory/zoo records. The objective of this chart is to provide a visual summary of broad trends within and among groups. (Carey and Judge 2000a, with permission from Odense University Press.)

observations (Gavrilov and Gavrilova 1991). This is a general principle of statistics—the likelihood of obtaining 10 heads in a row from flipping a fair coin is greater from a million trials (flips) than from 25 trials. Carey (1993) demonstrated this simple principle using the mortality data on 1.2 million *C. capitata* shown in figure 3.1 using different sample sizes of initial numbers of flies and replicating each sampling trial 100 times. The mean maximal age for sample sizes of 25, 100, and 1,000 flies was 43.5, 53.0, and 74.5 days, respectively. The main point is this—the putative maximal age for species in which the life spans of large numbers of individuals have been observed will be significantly greater than the corresponding figure for a species with the same longevity but represented by a few dozen individuals (Gavrilov and

Gavrilova 1991). The extent of the sample-size effect is also influenced by the shape of the survivorship curve. The more rectangular the survivorship the less variation in maximum observed life span with sample size (Calder 1996).

8.4.3. *Life Span Is Indeterminate*

Maximal length of life remains as one of the most compelling concepts in demography and gerontology. The validity of this concept is viewed by many as self-evident because different species exhibit different life expectancies; all individuals eventually die before the age of infinity; and, therefore, each species must possess unique and finite maximal ages. Kannisto (1988, 1991) noted that the problem with this idea is that our knowledge of the nature of mortality makes it difficult to accept the notion that there is a single age that some individuals may reach but that none has any chance of surviving. He views the only valid alternative as the existence of an asymptote to which the probability of dying tends and that may or may not be near 100%. Manton and Stallard (1996) noted that declines in the age-specific rates of increase of mortality for male and female cohorts in the United States are inconsistent with a fixed life span limit (Manton et al. 1991). Wilmoth and Lundström (1996) state, "[W]e have established the important empirical fact that the upper limits of the human age distribution has been rising steadily during the past century or more and shows no sign thus far of possessing a fixed upper bound." In general, I can conclude from our studies that it is possible to estimate *C. capitata* life expectancy, but that these flies, and most likely other species as well, do not appear to have a characteristic life span. The concept of an *indeterminate* life span implied by the medfly data is fundamentally different from the concept of a *limitless* life span.

8.4.4. *Reproduction Is a Fundamental Longevity Determinant*

Most organisms, from yeast and plants to invertebrates, to birds and mammals, suspend reproduction during periods unfavorable for reproduction by entering a different physiological mode. Such waiting strategies for prolonging survival while maintaining reproductive potential have been extensively documented in the physiological, ecological, and natural history literature. For example, when food is scarce, yeast enter a stationary phase, tardigrads form tuns, nematodes go into a dauer stage, mollusks and earthworms undergo a quiescence, fruit flies experience a reproductive diapause, long-lived queens in ants and wasps hibernate, some fish reabsorb their ovaries, amphibians and reptiles aestivate, mice retard their ovariole depletion, some birds (hummingbirds and swifts) become torpid, and plants suspend their physiological and reproductive activities (Audesirk and Audesirk 1996). Recent research on *C. capitata* aging (Carey, Liedo, Müller, Wang, and Vaupel 1998b) revealed that female medflies may experience two physiological modes of aging with different demographic schedules of fertility and survival. These include a waiting mode in which both mortality and reproduction are low, and a reproductive mode in which mortality is low at the onset of egg laying but accelerates as eggs are laid. *C. capitata* that switch from waiting to reproductive mode due to a change in diet (from sugar-only to full-protein diet) survive longer than those kept in either mode exclusively. The switch from waiting mode to reproductive mode initiates egg laying and reduces the level of mortality below current rates but increases the rate of aging. Understanding this relationship between longevity and reproduction in medflies is important because it links the reproductive fate of individuals with environmental conditions and points toward important causal mechanisms that may be related to and mediated by the rate of ovarian depletion and/or gonadal activity (Carey et al. 1998b).

8.4.5. The Heritability of Individual Life Span Is Small

Life span heritability is defined as the proportion of the variance among individual ages of death that is attributable to differences in genotype (Futuma 1998). Contrary to popular myth, parental age of death appears to have minimal prognostic significance for offspring longevity (McGue et al. 1993). Finch and Tanzi (1997) noted that the heritability of life span accounts for less than 35% of its variance in short-lived invertebrates (nematode, fruit flies), mice and humans (table 8.1). Although McGue et al. (1993) found evidence for genetic influences, environmental factors clearly accounted for a majority of variance in age at death. For example, these researchers reported that the average age difference at death for twins was 14.1 and 18.5 years for identical (monozygotic) and fraternal (dizygotic) twins, respectively, and 19.2 years for two randomly chosen individuals. The study by Herskind et al. (1996) followed more than 2800 twin-pairs with known zygosity from birth to death. This study showed that about 25% of the variation in life span in this population could be attributed to genetic factors. Generally, traits that are most essential to the survival of an organism including survival itself, show little heritability due to strength of selection and fixation (Strickberger 1996).

8.5. Biodemographic Principles and the Human Primate

Most of the biodemographic principles concerning senescence, mortality, and longevity presented in the previous section are general and thus apply to

TABLE 8.1
Heritability of Life Spans

Species	*Heritability*[a]
Nematode, *C. elegans*	34%
Fruit fly, *Drosophila*	6–9%
Mouse	29%
Human twins	23–35%

Source: Finch and Tanzi 1997.
[a]defined as the proportion of the variance among individual ages at death that is attributable to differences in genotype.

a large number of species. There are also actuarial characteristics in all species, including humans, that are specific to that species or a narrow group to which a species belongs. Such species-level characteristics are superimposed on the more general patterns. For example, the general mortality patterns in humans includes a decline after infancy, increases through the reproductive life span (the overall U-shaped trajectory), and a sex differential (Carey and Judge 2000b). However, the specific level pertains to details of the mortality experience unique to humans, including the actual probabilities of death by age, inflection points of age-specific mortality, the cause-specific probabilities of death, and the age-specific pattern of the sex differential. The observed actuarial patterns are a combination of the evolutionary components of the trajectory (which will be common to a large number of species with overlapping life history characteristics) and the proximate age and sex-specific factors contributing to mortality and survival under certain conditions.

A variety of life history traits largely unique to humans are widely documented in anthropology and human biology texts. These include bipedalism, large brains, complex language, tool use, and a prolonged juvenile period. However, the extraordinary absolute longevity of humans, as well as longevity relative to body size, is a life history trait that is not fully recognized and appreciated. The purpose of this section is to identify and describe 3 biodemographic principles that link our primate evolutionary past with modern human longevity. A substantial part of this section is based on results presented by Carey and Judge (2000b) and Judge and Carey (2000).

8.5.1. Body and Brain Size Predict Extended Human Longevity

Brain size is correlated with both body size and life span in mammals as a whole and within the Primate order. Relative brain size and relative life span (residual brain and life span after controlling for body mass) are highly correlated (Austad and Fischer 1991, where $n = 73$ species) (Hakeem et al.

1996, $n = 72$ species). Judge and Carey (2000) examined longevity records for 133 species of primates relative to adult female body size and adult brain size, and placed human life span in context relative to extant primates and relative to estimates for early (extinct) hominids. The great apes have absolutely long lives that slightly exceed the life span predicted by body and brain size. However, the closest relatives of humans (gorillas and chimpanzees) are exceeded in their positive deviation from the expected life span by 5 other Old World primate genera. No Old World nonhuman genus approaches the positive deviation from expected life span demonstrated by New World monkeys of the genus *Cebus* (i.e., Capuchin monkeys). *Cebus* exhibit life spans that rival those of chimpanzees even though chimps are roughly 15 times larger. The 25-year life span predicted by *Cebus* body and brain size is much exceeded by the 45–50 year life spans actually observed. *Cebus* are the most geographically widespread of New World monkeys, show convergent evolution in social structure to Old World monkeys (Kinzey 1997), and have a fruit-based diet supplemented by invertebrate and vertebrate prey.

8.5.2. *Long-lived Monkeys Have Life Spans Proportional to Human Centenarians*

Centenarian humans are not out of the scope of primate longevity, especially given the large numbers of human observations (i.e., high numbers increase the probability of sampling the extreme right tail of the distribution). *Cebus* monkeys exhibit *relative* life span potentials similar to humans and are convergent in traits such as a relatively large brain, generalized ability to exploit a wide range of ecological niches over a broad geographical distribution, fruit-based omnivorous diet, and polygynous mating systems. While *Cebus* are female philopatric (females remain in their natal groups while males disperse), whether human ancestors were male or female philopatric is unresolved. If human ancestors had the potential for 72–90 year life spans for 1–2 million years, one might wonder why prolonging life span to 100 years under modern conditions of ecological release has not been easier?

8.5.3. *Post-reproduction Expected from Primate Patterns*

Hammer and Foley (1996) incorporated body and raw brain volume estimates from fossil crania to predict early hominid longevity using a multivariate OLS regression of the log body weight and brain volume. Estimates based on regressions of anthropoid primate subfamilies, or limited to extant apes, indicate a major increase in longevity between *H. habilis* (52–56 years) to *H. erectus* (60–63 years) occurring roughly 1.7 to 2 million years ago. Their predicted life span for small-bodied *H. sapiens* is 66–72 years. From a catarrhine (Old World monkeys and apes) comparison group, our prediction is 91 years when contemporary human data are excluded from the predictive equation (figure 8.2). For early hominids to live as long or longer than predicted was probably extremely rare; the important point is that the basic Old World primate design resulted in an organism with the potential to survive long beyond a contemporary mother's ability to give birth. Notably, Hammer and Foley's predicted life span of *Homo habilis* exceeds the age of menopause in extant women by 7–11 years and that of *H. erectus* exceeds menopause by 15–18 years. This suggests that post-menopausal survival is not an artifact of modern life style (Washburn 1981) but may have originated between 1 and 2 million years ago coincident with the radiation of hominids out of Africa.

Williams (1957) first suggested that menopause may be the evolutionary result of a human life history that requires extended maternal care of offspring. Diamond (1992) noted that menopause probably resulted from two distinctly human characteristics: (1) the exceptional danger that childbirth poses to mothers; and (2) the danger that a mother's death poses to her offspring. Perinatal mortality increases with maternal age and the death of an older mother endangers not only her current infant but also those past infancy who still dependent on her for food, protection, and other forms of care. However, more recently Hawkes et al. (1998) have argued that it is post-reproductive longevity that has evolved rather than an early cessation of female reproduction; the reproductive spans of human and other ape females are not appreciably different. Rather, kin selection for older relatives subsidizing the reproduction of younger female kin may have been a primary mechanism extending human life span (the "grandmother hypothesis"). This subsidization also allowed humans a later age at maturity and, as a result, a longer period of time for growth and learning.

8.6. Biodemography of Human Development, Reproduction, and Genetics

8.6.1. Developmental Stages and Mortality

My objective in this section is to link trends in mortality at the early stages of the human life course with development (rate of increase in height) from birth through sexual maturity. Anthropologists consider six stages of the human life cycle (Bertalanffy 1960; Bogin 1990, 1997; Bogin and Smith 1996; Daly and Wilson 1983; Hawkes et al. 1998):

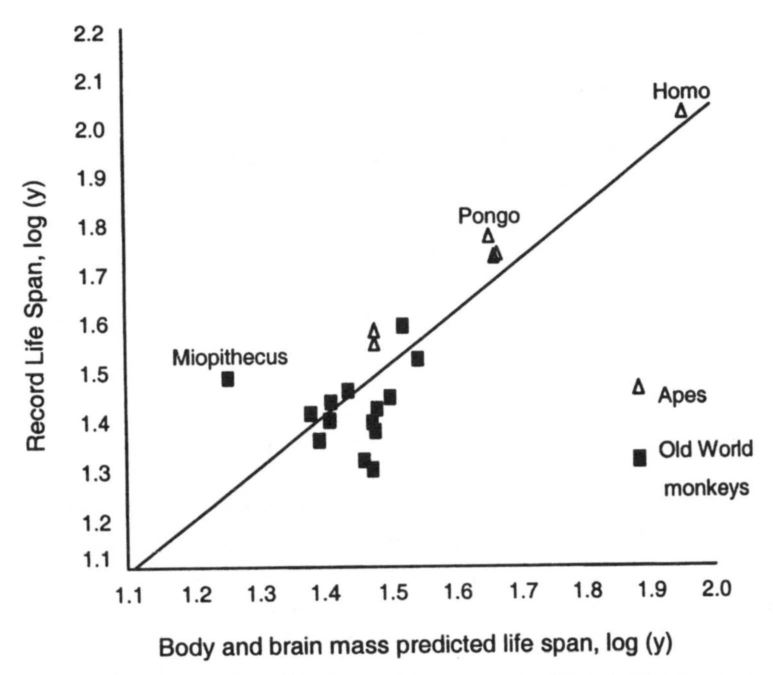

FIGURE 8.2. Observed and predicted record life spans for Old World primate genera. Predicted values are based on a multivariate linear regression, including adult female body mass and brain size as independent variables (redrawn from Judge and Carey 2000a).

1. Infancy (0–3 yrs)—when maternal lactation provides all or some nourishment to the offspring via lactation.

2. Childhood (3–7 yrs)—period following weaning when offspring still depends on older individuals for feeding assistance and protection.

3. Juvenile (7–10 yrs)—stage when the first permanent molars appear and when brain-maturation milestones occur but when others still provide care for the offspring.

4. Adolescence (10–19 yrs)—begins with the onset of puberty and ends with completion of growth spurt.

5. Reproductive adulthood (19–45 yrs)—attainment of adult stature, completion of dental maturation, achievement of social maturity and parenthood, completion of reproduction.

6. Post-reproductive (female > 45)—period of very low reproduction by age 45 followed by menopause when all primordial follicles are depleted and menstrual cycling shuts down.

The relationships between growth rate (a proxy for life-cycle stage) and relative female death rates from 1 to 25 years for U.S. females (1900–2000) (Bell et al. 1992) are shown in figure 8.3 (mortality for each year normalized relative to mortality at year 1). Relative mortality rapidly declines in the second year, continues to decline but at a less rapid rate through childhood and reaches a minimum in the prepubescent juvenile (Azbel 1996). It rises rapidly throughout adolescence concomitant with growth rate, and then slows during early adulthood. Understanding the details of the relationship between mortality and pre-adult developmental stages is important because changes in mortality in the early stages have a profound impact on the overall shape and trajectory of the mortality curve.

8.6.2. The Cost of Reproduction in Women

Cost of reproduction refers to the concept that an increment in reproduction at some age may result in a decrement in expected reproduction and an increase in mortality at later ages. It is derived from the antagonistic pleiotropy theory of evolutionary biology in which, according to Williams (1957), "selection of a gene that confers an advantage at one age and a disadvantage at another age will depend not only on the magnitudes of the effects themselves, but also on the times of the effects. An advantage during the period of maximum reproductive probability would increase the total reproductive probability more than a proportionately similar disadvantage later on would decrease it. So natural selection will frequently maximize vigor in youth at the expense of vigor later on and thereby produce a declining vigor (senescence) during adult life." Whereas a large literature exists in population and evolutionary biology showing unambiguous trade-offs between reproduction and mortality in many invertebrates (Bell and Koufopanou 1986) and some vertebrates (Reznick 1985), only a modest literature exists on reproductive trade-offs in human females and the results are mixed. For example, Westendorp and Kirkwood (1998) reported that among women who lived to 60 or more, the number of progeny was negatively correlated with age at death. This result was basically due to the somewhat lower fertility of women who lived beyond 80 years and the effect was more pronounced in periods of higher fertility.

Perls et al. (1997) found that 20th century Massachusetts women who lived to 100 years were three times as likely to have given birth in their 40s as were women who survived only to age 73. They concluded that "later menopause as well as pregnancy after age 40 might be associated with extreme longevity." The ability to reproduce later in life was not traded off with longevity; however, they provided no information on the intensity of reproduction for the two groups beyond the fact that they did not differ in

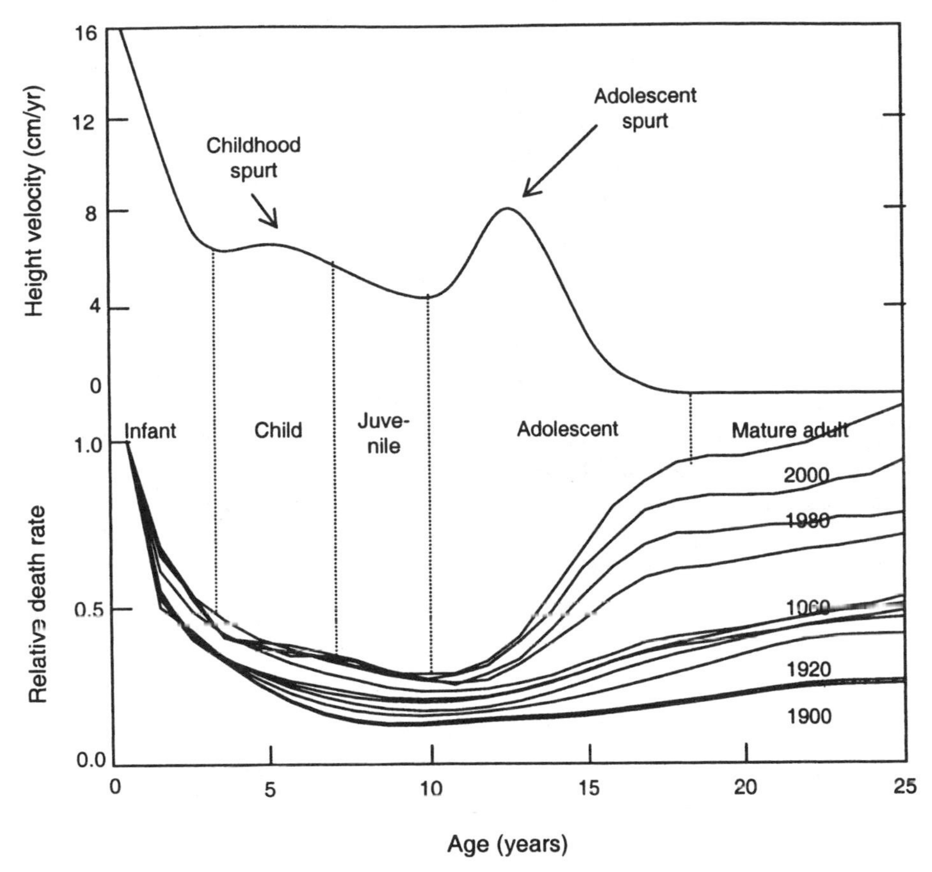

FIGURE 8.3. Relationship between human growth rate and stage-specific mortality. Top: female growth rate (height velocity). Bottom: relative age-specific mortality, μ_x (normalized at μ_1) (Bell et al. 1992).

rates of married childlessness. The argument by Perls and coworkers (1997) that late menopause may be a biomarker for extended longevity is strengthened by the findings by Snowdon et al. (Snowdon et al. 1989), who reported an age-adjusted odds ratio of death in Seventh-day Adventist women with natural menopause before age 40 was 1.95 compared to the reference group of women reporting natural menopause at ages 50 to 54.

Bourg et al. (1993) examined the reproductive life of over 5,000 French-Canadian women living in the 17th–18th centuries and also could find no evidence for a trade-off between fertility and longevity. My colleagues and I obtained the data set on which these researchers based their conclusions and also could find no relationship between timing and/or level of reproduction and longevity (figure 8.4). However, women who lived to menopause (defined as 50 years) had slightly longer interbirth intervals (by 2.0 and 2.4 months after daughters and sons, respectively). The number of sons born during the reproductive period was weakly negatively related to post-menopausal longevity. Facultative reproductive strategies may mask trade-offs between reproductive effort and longevity.

8.6.3. Extreme Longevity in Families and Close Kin

Despite studies demonstrating that genetics account for a relatively small fraction of the total variance in longevity (from a quarter to a third), one of the most enduring and widespread notions in gerontology is the longevity resemblance of kin, particularly between parents and offspring and among siblings. That is, the likelihood of living to the older ages is substantially increased if one's parents were long-lived. In this section I summarize the findings from studies on kinship resemblance for longevity and how this can be reconciled with the genetic finding in both humans and nonhuman species that the heritability of life span is somewhere between .25 and .35. There are essentially three approaches used to examine kinship resemblance of longevity.

PARENT-OFFSPRING (FAMILIAL) RESEMBLANCE

Pearl (1931) was one of the first researchers to examine this relationship and reported small but significant parent-offspring correlations. This modest familial component has also been established in several other subsequent studies (Abbott et al. 1978; Bocquet-Appel 1990; Gaesser 1999; Swedlund et al. 1983). However, as noted by Ljungquist et al. (1998), these studies are plagued by the problem of conflation of shared genetics and environments and so cannot provide conclusive evidence for a genetic component of longevity.

TWIN REGISTRIES

The use of twin registries has allowed more quantitative analyses of familial resemblance. Comparing the correlations between monozygotic (genetically identical) twins, fraternal twins, and non-twin siblings begins to disentangle the effects of shared genes from those of shared family environments. Investigations of the relationship between age of death between like-sex fraternal (dizygotic) twins and between identical (monozygotic) twins both reared together and reared apart reveal that longevity is moderately heritable (Ljungquist et al. 1998; McGue et al. 1993). McGue et al. (1993) reported that the average absolute difference in age at death between two members of a mono-

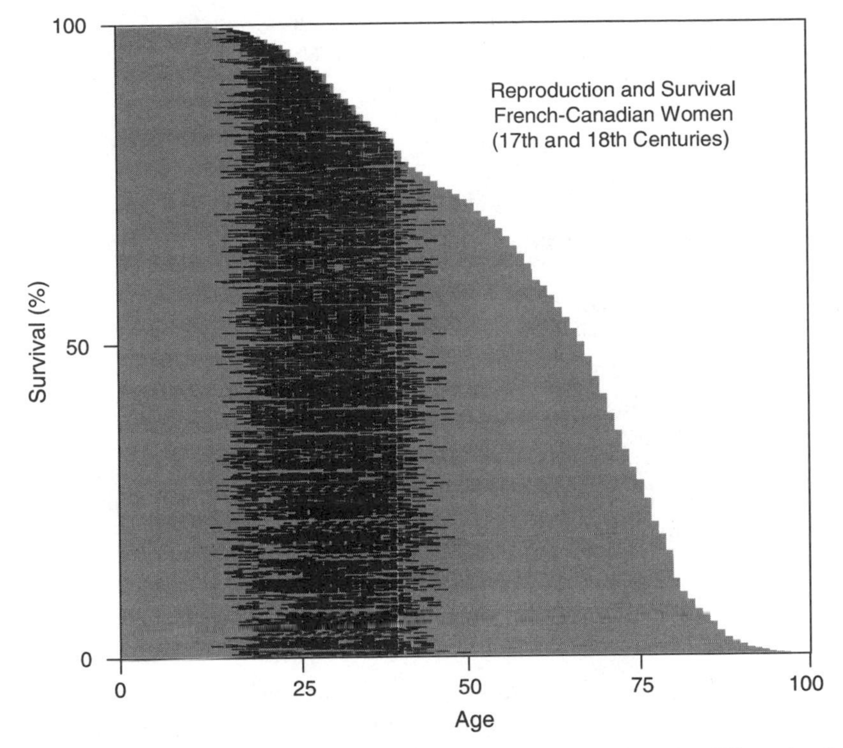

FIGURE 8.4. Event history reproductive chart (Carey et al. 1998a) for 500 randomly selected French-Canadian women rank-ordered by longevity after age 13. Each individual is represented by a horizontal "line" proportional to her life span. Each year an individual gave birth her life course is coded black, otherwise it is coded gray. The vertical dashed white lines delineate the 30-year span from age 15 (left-most line) to age 45 (right-most line).

zygotic twin pair, two members of a dizygotic twin pair, and two randomly selected same-sex individuals was 14.1, 18.5, and 19.2 years, respectively.

COMPARATIVE ANCESTRY OF CENTENARIANS AND SHORTER-LIVED GROUPS

This comparative approach can elucidate the familial resemblance in extreme longevity. Perls et al. (1997, 1998) found that the siblings of the centenarians whom he and his coworkers studied had a four times greater chance of surviving to their early nineties than did siblings of a comparison group that died at age 73. Robine and Allard (1998) identified immediate ancestors of the oldest person on record, Jeanne Calment from France (died at 122 years 164 days), and found that twenty four percent (13/55) of her ancestors lived longer than 80 years, versus 2% in a reference family (1/50). Her immediate ancestors lived an average of 68.2 years versus 57.7 years for controls in the reference family. Robine and Allard noted that genetic inheritance could be explained by the absence of alleles predisposing individuals to various life-threatening degenerative diseases and hypothesized that Jeanne Calment lived in an environment to which she was well adapted.

8.7. Proximate Determinants of Human Longevity

Christensen and Vaupel (1996) noted that the factors determining human longevity might be expected to be well understood but, surprisingly, knowledge about the determinants of longevity—particularly of extreme longevity—is still sparse. In this section I outline and briefly discuss three broad categories of factors that influence health and longevity. Although there are many sources of information on the proximate determinates of longevity (Hayflick 1994; Rowe and Kahn 1998), I include this information in this chapter for both completeness and continuity.

8.7.1. Socioeconomic Factors

In most populations, people with more schooling, higher income, and more prestigious occupations enjoy better health and longer life. Mortality is inversely related to educational attainment—the average risk of death decreases markedly with increasing educational attainment. Age adjusted death rates in the United States for those with less than 12 years of education was 17% higher than the rate for those with 12 years of education, and 2.4 times higher than those with 13 years or more of schooling (Hoyert et al. 1999).

Rowe and Kahn (1998) note that humans are not meant to lead solitary lives. Social integration measured by marital status, contacts with family and friends, church membership, and membership in other organized groups was a significant predictor of longevity—people who lack social ties die earlier (Berkman and Syme 1979). In 1997, the never-married group of persons in the United States had an age-adjusted death rate 68% higher than that of the ever-married and 2.0 times the rate for the currently married. Age-adjusted death rates for widowed and divorced persons were 80% and 74% higher, respectively, than for those who were currently married at the time of death (Hoyert et al. 1999; Waite 1995). Hummer et al. (1999) reported that religious attendance is associated with U.S. adult mortality in a graded fashion with those attending church regularly experiencing a 7-year advantage in life expectancy at age 20. While selectivity clearly accounts for part of the difference, religious attendance also increases the social ties and hence enhances coping resources during times of stress (Ellison 1991).

8.7.2. Physical Fitness, Exercise, and Nutrition

One of the most surprising recent findings concerning the relationship between physical fitness and longevity is that, contrary to the conventional wisdom and oft-cited public health policies, thinness that is not attributed to smoking or disease, is not related to reduced mortality (Gaesser 1999) and that being overweight actually had a protective effect for some types of diseases such as lung cancer. The majority of studies of the relationship between body weight and mortality do not support the contention that greater longevity is enjoyed by individuals with lower-than-average body mass indices (BMI). Gaesser (1999) noted that high BMI could be a symptom of a sedentary lifestyle and low fitness level. Hayflick (1994) also reported that, contrary to popular belief, the lowest death rates among middle-aged people occur not in the leanest segment of the population but among those with weights ranging from the midpoint to at least 20 percent over the midpoint. A meta-analysis of published studies on the relationship between body weight and mortality found a U-shaped relationship between body mass index (BMI) and mortality for 50-year-old, non-smoking men followed for 30 years (Troiano et al. 1996). For women (followed for 10 years) there was little relationship between BMI and mortality. The most important and consistent factor related to longevity and fitness was not weight, per se, but exercise. Hayflick (1994), Buchner et al. (1992) and Rowe and Kahn (1998) all noted that although exercise does not slow or otherwise alter normal aging processes it may increase longevity by modifying disease processes (e.g., cardiovascular disease).

Surprisingly little is known about the relationship between nutrition and longevity beyond the straightforward conclusion that both nutritional intake and status are known to influence life span and healthy life expectancy (Casper 1995). Although caloric restriction in laboratory rodents has a positive impact on longevity (C. K. Lee et al. 1999; Masoro 1988), its positive effect on human longevity has yet to be demonstrated. Various nutrients have been linked to prolonged health and indirectly to longevity, generally through their antioxidant activity (Ames et al. 1993).

8.7.3. Behavioral Factors: Smoking and Alcohol Use

Cigarette smoking is the best known behavioral factor influencing longevity and has now been positively associated with some 40 causes of death and negatively associated with eight or nine (Doll 1999). Smoking can increase the risk of death 10-fold and altogether doubles the annual risk of death at all ages combined in both sexes. Thun et al. (1997) reported that alcohol consumption has both adverse and beneficial effects on survival. The overall death rates were lowest among men and women reporting about one drink daily. Mortality from all causes increased with heavier drinking, particularly among those adults under age 60 with a lower risk of cardiovascular disease.

8.8. Longevity Gains Are Self-reinforcing

Improved health and increased longevity in societies may set in motion a self-perpetuating system of longevity extension. This positive feedback relationship is based on the demographic tenet that increased survival from birth to sexual maturity reduces the number of children desired by parents, *ceteris paribus* (Bourgeois-Pichat 1967; Ryder 1959). Because of the reduced drain of childbearing and rearing, parents with fewer children remain healthier longer, raise healthier children with higher survival rates, which, in turn, fosters yet further reductions in fertility. Greater longevity of parents also increases the likelihood that they can contribute as grandparents to the fitness of both their children and grandchildren. And the self-reinforcing cycle continues.

In an essay on the formation of human capital, Abramovitz (1989) noted that the decline in mortality rates during the early stages of industrialization in the United States was probably one of the forces behind the expansion of educational effort and growing mobility of people across space and between occupations. Whereas previous conditions of high mortality and crippling morbidity effectively reduced the prospective rewards to investment in education during the preindustrial period (Landes 1998), prolonged expectancy for working life span must have made people more ready to accept the risks and costs of seeking their fortunes in distant places and in new occupations. The positive feedback of gains in longevity on future gains involves a complex interaction among the various stages of the life cycle with long-term societal implications regarding the investment in human capital (Abramovitz 1989), intergenerational relations (Kaplan 1997), and the synergism between technological and physiological improvements (so-called "technophysio evolution" (Fogel 1994; Fogel and Costa 1997). In other words, long-term investment in science and education provides the tools for extending longevity, which, in turn, make more attractive the opportunity cost of long-term investments in individual education, and thus help humans gain progressively greater control over their environment, their health, and overall quality of life.

8.9. Summary

My broad objective in this section was to develop a framework for the emerging field of biodemography. I progressed through a hierarchical sequence between as well as within each of the two main sections. In the first main section I described sets of general principles on senescence, mortality, and longevity derived from research on model systems and various longevity data sets on both human and nonhuman species. In the second main section I attempted to situate human longevity within the conceptual framework of biodemography by focusing on data that were expressly related to humans. I demonstrated allometric longevity relationships in nonhuman primates, and the relationship of human longevity to development, reproduction, and kinship. I then outlined the contemporary proximate determinates of longevity. Lastly, I described how gains in longevity in modern societies can be self-perpetuating.

In general, I believe that the biodemographic principles and concepts outlined in this paper are useful to demographers concerned with human populations in three contexts. *First*, the principles provide a scientific *coherence* that is lacking in conventional texts on demography and actuarial science. Biodemography has the potential for integrating biology into the pedagogical framework of classical demography in much the same way as basic biology is integrated into biomedicine. The focus on humans is retained but the epistemological foundations are strengthened, the biological scope is expanded, and the demographic perspectives broadened. *Second*, the biodemographic principles provide *explanations* for life table patterns in human populations that are not evident in the absence of broader biological concepts. For example, the principles link senescence and sexual reproduction. They suggest explanations of sex differentials in life expectancies, why older individuals may grow old more slowly, whether life-span limits exist, whether post-reproductive life is common or rare, and if and how post-reproductive life spans in other species increase fitness. *Third*, biodemographic principles can be used as a more secure foundation for qualitative *predictions* in the context of old-age mortality, upper limits of life span (or lack thereof), and the magnitude and sign of the gender gap (Friedland 1998; Judge and Carey 2000; Tuljapurkar and Boe 1998; Wilmoth 1998).

Wilson (1998) noted that human beings may be unique in their degree of behavioral plasticity and in their possession of language and self-awareness, but all of the known human systems—biological and social—taken together form only a small subset of those displayed by the thousands of living species. I believe that the integration of biology into demography will provide a deeper understanding of demographic processes and will allow us to determine what patterns are common to a broad range of organisms and thus which demographic patterns are uniquely human.

9

A General Theory of Longevity

SACHER (1978) noted that the approach to one basic question in the biology of aging "Why do we grow old?" is guided largely by a research paradigm involving model systems designed to compare ontogenetic functions in old and young animals. One of his primary concerns with this aging-oriented approach to understanding longevity was that the far-reaching physiological, cellular, and molecular correspondence between model systems (mice) and humans concealed a paradox: if model species are so similar to humans in molecular makeup that they can serve as models, why do mice age as much in 2 years as humans do in 80 years? This paradox motivated him to ask his second question: "Why do we live as long as we do?" This question cannot be answered within the framework of ontogenetic research on aging, but rather requires the development of an evolutionary-comparative paradigm concerned with longevity. I use life span in the sense articulated by Wilmoth and the participants of the Santorini conference on longevity (Goldwasser 2001) and use longevity in a general manner to more specifically focus on prolonged or increasing life spans.

Despite the arguments by Sacher (1978) and others (e.g., Hayflick 2000) in support of developing a longevity-oriented theory of the finitude of life, no such theory has ever been published. My objective in this chapter is to describe a general theory of longevity extension in social species, particularly as it applies to humans. These central concepts derive from patterns identified in vertebrate life-span data (Carey and Judge 2000a), insect life spans (Carey 2001b), and humans (Kannisto et al. 1994; Tuljapurkar et al. 2000; Vaupel et al. 1998; Wilmoth et al. 2000), environmental factors associated with the evolution of extended longevity in insects (Carey 2001a; Carey 2001b), and the preliminary idea that longevity extension in humans is self-reinforcing as first proposed by Carey and Judge (2001b). This theory also builds on and extends work by Fogel (1993, 1994; Fogel and Costa 1997) on what he and his coworkers refer to as "technophysio evolution"—a theory explaining the decline in morbidity and mortality since 1700.

A theory of longevity that extends beyond the classical evolutionary theory of aging is important for several reasons. *First*, whereas senescence is a by-product of evolution (Medawar 1955), life span is an evolved life-history trait that results from positive selection. *Second*, unlike the evolutionary theory of senescence, which is based solely on individual natural selection

(Williams 1957), this theory includes processes of sexual selection and kin selection, bringing life history theory more fully to bear on questions concerned with the latter portion of the life cycle. *Third*, longevity-oriented theory allows consideration of behaviors that are characteristic of older individuals, including divisions of labor and intergenerational transfers (Beshers and Fewell 2001). *Fourth*, mortality factors unrelated to aging (accidents, acute diseases, socioeconomic factors) can be considered, and therefore a longevity-oriented theory fosters a greater integration of demographic (Hauser and Duncan 1959) and gerontological (Finch 1990) pedagogies.

9.1. Comparative Demography of Longevity

My purpose in this section is to situate human longevity in a broad biological context. I first describe a general classification system concerning factors that favor the evolution of extended longevity and then provide a brief sketch of longevity extension in social wasps, including a discussion of its relevance to humans.

9.1.1. A General Classification Scheme

The literature on aging and longevity includes descriptions of a small number of life-span correlates, including well-known relationships between longevity and both body mass and relative brain size (Austad 1997; Comfort 1961; Hakeem et al. 1996) and the observation that animals that possess armor (e.g., beetles, turtles) or capability of flight (e.g., birds, bats) are often long-lived (Austad 1997; Kirkwood 1992). The evolutionary theory of senescence suggests that animals better able to escape stochastic mortality sources such as predators (e.g., via armored defenses or better escape mechanisms) live longer and thus the force of selection at older ages is increased and the evolution of longer presenescent life span is possible. But major inconsistencies exist within even this small set of correlates. For example, there are several exceptions regarding the relationship of extended longevity and large body size (e.g., bats are generally small but most species are long-lived) and this positive relationship may be either absent or reversed within orders. Likewise, the observation that flight ability and extended longevity are correlated does not provide any insight into why within-group (e.g., birds) differences in life span exist, nor does it account for the variation in longevity in insects, where adults of the majority of species can fly.

An alternative approach for identifying broad correlates of longevity emerged from an examination of several large-scale databases containing the maximum recorded life spans of both vertebrate and invertebrate species

(Carey 2001a, 2001b; Carey and Judge 2000a). Across a wide taxonomic spectrum, many long-lived species appeared to cluster within 1-of-2 general ecology and/or life history criteria: (1) species that live in either unpredictable environments (e.g., deserts) or where food resources are scarce (e.g., caves, deep water); or (2) species that exhibit extended parental care, and/or live in groups with complex or advanced social behavior. This led to a classification system regarding the life-span determinants of species with extended longevity (table 9.1) that I believe is general and applies to a wide range of invertebrate and vertebrate species:

1. Environmentally Selected. This category includes animals whose life histories evolved under conditions where food is scarce and resource availability is uncertain or where environmental conditions are predictably adverse part of the time. The extended longevity of animals in this category evolved through natural selection.
2. Socially Selected. This category includes species that exhibit extensive parental investment, extensive parental care, and eusociality. The extended longevity of animals in this category results from natural, sexual, and kin selection.

This classification system places the relationship of life span and two conventional correlates, relative brain size and flight capability, in the context of life history. That is, brain size is related to the size of the social group and the degree of sociality (Dunbar 1992; J. M. Smith and Szathmary 1999), which, in turn, is linked to extended life span. And intensive parental care is linked to flight capability in birds and bats, which, in turn, is also linked to extended life span. No system of classification is perfect and the one presented in table 9.1 is no exception—the categories are not mutually exclusive and therefore some species could be placed in either or both categories. However, this classification serves as a practical and heuristic tool for considering the evolution of animal life spans. In particular this system provides a general background for closer examination of specific human attributes and sets the stage for addressing questions of process.

9.1.2. Evolution of Extended Longevity in Wasps: The Interactive Role of Sociality

Classifying long-lived species by life history characteristics is important because it suggests that the evolution of extended longevity is fundamentally different in the social versus solitary species; the extended longevity of species in these different categories was a response to different types of environmental and socioecological problems. In this section, I consider the coevolution of longevity extension and sociality in social wasps in a sequence of grades (Carey 2001b; Carey and Gruenfelder 1997; H. E. Evans 1958) that can be seen logically as evolutionary steps (H. E. Evans 1958; Wilson

TABLE 9.1
Categories of Factors That Favor the Evolution of Extended Life Span in Insects, Arachnids, and Vertebrates: Selected Examples Species or Groups. Life Span (Years) in Parentheses

Selection Factors[1] *Taxonomic Group*	*Examples*[2]
Environmentally Selected Life Spans (conditions of resource uncertainty and/or scarcity)	
Insects/acarina	*Heliconius* butterflies (0.9); cave beetles (4); orchard bees (1); paper wasp queens (1); soft ticks (20); treehole mosquitoes (1); flour beetles (4); monarch butterfly (1); African locusts (2); tarantulas (10); bed bugs (1)
Mammals	Gray seal (46); dromedary camel (40); horseshoe bat (30); little brown bat (30); flying fox (31); African wild ass (47); white rhinoceros (45); short-nosed echidna (50); wombat (26); gray seal (46); crab-eater seal (36); ringed seal (46); Caspian seal (50); kangaroo rat (10); Greater Egyptian gerbil (8); hairy armadillo (20); rock hyrax (14)
Birds	Great blue heron (23); marabou (41); southern ground hornbill (70); sandhill crane (22); whooping crane (40); Cheriway caracara (26)
Reptiles & Amphibians	American toad (30); Japanese giant salamander (55); Eastern hellbender (25); greater siren (25); common boa constrictor (39); American alligator (56); Chinese alligator (38); West African dwarf crocodile (42); salt water crocodile (42); slowworm (54); Mexican burrowing python (33); loggerhead turtle (33); common snapping turtle (47); alligator snapping turtle (59); stinkpot (55); Eastern indigo snake (25); San Diego gopher snake (20); Aldabra tortoise (152); tuatara (77)
Fish	White sturgeon (100); Beluga sturgeon (118); European eel (88); carp (38); wrasse (53); sole (45); rockfish (140); Bunnylake brook trout (24); arctic char (41); spur dogfish (60)
Socially Selected Life Spans (kinship, cooperation, parental care, monogamy, and helpers)	
Insects/acarina	Tsetse fly (1); dung beetles (2); *Bembix* spp—progressive provisioning wasp (1); *Polistes* spp; apo-

TABLE 9.1 *(continued)*

Selection Factors[1] *Taxonomic Group*	*Examples*[2]
	sematic Saturnid butterflies (1); bumblebees (1); ant queens (30); termite queens (30)
Mammals	Bowhead whale (40); blue whale (65); pilot whale (65); sperm whale (70); golden jackal (16); spotted hyena (40); naked mole rat (21); black legged mongoose (16); humans (122); white-faced capuchin (47); squirrel monkey (27); gray-cheeked mangabey (33); yellow baboon (45); chimpanzee (60); orangutan (59)
Birds	African grey parrot (73); golden-naped parrot (49); red and blue macaw (64); trumpeter swan (33); mute swan (27); ring-billed gull (32); scrub jay (16); Royal albatross (58); northern fulmar (48); Manx shearwater (30)
Reptiles & Amphibians[3]	Rosenberg's tree frog (3.5)
Fish	basking shark (32); dusky shark (30); hound shark (28); nurse shark (25); whale shark (70)

Source: Carey (2001b); Carey and Judge (2000).

[1]The two general categories "Environmentally Selected" and "Socially Selected" are not mutually exclusive. Environmentally selected life-span extension may allow overlapping generations and sociality; sociality may allow expansion into otherwise inhospitable environments.

[2]Scientific names given in Carey (2001b); Carey and Judge (2000a).

[3]Care of the mass of developing eggs and/or the hatchlings is exhibited in some members of all 3 major groups of amphibians (Wake 1998); however no species of either amphibian or reptile are truly social and only a small number of reptilian species such as crocodilians and pythons exhibit any level of parental care (protection).

1971). Comprehending this evolutionary progression from short-lived, solitary species to long-lived, eusocial species provides important perspectives that foster understanding of the process of longevity extension in a broad array of social species.

This schema is summarized in table 9.2. Basically, nesting behavior and sociality result in improved microenvironmental conditions that foster greater survival. This survival improves conditions for increased provisioning and a more intensive social organization, which then generates a positive feed back and longevity self-reinforcement. Incipient sociality creates conditions for the evolution of incremental increases in longevity, which, in turn, create conditions for the evolution of more complex and innovative social structure. This positive feedback relationship is based on the demographic relationship between increased offspring survival and reduced birth rate, *ceteris paribus.*

TABLE 9.2
Evolutionary Changes in Wasp Longevity and Social Complexity

Stage	*Description*	*Key Concepts and/or Events*	*Life Span (Days)*
I[a]	Solitary	Parasitoids; host as incubator of parasitoid larvae	14–60
II[b]	Nest as locus	Nest provides a protected microenvironment for female and brood; thus different selective factors operate; selection reduces senescence rates and increases life span	30–90
III[c]	Extensive parental care	Extensive parental care increases life span to 1 year; larval and pupal mortality reduced to near zero; this fosters reduction in birth rate; allows females to invest more resource for their own maintenance and for rearing of their offspring	60–365
IV[d]	Colony and queen concepts	Female's life further prolonged and overlaps with progeny that remain in nest to form extended family (helpers); colony, rather than the individuals within it, begins to become the unit of selection with the fate of the queen inextricably tied to fate of colony; a new level of individualism emerges (colony) and thus life of queen becomes adaptive to the long-term colony needs	180–1,000

Sources: H. E. Evans (1958); Wilson (1975).
[a]Example: spider wasp, *Natocyphus.*
[b]Example: sand-nesting wasps, *Haploneurion; Amophila.*
[c]Example: progressive provisioning *Gorytes* spp.; and primitive paper wasps, *Stenogaster spp.*
[d]Example: advanced eusocial wasps, including yellow jacket, *Vespula spp.*

Because of the reduced costs of reproduction and parental care, female wasps with fewer offspring remained healthier longer and raised healthier offspring with higher survival rates that fostered yet further reductions in mortality and reproduction. Greater longevity of parents also increased the likelihood that they can contribute as grandparents to the fitness of both their offspring and grand-offspring. Inasmuch as the principles of social evolution are general (Wilson 1975), it follows that the same pattern of evolution of longevity extension in social species will also be general. The main point is that the ongoing process of longevity extension in social species adds age classes that increase the fitness of other age classes. This creates a dynamic that changes the rhythm and synchrony of life-cycle events that alter the qualitative properties of the life history.

9.2. Foundational Principles

The purpose of this section is to describe three basic principles concerning longevity that emerge from the wasp example and will serve as the foundation for the concept that longevity extension is self-reinforcing in social animals. The underlying concept is that sociality initially evolved through natural selection to increase survival and/or reproduction through sharing and helping (Bonner 1980; Wilson 1975); that, as Wilson notes (1975), the proximate rewards of cooperation (including status, power, sex, access, comfort, and health) are reflected in the universal bottom line of fitness, including greater longevity and reproduction. However, these conditions then became both cause and consequence of extended adult longevity as an outcome of natural and kin selection. The evolutionary theory of aging (Principle #1) serves as the basis by which adult longevity is initially extended, which then sets the stage for intergenerational transfer (Principle #2) and division of labor (Principle #3), both of which underlie the self-reinforcing dynamic of prolonged life span.

9.2.1. Principle #1: Evolutionary Theory of Aging

Medawar (1957) proposed that if deleterious hereditary factors are expressed at some intermediate age and if the age of this expression is both variable and heritable, then selection will weed out earlier expressions, more effectively than later expressions delaying the average age of expression and increasing longevity. As the force of selection is reduced by the declining reproductive value of increasingly older individuals, those deleterious traits will accumulate resulting in a mosaic and variable pattern of age specific infirmity and thus senescence (Kirkwood 1997). This argument requires that reproductive value (Frank 1998) decrease with age, and this will be generally true due to stochastic sources of mortality even in the absence of physiological deterioration. As more individuals live longer (especially due to reductions in early-life mortality), the force of selection increases at later ages weeding out later-expressing deleterious alleles, selection for somatic maintenance is prolonged, and senescence is delayed (Charlesworth 1994;

Medawar 1957; Roff 1992), resulting in extended life spans. Williams extended this line of thinking by noting that pleiotropic genes may have both beneficial and deleterious effects, and, to the extent that beneficial effects precede deleterious ones, the genes may be selected for even in the face of their positive effect on mortality later in life (Williams 1957).

9.2.2. Principle #2: Intergenerational Transfers

Investment of resources in reproduction can be extended to investment of resources in offspring and other relatives after birth. Increased per capita investment in offspring *decreases* juvenile mortality, *increases* the health and well-being of offspring and thus *improves* adult health and survival. At the extreme, older parents can "bankroll" reproductive offspring and thus increase their production of grandchildren (Hill and Kaplan 1999; Kaplan et al. 2000). In systems where different age-classes interact, natural selection will favor net transfers from old to young because the reproductive value of young individuals is greater (Frank 1998). However, intergenerational transfers can flow up as well as down generations (Lee 1997). If investment from older to younger individuals decreases mortality, then selection for increased longevity results, and that investment can flow for a longer period. Selection for longevity also increases the return on higher levels or prolonged periods of investment. Intergenerational transfers are most prolonged in highly social species (Wilson 1971, 1975). While transfers of genetic information and resources from parents to offspring contribute to the evolution of longevity, additional forms and pathways of intergenerational transfer can elaborate and increase the rate of impact on longevity extension. Social learning and community-level organization increase the pathways and volume of intergenerational transfers of information and the benefits that accrue from innovation (Boyd and Richerson 1985) and should be associated with more extreme longevity extension compared to related species with fewer transmission paths. In humans the transmission of longevity promoting genes and resources are augmented by culturally transmitted information, or "memes" (Dawkins 1976).

9.2.3. Principle #3: Division of Labor

Division of labor distributes related tasks among members of a population in a coherent system of symbiotic relationships (Petersen and Petersen 1986; A. Smith 1776). It is based on specialization of function—individuals use their differences in skill and/or resources to best advantage (Abramovitz 1989) and the energy-returns per unit of labor increase. The concept is fundamental to economics and business, as well as biology and anthropology because it is the basis for increasing efficiency at all levels of organization—from enzymes and cells to organisms and societies. In addition to increased efficiency, division of labor: (1) accentuates individual differences, (2) produces new specialization through finer and finer task division, and (3) increases interdependencies—the entire colony will die if the queen dies, families are at great risk if a parent is lost (Wilson 1971). All of these outcomes have a direct bearing on longevity extension because they reduce innovation costs, increase the efficiency and thus the net resource return to the (kin) group, and increase the behavioral repertoire of the colony or community as a whole. The concept of a complex, integrated society is inextricably linked to the dual concepts of division of labor and individual continuity, and so, with individual specialization and longevity.

9.3. Model of Longevity Extension

My objective in this section is to describe a general model of longevity extension in social species that builds on the three fundamental principles described in the previous section and on basic demographic and life history tenets. A primary tenet is the cost of reproduction (Bell and Koufopanou 1986; Partridge and Harvey 1988; Reznick 1985): allocation of energy to reproduction removes its availability for somatic growth and repair and thus is reflected in increased mortality. The second tenet is that demographic/epidemiologic transition theory is based on declines in mortality that precede declines in fertility (Caldwell 1976; Demeny 1968; Keyfitz 1985; Notestein 1945). Changes in patterns of disease result in shifts in ages of deaths as well as changes in the age-sex structure of the population (Omran 1971; Preston 1978; Preston et al. 1978; Ross 1982). The three principles presented in the previous section serve as the foundation for demographic shifts, including mortality reduction in infants, reduction in numbers of births, improvement in parental health, and increase in offspring quality. Although I draw many of the concepts from the demographic literature, I believe that the model is general and thus applies to a wide range of both vertebrate and invertebrate species.

9.3.1. Reduced Infant Mortality

All else being equal, increased survival from birth to sexual maturity improves energetic efficiency to the parent by decreasing reproductive waste (offspring who die prior to adulthood but after parental investment) and allows parents either to produce more young or to invest more per capita in existing young (Clutton-Brock 1991; Preston and Haines 1991; Szirmai 1997). Inasmuch as early reproduction in increasing populations will have a greater effect on fitness than later reproduction (Futuma 1998), the former will outcompete the latter. But, even in increasing populations, increased *per capita* investment can be selected if the age-specific increase in fitness of those offspring more than offsets losses due to numbers of young foregone. Increased resources available for reproduction can have any of the following effects depending on which age classes can access them: (1) increase number of young, (2) increase survival of young, (3) delay maturity of young by allowing the parental generation to support overlapping sets of offspring, and/or (4) reduce the cost of reproduction to mothers—"grandmothering" (Hawkes et al. 1998). The first will have no necessary effect on longevity while the latter three outcomes are conducive to longevity reinforcement through increased size, improved health, and better developed resource acquisition skills.

9.3.2. Demographic Transition

A general observation from studies on the comparative life histories of both invertebrate and vertebrate species is that an increase in offspring survival is followed by a decrease in fecundity (Clutton-Brock 1991; Ricklefs 1979). Most demographers and sociologists agree that one of the basic causes of a general decline in fertility is a reduction in mortality that reduces the number of births necessary to have any "desired" number of children (Becker 1991; Montgomery and Cohen 1998; Preston 1978). Parents (especially mothers) with fewer offspring remain healthier due to reduced energy allocation to early stage parental effort (especially pregnancy and lactation in contrast to effort expended on older children), and healthier young with even higher survival rates at both young and old ages result (Walle 1986). The decline in fertility may also be understood as a balance between growing family size and declining economic advantage of family (Buikstra and Konigsberg 1985). The transition from simple agriculture to industrial economies devalues children as early producers. Whereas a young person on a farm becomes a useful member of the labor force at an early age (Cain 1977; Kaplan 1996), the young person in industrial or postindustrial society requires an extensive and expensive education (Ryder 1959). Kaplan's model of embodied human capital provides important insights into the trade-offs inherent in providing necessary skills to offspring versus producing additional offspring and suggests that the inherent mechanisms of demographic transition characterized pre-transition humans as well. There is some empirical evidence that mothers in traditional horticultural/pastoral societies appear to produce optimal numbers of offspring to maximize their subsequent reproductive success (Borgerhoff-

Mulder 1998). Where heritable resources influence reproductive success there appears to be some trade-off between numbers of offspring produced and the per capita wealth that is inherited (Borgerhoff-Mulder 1998).

9.3.3. Improvement in Parental Health and Survival

Parents experiencing fewer births but unchanged reproduction remain healthier. Consequently they experience higher survival and can invest more of their resources in a smaller number of "high quality" offspring—healthier, larger, more competitive (Clutton-Brock 1994), skilled, and innovative (Preston and Haines 1991; Richard 2001). That healthier mothers produce healthier infants (Wood 1994) and that excessive maternity demands deplete maternal health are well documented (Hrdy 1999). In addition to intergenerational effects, reducing numbers of reproductive attempts without sacrificing reproductive output will improve the health and survivorship of first-generation mothers. The fitness benefits of this trajectory are even greater in species with long and costly post-infancy dependency. For example, improving the health and nutrition of girls and women and reducing maternal depletion improves the health of subsequent generations because healthy, better educated mothers can provide high-quality care over an extended period, including investments such as prolonged breast-feeding, increased vigilance, and longer education. Improved survivorship and competitive ability increases the number of productive adult (parent and grandparent) years relative to the number of juvenile/sub-adult years as well as the per year return—years of additional productivity that can further expand intergenerational transfers (Hawkes et al. 1998; Kaplan et al. 2000).

9.3.4. Increase in Offspring Quality

A second result of a decrease in the number of offspring can be an increase in the average quality of a smaller number of offspring because the per capita amount of both depreciable (e.g., food, nutrients) and nondepreciable (e.g., vigilance, teaching) care will increase (Clutton-Brock 1991). For example, in humans a decrease in the number of children will increase the average quality of a smaller number of offspring in two respects. *First*, the nourishment that a baby receives from its mother mediates its mortality risk due to infectious diseases early, and nutrition and exposure to infection after birth determine its susceptibility to disease in later life (Barker 1994; Elo and Preston 1992). Thus healthy mothers may prevent the early onset of the degenerative diseases of old age that are linked to inadequate cellular development early in life. The increase in childhood survival then fosters even further reductions in fertility and improvements in offspring quality. *Second*, parents can invest heavily in one or a small number of offspring for a longer period. Development of the brain, of language as a means of transmitting social learning, and of group size are characteristic of hominid evolution (Hammer and Foley 1996). Language as a mechanism for social (and therefore less hazardous) learning allowed humans not only to transmit information across generations via reproduction with variations that arise independent of real-world problems, but also to transmit information directed toward solving ecological problems. Cultural transmission increased the production of innovation, the rate at which innovations dispersed, and allowed innovations to be directly targeted toward hazards of the environment that threatened survival.

Contemporary studies demonstrate that family size is negatively related to high school graduation (Blake 1989a, 1989b), but this effect is mediated through resources (social economic status). These are modern incarnations of processes ongoing since the first information revolution in early *Homo sapiens*. Education is a measure of offspring "quality" that impacts longevity in two respects. At the individual level, the relationship between educational attainment and longevity is well documented (Hoyert et al. 1999). Even if the correlation between these two is mediated by intelligence (as measured by IQ, which is susceptible to poor prenatal and postnatal environments, including nutrition and social environment). At the societal level, economic growth can be attributed to advances in science and technology that are only possible through education and training of highly skilled specialists. The focus of contemporary specialists on biomedical research (Cruse 1999; Nutton 1996; Szirmai 1997; Thomas 1988), for example, is as natural as was the focus of *Homo erectus* on the use of fire to cook, warm, and deter predators.

9.3.5. Incremental Increase in Longevity as Cause and Consequence

The culmination of the effects of the four factors previously described yields an overall increase in longevity that is related to the extent and speed of resource, quality, and productivity changes. Reduced litter size, mortality, and increased parental investment are associated with a broad array of organisms with prolonged life spans. As generational overlap extends, healthy adults are longer lived and thus become healthy grandparents who can contribute to offspring rearing and further increase the subsidies discussed above. In humans this process is exaggerated through the ability to transmit symbolic information across and within generations and at a tempo substantially faster than generation times. Solutions to environmental or physiological hazards are transmitted broad band across all age classes of the population rather than solely to younger age classes through genetic modes. Incremental increases in longevity increases survival at early ages through increased subsidization of development. This is a new and expanded way of viewing the "contract between the generations," which is invoked for economic transactions across age classes by economists (Szirmai 1997). When intergenerational transfers occur at the pace of generation times (via natural selection) the process is self-promoting but at a slower rate of increase than when multiple means and directions of information and resource flow are possible (Boyd and Richerson 1985). In humans the cognitive ability to target innovation to specific problems can result in technologies that further improve health and do so much more rapidly than could be accomplished by natural selection on random variation (Bonner 1980; Durham 1991). Better-educated, more innovative children then increase the rate of technological innovations (Chant 1989) that contribute to health and longevity (Cruse 1999; Nutton 1996) and higher adult survival to older and older ages. Innovation is spurred by the increased return through prolonged later life to the increased number of surviving older adults who can benefit. Furthermore, expectations of increased benefits accrued through longer life may promote longevity itself (Becker and Murphy 1992).

9.4. Model Application

9.4.1. Human Longevity in Biodemographic Context

A graphical perspective of the primate life span relative to life spans of other mammalian orders reveals two relationships that provide important context. First, the life spans of nearly all primate species are greater than those predicted by body mass alone. Chiroptera (bats) are the only other major (that is, with more than a few species) group that uniformly exhibits greater life spans than those predicted from body mass alone (figure 9.1). Primates are long-lived mammals with life spans of most species ranging from 1.25 to 1.5-fold greater than that predicted from body mass alone (Judge and Carey 2000). Humans, in turn, are long lived for a primate (figure 9.2), with a life span over 2-fold greater than that predicted from body mass. The inclusion of primate brain mass in predictive regressions increased the human predicted life span into the region of 8 to 9 decades (depending on comparison group) (Judge and Carey 2000). Anthropoid primates are relatively large brained and social, and have complex forms of communication. Language, as the coevolutionary product of social and biological processes, changed the nature of information transfer across and within generations (Deacon 1997). Potential life span appears to have increased rapidly in *Homo*, with predicted life spans into 7 decades for *Homo erectus*, and between 8 and 10 decades for *Homo sapiens*, which has a current record of 122 years (Allard et al. 1998)—over twice the life span of any other documented primate. Evidence

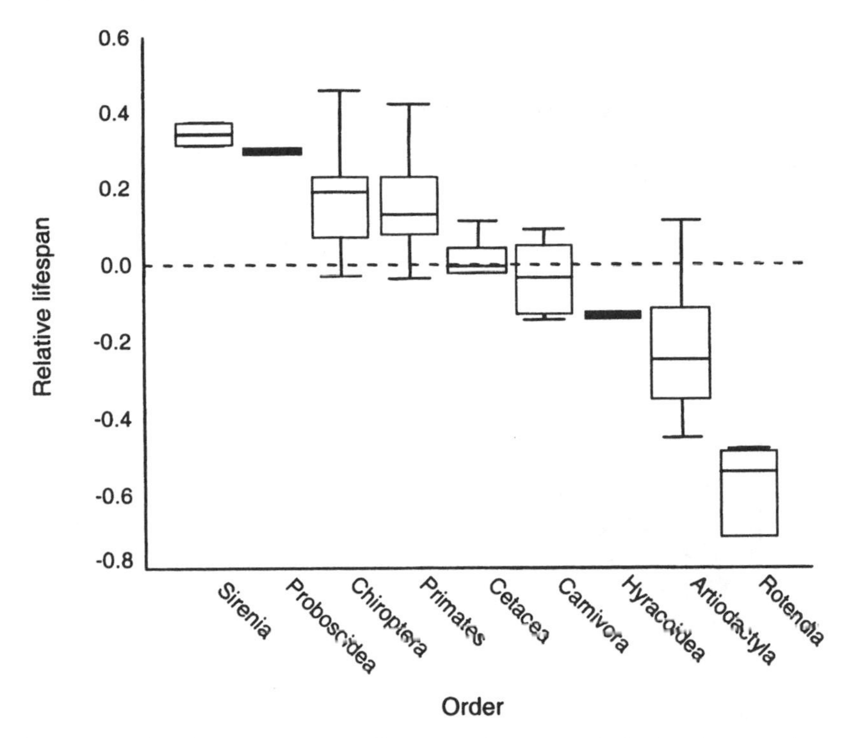

FIGURE 9.1. Relative life span for 9 mammalian orders. Relative life span is the record life span less the life span value predicted from mammalian regression of life span on body mass. Predictions were calculated using ordinary least squares regression with log transformed values and case deletion. Sample comprises average logged values of species for 62 families of mammals in 9 orders. Means are indicated within boxes (Carey and Judge 2001a).

of extensive intergenerational subsidization is widespread and has given rise to hypotheses regarding the role of grandmothers (Hawkes et al. 1998), for which corroborative data are mixed (Hill and Hurtado 1991), and for parental investment more generally (Kaplan 1996). I believe that this model helps to explain the role of intergenerational transfers in the evolution of longevity and suggests that the specific transfer pathways may differ under varying ecologies.

9.4.2. Primate Origins of Longevity Extension

Judge and Carey (2000) combined the primate life span and body mass data presented in figure 9.2 with brain size to develop a regression equation for predicting the evolved life span of both modern humans and ancestral homi-

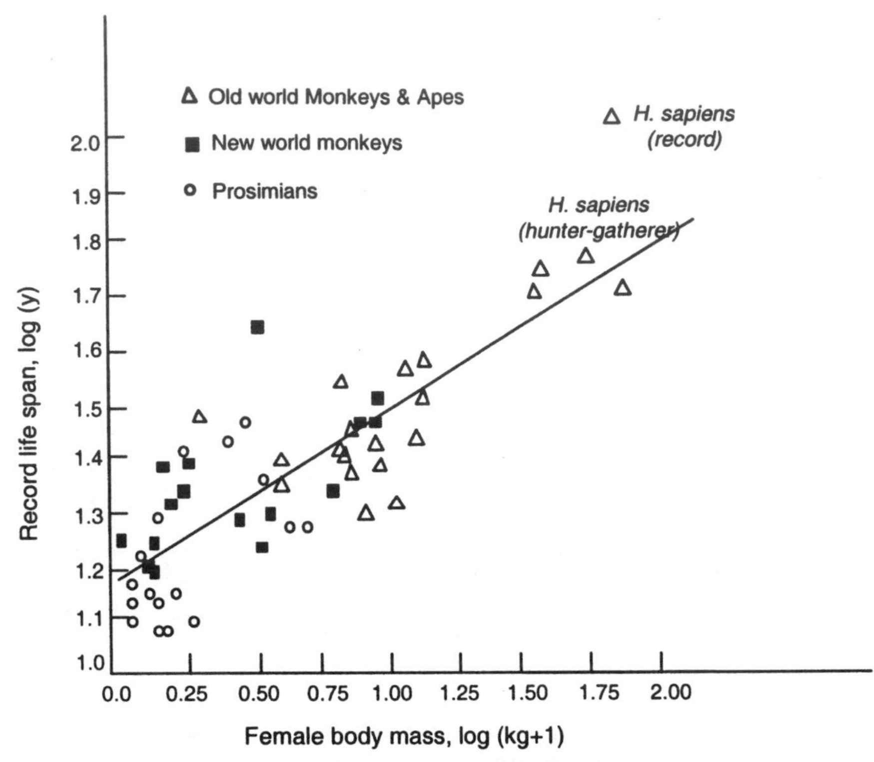

FIGURE 9.2. Record captive life spans by body size for primate genera. Mean record captive life spans (years) by female body mass (kg + 1) for 53 genera of primates. The left axis and abscissa are the log-transformed values used in ordinary least-squares regressions yielding a positive relationship with $r = 0.817$, $p < 0.001$, R^2 adjusted = .660); the right axis provides values in years. The statistical outlier in the midrange of body mass is the New World monkey genus, *Cebus*, with a mean life-span record of 46 years (Carey and Judge 2001a).

nids (table 9.3). The analysis revealed that humans, both archaic and modern, have a primate morphology that predicts a life span of 72 to 90 years—well beyond the age of menopause in modern females. These predictions also exceed other estimates (Angel 1969; Buikstra and Konigsberg 1985; Washburn 1981). With a primate longevity advantage, reduction of litter size to one and shifts to a higher quality diet that required more skill to obtain (Kaplan 1996), hominids appear to have broken into the longevity feedback loop. Fossil evidence indicates that *H. erectus* demonstrated larger stature (McHenry 1994) and slower dental development, suggesting longer inter-birth intervals than earlier hominoids. This suggests overall increases in per offspring parental investment rather than translation of higher quality diet

TABLE 9.3
Evolved Life Span of Ancestral Hominids and Modern Humans

Hominid Species	*Life Span (Years)*	*Incremental Change*	*Cumulative Change*
Australopithecus afarensis	46.6	8.4	8.4
Homo habilis	55.0	7.0	15.4
H. erectus	62.0	10.9	26.3
H. sapiens (pre-historical)	72.9	49.1	75.4
H. sapiens (contemporary)	122.0		

Note: Estimates of longevity for fossil hominids (Judge and Carey 2000), based on hominoid body-size relationships, range from 42 to 44 years for *Australopithecus* to 50 years for *Homo erectus* (McHenry 1994). Incorporation of brain mass increased estimates for *Homo habilis* from 43 years to 52–56 years and for *Homo erectus* from 50 years to 60–63 years.

into more offspring. Greater adult life expectancy and generational overlap is required for the intergenerational transfers of resources, knowledge, and skills in later *Homo*. Reductions in juvenile and subadult mortality, through adult subsidization of juvenile diet, probably increased life expectancy at birth as well as the age at sexual maturity. Thus, additional investment in individual young by parents was more likely to have a fitness payoff (Kaplan et al. 2000).

9.4.3. Prehistoric and Historical Patterns of Longevity Change

Human life spans that approach and sometimes exceed the "design capacity" indicated from primate allometry are not unusual in contemporary Western societies and may have been more common in early hunter-gatherer societies than is generally realized. Studies of extant hunter-gatherers find that a significant proportion of those reaching adulthood lived into their seventh decades (Hill and Kaplan 1999). As more people reach older ages, investments in health and well-being are more likely to yield returns, in terms of prolonged working life span, to more individuals. Historical developments substantially reduced early mortality—revealing design capacity longevity—and subsequently allowed technological attenuation of early senescing systems (e.g., eye glasses, dentures) (Fogel and Costa 1997; Landes 1998). Only with the last quarter century have humans pressed beyond the "design capacity for longevity" and extended that capacity with large-scale systemic innovations such as joint replacement and organ transplants (Nutton 1996; Smith 1993; Thomas 1988). The self-reinforcing process of increasing life expectancy increases age-specific productivity, and thus both demand and resources for health improvements should occur only to the extent that productive life span is increased along with absolute life span.

9.4.4. Health, Wealth, and Longevity

It is intuitively obvious that wealth is conducive to longevity—wealthier nations have higher life expectancies and larger proportions of very elderly citizens (Bloom and Williamson 1998). However, at the population level, improvements in longevity also are conducive to improvements in wealth and thus create a so-called "virtuous spiral." Four main categories of relationships may account for this relationship (Bloom and Canning 2000):

1. Productivity. Longer-lived populations tend to be healthier and thus have higher labor productivity because workers are physically more energetic and mentally more robust for longer. Societies in which a large fraction of the population attains older ages can have a greater division of labor.

2. Teaching, learning and education. Long-lived people are repositories of knowledge, wisdom, and technical expertise and therefore are teachers and educators. Healthier people who live longer have stronger incentives to invest in developing their skills because they expect to reap those benefits over a longer period. Increased technical training and schooling promotes greater productivity and higher quality output and thus increases higher income.

3. Investment in physical capital. Improvements in health and longevity create an incentive to plan for the future, including the need for savings, which, in turn, promotes greater investment. Investment promotes employment, and workers will thus have access to more resources as their incomes rise. A healthy, long-lived, and educated workforce acts as a strong magnet for foreign investment.

4. Demographic dividend. The transition from high to low mortality rates, particularly in the young, typically trigger declines in fertility rates. This, in turn, increases the proportion of the population of working age that increases productivity. In addition, populations with large cohorts of older (retired) persons actually stimulate economic growth, presumably because they work or enable others to work by minding their children, and many continue to impart their accumulated knowledge to others.

9.5. Implications of Longevity-oriented Theory

I believe that the theory described in this chapter has several important implications. First, expansion of the theoretical foundations of the finitude of life provides a new framework for discussing the future of human life span (Hayflick 2000; Manton and Stallard 1996; Vaupel et al. 1998; Wilmoth et al. 2000). In particular, it shifts discussion from the issue of mortality reduction, based on putative limits imposed by the evolutionary theory of aging, to discussion of a more inclusive process of longevity extension of which reduction in aging rate is one component.

Second, a theory of longevity raises questions about the interpretation of results of studies on the biological mechanisms of aging using model species such as fruit flies, laboratory rodents, and nematode worms, all of which are solitary (nonsocial) species. Inasmuch as the theory suggests that extended longevity in solitary species evolved under different ecological contexts than in social species, the underlying mechanisms of aging between animals in these two broad categories may be different and assumptions about age-specific reproductive value are certainly different. For example, the aging response to caloric restriction (Sohal and Weindruch 1996; Weindruch 1996) in solitary species that must survive independent of a social group may be fundamentally different than the mechanisms in social species with evolved behaviors for helping, sharing, and food storage. Whereas aging-oriented research necessarily focuses on lower levels of biological organization such as the molecule and cell, longevity-oriented theory focuses on the level of organization considered by many biologists to be the quintessence of biological relevance—the whole organism. All discoveries related to the rate of aging at lower levels of biological organization must ultimately be tested at the level of the individual.

Third, the theory suggests that life span as a measure of length of life should be considered more central to life history theory than has been the case. Life span as a concept is either absent from essentially all of the mainstream texts in ecology (Begon et al. 1996), evolution (Futuma 1998), and population biology, or is discussed as an outcome variable of size and mortality (Roff 1992). Classic life history theory treats life span as an outcome of the relative mortality of juvenile and adult stages and trade-offs between early and late reproduction (Stearns 1992). Evolution of life span per se is considered only in the context of the evolution of repeated reproduction (Stearns 1992). Inasmuch as life history theory is concerned with trade-offs, questions concerning trade-offs between extraordinary longevity and other traits must certainly constitute issues central to this area of inquiry.

Fourth, a longevity-oriented theory concerns human longevity in the broader context of its life history (Hill and Kaplan 1999) and culture (Durham 1991). Whereas most anthropologists consider many traits such as covert ovulation, menopause, and speech as unique, and bipedalism, brain size, and the use of tools as exceptional, extended longevity remains unappreciated. Human life span should be considered nearly as extraordinary as human brain size. Life span receives scant attention in life history context and yet is 2.5-fold greater than would be predicted from body size alone (Judge and Carey 2000), whereas the brain, which receives considerable attention, is 3-fold greater than is predicted from body size alone (Diamond 1992).

9.6. Human Life Span Extension: A Framework for the Future

The extensions of human longevity occur through the same basic processes as in other organisms. The difference is not in the relationships of reproductive value, intergenerational transfer, or specialization, but in the mechanisms by which those processes are implemented. Intergenerational transfers are expanded beyond genetic and immediate resources to include the information necessary to increase the production of resources, highly transferable and partible symbols of resources, and information about ecological problems that allow targeted adaptive responses. Specialization is developed beyond that in other species through these same unusual mechanisms.

The theory presented here extends Fogel and colleagues' concept of technophysio evolution (Fogel 1993, 1994; Fogel and Costa 1997), which states that mortality reductions in human populations occur because our species has a level of control over our environment that is unprecedented among all animals. The model demonstrates the link between the Fogel model and Becker's "demand for children," human capital, and specialization (Becker 1991; Becker and Murphy 1992). I argue then that advances in human longevity are the result of some special conditions within a broader evolutionary model applicable to animals more generally. This self-reinforcement model of longevity extension in humans integrates evolutionary (sociobiological) and economic processes that collectively apply to longevity extension in many social species. In other words, sociality evolved in many groups in response to the advantages of controlling the physical and biological environment. Advantages include protection from predators, temperature and humidity controls, and food acquisition, storage and production (Hill and Kaplan 1999; Turner 2000). Generalizing the concept reveals that humans are simply one of a number of social species in which longevity extension has occurred and is continuing to occur as a result of basic principles linking sociality and longevity extension. The theory expands Fogel's time horizon from 300 years of modern technological development to the entire history of human technological development—on the order of some 2 million years. Early technological developments influenced longevity through environmental modifications, including the use of stone tools, control of fire, shelter construction, invention of farming, and the emergence of bureaucracies (Foley 1992; Zvelebil 1994). Subsequent technological changes in communication and organizational efficiency included the invention of writing and improvements in information transfer, improvements in transportation, industry, and sanitation (Cruse 1999; DeGregor 1985), and continuing scientific specialization, such as cloning, genomics, and molecular techniques in biomedicine (Collins 1998, 1999; Gurdon and Colman 1999; Silver 1997). The cost of coordinating specialized workers may constrain the division of labor and specialization (Becker and Murphy 1992). The declining costs of coordination associated with the explosion of information technology over the past century may have contributed to increased specialization and greater rates of scientific progress, and contributed to the vast increases in observed human life spans over the same period.

One of the most important contributions of a longevity-oriented theory is to reframe the way scientists consider the future of human life expectancy. The more conventional aging-oriented question is "How low do mortality rates have to go to substantially increase longevity in the future?" (Fries 1980; Olshansky et al. 2001), whereas the longevity-oriented question is "To what extent will future gains in longevity (or improvements in mortality) extend life expectancy further?" The aging-oriented question assumes that the age-specific improvements in survival impact only older ages, but the self-reinforcing model argues that age-specific decrements in late life mortality can also have a positive effect on early survival and productivity through selection, intergenerational transfers of resources and information, and increasing innovation from specialization. The longevity-oriented question recognizes a built-in dynamic whereby the effects of reducing mortality at one age are spread across multiple ages and thus amplify longevity extension effects. The longevity-oriented approach also suggests that the present trajectory of science and technology, including organ cloning, xenotransplantation, and molecular medicine will lead to (1) sweeping transformations in health, and (2) to what Wilson (1998) refers to as "volitional evolution"—the decommissioning of natural selection so that individuals can make choices conducive to engineering a long life for themselves and their children (Baltimore 2001). As de Grey (2000) notes, it is certain that human scientific knowledge and consequent technological prowess will continue to advance at a non-negligible rate for as long as civilization survives. It is thus likely that human longevity will continue to advance and to have lasting and transformational effects on society.

9.7. Summary

To understand human longevity we must look not only to our ability to control our environment, but also to our phylogenetic and historical legacy (Coe 1990; Foley 1992, 1995). Parental care is part of our mammalian heritage as is the tendency for females to be more risk-averse and males to be more risk-prone. Our anthropoid heritage produced long gestation times, single infant births, and long interbirth intervals consonant with high levels of maternal effort, and sociality is a firmly established core heritage. Brain expansion and use of manufactured tools were linked to increased longevity in the genus *Homo*. With the fully modern brain, anatomy, and behavioral repertoire of *Homo sapiens* came an awareness of our own mortality and directed attempts to escape it. Our modern human heritage includes living in larger groups through the development of agriculture and systems of production (Diamond 1998). Humans were gradually able to order their experience more consciously in accordance with a plan. They replaced the concept of life as an exercise in endurance with the thought of life as a seeking and a journeying (Southern 1953, p. 222). We are now in the era of emerging ability to control our actuarial destiny in response to the desire in humans throughout history to live comfortably and to delay death (Holliday 2001; Preston et al. 1978).

10

Epilogue

A CONCEPTUAL OVERVIEW OF LIFE SPAN

HISTORICALLY the life span concept has typically been considered in the narrow context of a life table–like parameter defined as "record age" or "maximum life span potential." However, in this chapter I examine the concept in broader contexts, including (1) abstract and fundamental contexts, noting that the concept applies only to clearly defined individuals with clear start and top boundaries that delineate its existence; (2) evolutionary, including senescence, elderly in nature, and longevity as an adaptation); (3) human, including predictions from primate patterns and a theory of life span extension; and (4) future, including demographic predictions and possibilities and so-called demographic ontogeny. At the end of the chapter I point out a conceptual paradox. On one hand, the life span concept is decoupled from death inasmuch as it implies living (life) rather than dying. On the other hand, the concept is inextricably linked to death because it implies life's ultimate age of closure.

10.1. Background

Life span is an evolved life history characteristic of an organism, which refers to the duration of its entire life course. Application of the concept is straightforward at both individual and cohort levels and specifies the period between birth and death for the former (individual) and to the average length of life or life expectancy at birth for the latter (including both real and synthetic cohorts). However, when life span is applied to a population or a species it requires a modifier to avoid ambiguity (Goldwasser 2001). Maximum observed life span is the highest verified age at death, possibly limited to a particular population, cohort, or species in a specified time period. The overall highest verified age for a species is also called its record life span. The theoretical highest attainable age is known as either maximum potential life span, maximum theoretical life span, or species-specific life span.

Maximum observed life spans (i.e., longevity records) are not synonymous with theoretical maximums for at least two reasons. *First*, maximum longevity is an inappropriate general concept because an animal dies before the age of infinity, not because it cannot pass some boundary age but because the probability of its riding out the ever-present risk of death for that long is infinitesimally small (Gavrilov and Gavrilova 1991). In other words, there is no identifiable age for each species to which some select individuals can survive but none can live beyond. Second, the record age of a species is heavily influenced by the number of individuals observed. That is, the longevity records for species in which the life spans of large numbers of individuals have been observed will be significantly greater than the corresponding figure for a species with the same longevity but represented by a few dozen individuals. For the vast majority of longevity records by species, the population at risk and therefore the denominator are completely unknown.

10.2. Abstract Perspectives

The life span concept should be considered within a larger biological framework concerned with *life duration*—a model where the end points of life for an *individual* do not necessarily correspond to conventional events of the birth and death. The life span concept does not apply to bacteria that reproduce by binary fission, to plant species that reproduce by cloning, or to modular organisms with iterated growth, such as coral or honeybee colonies. The *state of existence* for an individual refers to one of several possible states of living, such as "normal state," metabolic reduction (e.g., hibernation) and/or arrest (e.g., frozen embryo, dormancy), its *time of existence* refers to the sum total of all stages of the individual's life, and its *extinction* refers to the point at which the individual ceases to exist. What constitutes an individual is conditional on the level of *individualism*. For example, the individual cell within a multicellular organism and the individual worker bee in a colony of honeybees are individuals on one level but are part of a greater whole on another. Death of a component part (e.g., through apoptosis) may or may not determine the death of the whole, although in each case the components (cells, worker bees) relinquish both their evolutionary heritage and their autonomy in favor of the higher organisms of which they are a part (Michod 1999).

The distinction between mortal macroorganisms and immortal microorganisms is unclear with certain organisms. For example, slime molds such as *Dictostelium discoideum* feed as millions of individual, potentially immortal single-celled amoeba in the forest soil and litter most of the time. However, under certain conditions these millions form swarms in dense, coordinated, visible, mortal masses to ooze about in the form of a small garden slug before sprouting into a "flower" for creating and disseminating spores (Bonner 1967; Murchie 1967). A different life span issue concerning the individual concerns the concept of replacement of constituent parts (i.e., cells). This is illustrated biologically by the hydra, *Hydra littoralis* (Campbell 1967), which replaces all of its cells over the course of a few weeks and thus can potentially live forever.

When a single reproductive event occurs at the end of the life course and results in the death of the individual, then life span is linked deterministically to the species' natural history. This occurs with the seed set of annual plants (grasses), in drone (male) honeybees as a consequence of the mechanical damage caused by mating, in many mayfly species when a female's abdomen ruptures to release her eggs after she drops into a lake or stream, and in anadromous salmon, which die shortly after spawning. Life span can be considered *indeterminate* for species (including humans) that are capable of repeated (iteroparous) reproduction. That life span is indeterminate in many species is consistent with everything that is known about the lack of cut-off points in biology—all evidence suggests that species do not have an internal clock for terminating life.

10.3. Death and Extinction

Death is the unique, singular, and fateful event that stamps life's closure (Hillman 1999). Death—that is to say dying—is a graded process in virtually all organisms starting with substages of infirmity (weakness) and morbidity (sickliness) and progressing to coma, breath stoppage, heart stoppage, brain death, cessation of metabolism, tissue degradation, and loss of DNA integrity. Because of this progression of death by stages, it is useful to distinguish between the death of the *organism as a whole*, which refers to somatic death (the organism as an integrated functional unit), and the death of the *whole organism*, which refers to death of the various parts (Kass 1983). Whereas the death of the organism as a whole is usually an event—an abrupt crossing-over from the live to the dead state—the death of different parts of the whole organism is a long, drawn-out process of degradation.

In broader biological contexts, the ways in which individuals can go extinct (as opposed to extinction by dying) as singular entities is through *fission* in bacteria, where one individual becomes two, or through *fragmentation*, as in fungi, where the filaments of the original fungal growth breaks into many pieces, each of which is essentially a 'seed' for new growth. An especially striking example is *splitting* in flatworms, where the original individual can be divided into parts, each capable of growing into a new worm (Anon 2001). In each of these cases, the original individual no longer exists; no part or division has precedent over the other.

These same concepts can also apply to humans if an individual embryo splits into two or more smaller embryos, each of which then develops into separate individuals (e.g., monozygotic twins). One individual existed before the split, but two or more new individuals exist afterward—the "original" individual no longer exists. A related concept is that of fusion of embryos into a *chimera*, in which two individuals become one. Indeed, since the 1950s more than one hundred natural-born chimeric humans beings have been identified by medical geneticists (Silver 1998). Each of these people emerged from the fusion of two embryos that resulted from the fertilization of two eggs that had been ovulated simultaneously by their mother. Thus what started out at conception as two individuals ended up being a single individual. In short, conceptual difficulties regarding life span may arise when the life of an individual is not bounded by either discrete birth and death processes or where the level of individualism is unclear.

10.4. Boundary and Perpetuity

It is useful to distinguish basic terms that are often used in the context of longevity and life span. *Mortal* refers to individuals destined to die, *immortal* refers to individuals living forever (however, the latter word also is used in the context of "immortal" germ lines), *eternal life* refers to endless life after death, *finitude* (or finite) refers to limits, *infinitude* (or infinity) is a general term referring to anything without limits, *determinate* refers to a fixed process or number, and *indeterminate* refers to an open-ended process or rate. The term *immortalization* is used to describe cell cultures that can be propagated generation after generation with no loss of viability (Blackburn 2000), such as the HeLa cell line (the first human cells to live indefinitely outside the body), which were derived from the cells of cancer victim Henrietta Lack in the 1940s (Murchie 1967; Skloot 2001) and immortalized B lymphocytes, which are used to produce virtually unlimited quantities of identical antibody molecules (Köhler and Milstein 1975). The term *preservation* refers to maintenance in the existing state. The bacteria trapped in 25–30-million-year-old amber (Cano and Borucki 1995) or in a 250-million-year-old primary salt crystal (Vreeland et al. 2000) and the mitochondral DNA isolated from a 29,000-year-old Neanderthal (Ovchinnkov et al. 2000) are all examples of preservation. *Cryopreservation* is a special type of preservation using ultra-low temperatures (−196° C) to preserve human sperm and embryos (Silver 1998; Trounson 1986). *Dormancy* is a general term used to describe one of a number of different evolved types of arrested growth ranging from *torpor* to *hibernation*. (Carson and Stalker 1947; Hairston and Caceres 1996; Mansingh 1971; Masaki 1980).

10.5. Evolution

10.5.1. Evolutionary Origins of Senescence

Bell (1988) established the deep connection between the two invariants of life—birth and death—by demonstrating that protozoan lineages senesce as the result of an accumulated load of mutations. The senescence can be arrested by recombination of micronuclear DNA with that of another protozoa through conjugation. Conjugation (sex) results in new DNA and in the apoptotic-like destruction of old operational DNA in the macronucleus (figure 10.1). Thus, rejuvenation in the replicative DNA and senescence of operational DNA are promoted by sexual reproduction.

When this is extended to multicellular organisms, sex and somatic senescence are inextricably linked (W. Clark 1996). In multicellular sexually reproducing organisms, the function of somatic cells (i.e., all cells constituting the individual besides the germ cells) is to ensure the survival and function of the replicative DNA—the germ cells. Prior to bacteria, the *somatic* DNA was the *germ line* DNA; prior to multicellular animals, the *somatic cell* and the *germ cell* were undifferentiated. Like the macronuclei in the paramecia, the somatic cells senesce and die as a function of their mitotic task of ensuring the survival and development of the germ cells. The advent of sex in reproduction allowed exogenous repair of replicative DNA (W. Clark 1996), while in multicellular organisms the replication errors of somatic growth and maintenance are segregated from the DNA passed on to daughter cells and are discarded at the end of each generation. Thus senescence is built into the life history concept of all sexually reproducing organisms; sex is an adaptation to circumvent mutational errors in somatic growth during reproduction.

10.5.2. Life Span as an Evolutionary Adaptation

In evolutionary biology an "adaptation" is a characteristic of organisms whose properties are the result of selection in a particular functional context. Just like the different bird beaks that are adaptations for exploiting different niches and that must be balanced with other traits such as body size and flight propensity, the longevity of an animal is also an adaptation that must be balanced with other traits, particularly with reproduction. The variations in the relationship between reproduction and longevity can only make sense when placed within the context of such factors as demographics, duration of the infantile period, number of young, and the species' ecological niche—the organism's overall life history strategy. Inasmuch as life spans differ by 5,000-fold in insects (2-day mayflies to 30-year termite queens), by 60-fold in mammals (2-year mice to 122-year humans), and by 15-fold in birds (4-year songbird to 60-year albatross), it is clear that life span is a life history adaptation that is part of the grand life history design for each species.

10.5.3. Evolutionary Ecology of Life Span

The literature on aging and longevity contains descriptions of only a small number of life-span correlates, including the well-known relationship between longevity and both body mass and relative brain size and the observa-

Conjugation in Paramecium

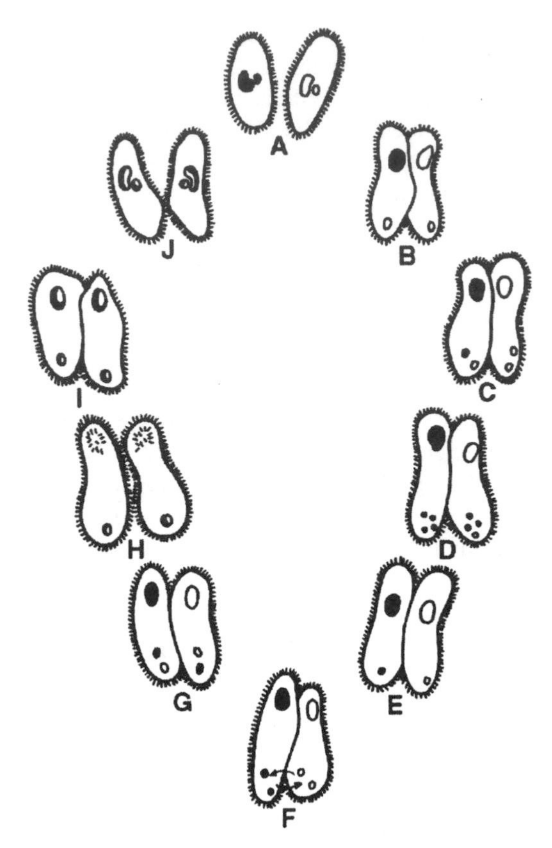

FIGURE 10.1. Schematic of sexual reproduction in the protozoan *Paramecium aurelia* (redrawn from Clark 1996, p. 70). A. Two genetically different protozoa, each with macronucleus (large oval in individual protozoa) and a micronucleus (small oval in protozoa). B. The two protozoa fuse in the first step of conjugation, and the macronuclei and micronuclei move to opposite ends of the cell. C. Each micronucleus divides once by meiosis. D. The daughter micronuclei each divide again, to produce four haploid micronuclei. E. Three of the four haploid microcuclei disappear. F. The remaining micronucleus divides once more, to produce two identical micronuclei. G. The two conjugants exchange one micronucleus. H. The two haploid micronuclei fuse to produce a single diploid micronucleus. I. The new micronuclei each direct the production of a new macronucleus; the old macronucleus begins to disintegrate. J. The two protozoa disengage, and the nuclei assume their starting positions in the cell. The exconjugates are now genetically identical to one another, but genetically different from either of the two starting cells. Each will go on to produce genetically identical daughters by simple fission.

tion that animals that possess armor (e.g., beetles, turtles) or capability of flight (e.g., birds, bats) are often long-lived. But major inconsistencies exist within even this small set of correlates (Carey and Judge 2000a; Carey and Judge 2001a). For example, there are several exceptions regarding the relationship of extended longevity and large body size (e.g., bats are generally small but most species are long-lived), and this positive relationship may be either absent or reversed within certain orders including a negative correlation within the Pinnipeds (seals and walruses) and no correlation within the Chiroptera (bats). Likewise, the observation that flight ability and extended longevity are correlated does not provide any insight into why within-group (e.g., birds) differences in life span exist, nor does it account for the variation in longevity in insects, where adults of the majority of species can fly.

A classification system regarding the life-span correlates of species with extended longevity that applies to a wide range of invertebrate and vertebrate species consists of two categories. The first is *environmentally selected life spans*. This category includes animals whose life histories evolved under conditions in which food is scarce and where resource availability is uncertain or environmental conditions are predictably adverse part of the time. Some of the longest-lived small and medium-sized mammals live in deserts where rainfall and, thus reproduction, is episodic and unpredictable; they include foxes, rodents such as gerbils and rock hyrax, and small equines and ungulates such as feral asses. The recent findings that the life span of bowhead whales, *Balaena mysticetus* (a solitary species of baleen whales), may exceed 200 years (George et al. 1999) is another important example of how environmental factors shape life span through *direct natural selection*. Bowhead whales live in the harsh environment of the Arctic ocean with low prey densities, and thus require great investment in fat storage, body mass, and thermoregulatory mechanisms. Cetacean biologists studying this species have suggested that these stressors led to slow growth, delayed maturity, and subsequently extended longevity to ensure reproductive success (George et al. 1999). Several additional examples of animal species whose life spans were environmentally selected are shown in figure 10.2.

The second category is *socially selected life spans*. This category includes species that exhibit extensive parental investment, extensive parental care, and eusociality (social strategy arising from the study of ants, bees, wasps, and termites that have overlapping generations, cooperative care of young, and a reproductive division of labor). The extended longevity of animals in this category results from natural, sexual, and kin selection and includes all of the social primates including humans. An example and description of a long-lived species (ant queen) whose life span was socially selected is shown in figure 10.3.

This classification system places the relationship of life span and two conventional correlates, relative brain size and flight capability, in the context of

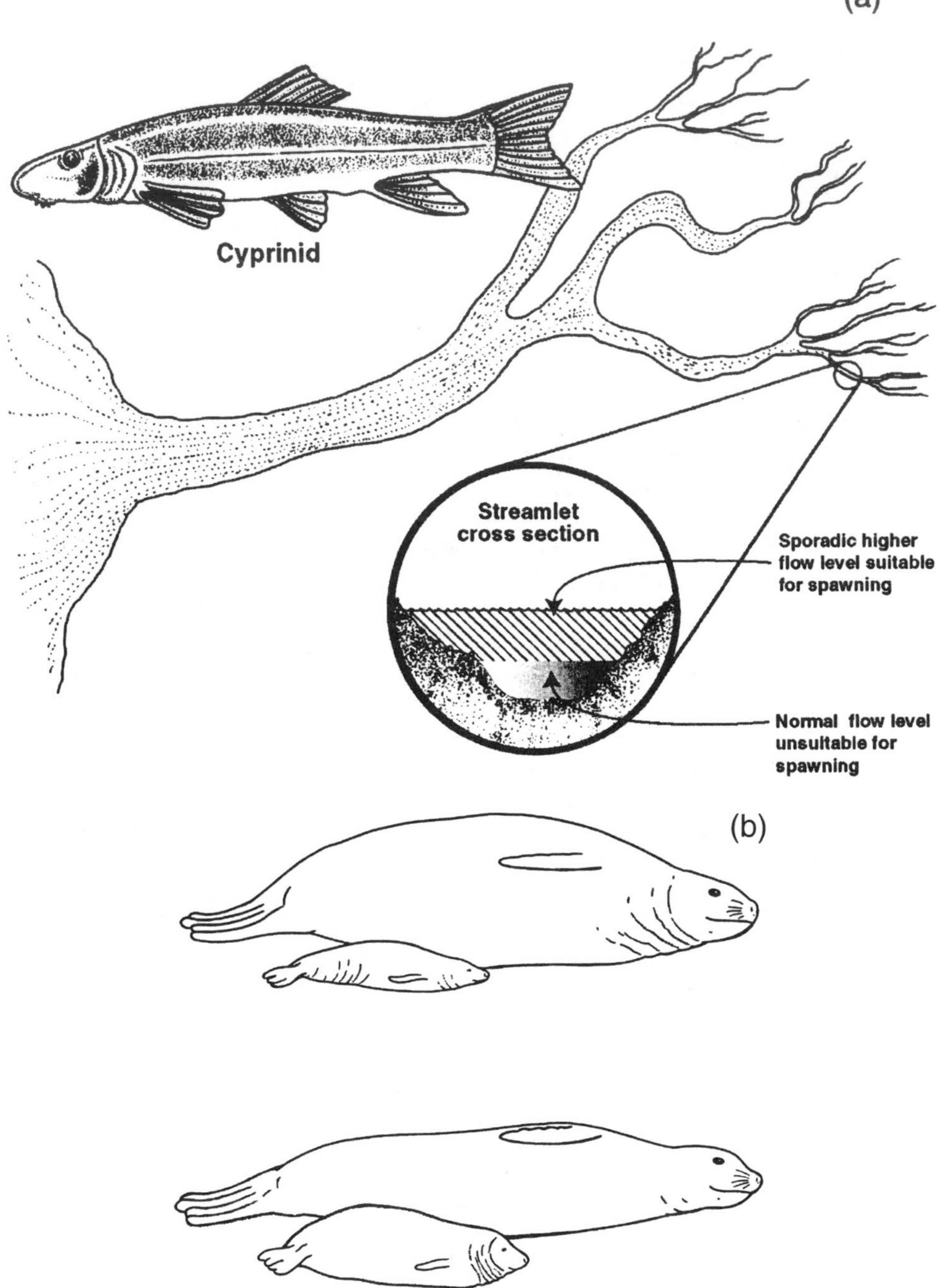

(a)
Cyprinid
Streamlet cross section
Sporadic higher flow level suitable for spawning
Normal flow level unsuitable for spawning
(b)

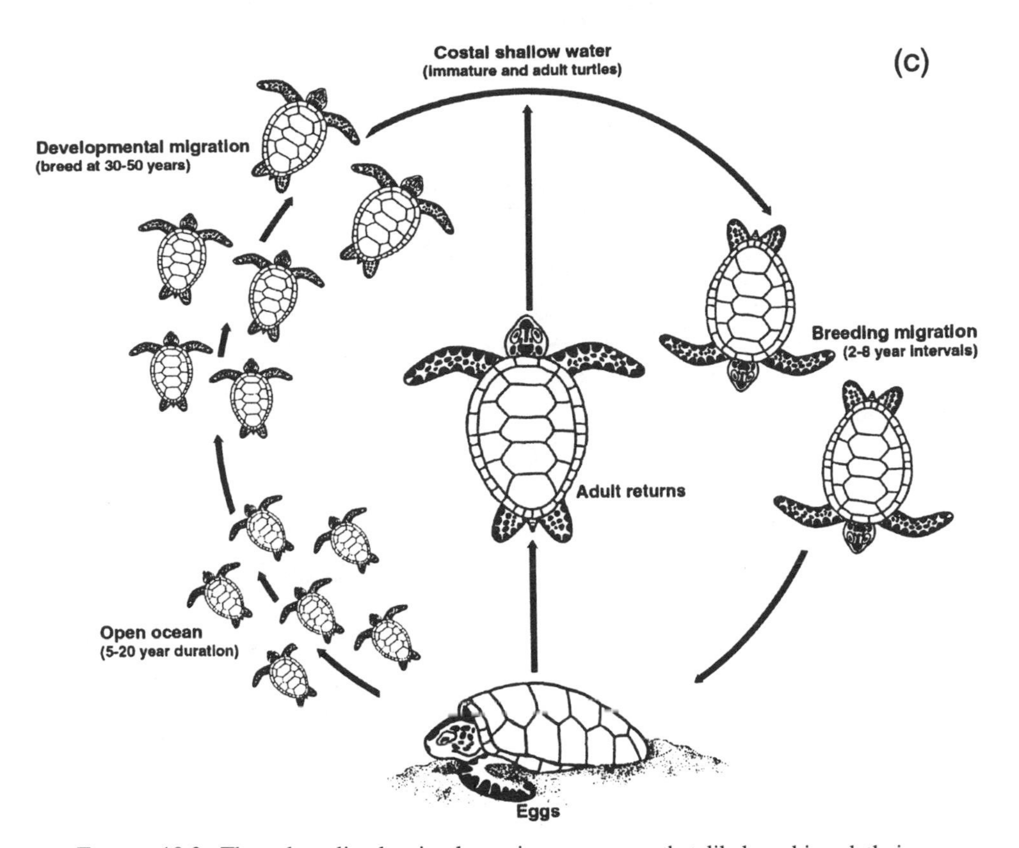

FIGURE 10.2. Three long-lived animal species or groups that likely achieved their extended life span due to direct natural selection. (a) Evolutionary ecology of extended longevity in cyprinids (suckers) in California. For spawning success, stream flow in the streamlets feeding into the main river must be both high and sustained since these serve as nurseries for the fry. However, many years are unsuitable for spawning because of the uncertainty of stream flow. Thus the uncertain environment selects for fish that live long enough to sustain the longest period between favorable spawning conditions. (b) During lactation the northern elephant seal mother loses weight while her pup gains correspondingly (redrawn from Riedman 1990). The top schematic shows the beginning, and the bottom shows the end of lactation period. The life span of most Pinnepeds (seals and walruses) is greater than predicted from their body mass alone. Their long life likely evolved to compensate for the high-risk environment of the young, which receive no parental care once they are weaned and must find food by themselves in the open ocean. This constraint means that seal mothers must invest heavily in a single young, transferring fat and protein to them over a relatively short span of time. If they were to have two pups, both would likely die shortly after weaning since neither would have achieved sufficient size to survive independently of their mother. Thus the long life of Pinnepeds is required to produce a series of singleton pups over a long span of time. (c) The life history of sea turtles requires a long life span because of their extraordinarily long developmental period, the long interval between egg laying bouts due to the time required to build up sufficient resources for producing new batches of eggs, the long distances that they must travel when migrating to their natal breeding grounds, and the high attrition of eggs, hatchlings, and immatures (redrawn from Miller 1997).

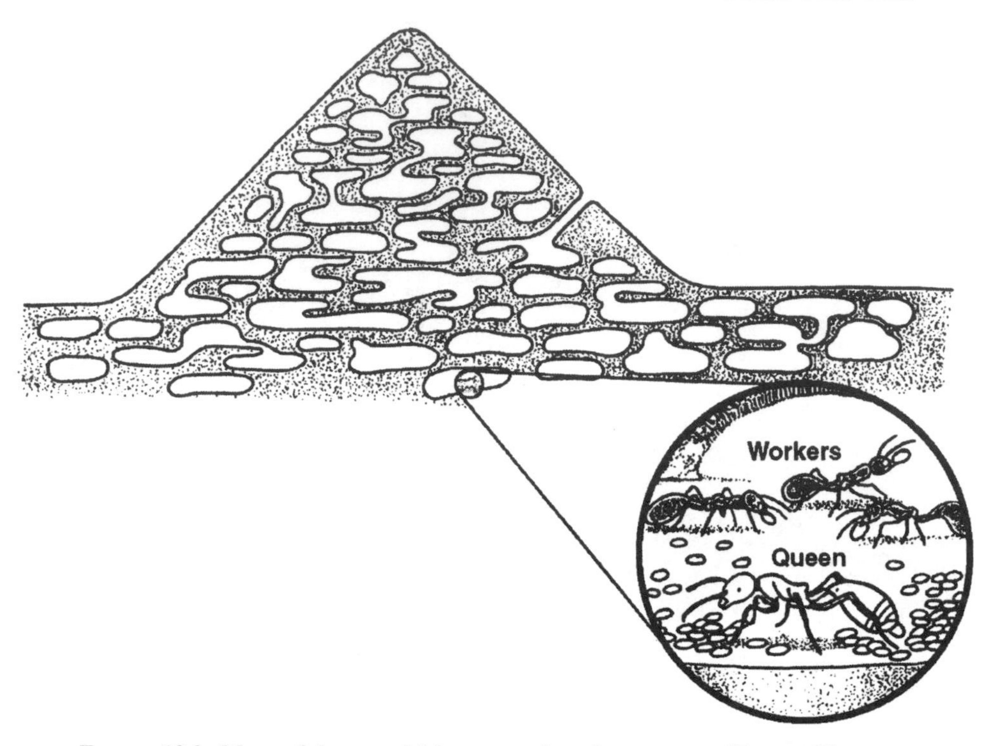

FIGURE 10.3. Many of the eusocial insects such as the ant queen illustrated here can live several decades (redrawn from Wheeler 1928). This life span evolved in a different context than the environmentally selected life spans of seals, cyprinids, and sea turtles shown in figure 10.2. Whereas the life spans of these groups evolved through direct natural selection, the life spans of social species such as the social insects and many of the highly social vertebrates, including humans, evolved partly as the result of kin selection and the coevolution of sociality and longevity.

life history. That is, brain size is related to the size of the social group and the degree of sociality that in turn is linked to extended life span. And intensive parental care is linked to flight capability in birds and bats, which is also linked to extended life span. For example, most bird species are monogamous, with both sexes helping in the rearing (e.g., one can protect the nest while the other collects food). The reproductive strategy of the majority of bat species is to produce only a single altricial (naked and helpless), relatively large offspring at a time—flight preempts the possibility of the female foraging for food while gestating multiple young. Thus bat maternal investment in a single offspring is substantial.

10.6. Roles of the Elderly

The long life span of in many nonsocial species is an adaptation designed to increase the likelihood of population replacement by their ability to survive through extended periods that are unfavorable to reproduction, to acquire scarce resources over a long period, or to repeatedly produce and invest heavily in singleton offspring. The elderly in social species play important roles in the cohesion and dynamics of groups, populations, and communities. In species such as elephants, whales, and primates, the elderly serve as caregivers, guardians, leaders, stabilizing centers, teachers, sexual consorts, and as midwives (Carey and Gruenfelder 1997). Older individuals in human populations are the culture bearers and the conscience of the societies of which they are a part. They provide economic and emotional support, help in child care, nursing, counseling, and homemaking, teach skills and crafts, conserve and transmit the traditions and values of culture groups, are valued voices in communal council, and provide historical continuity for younger members (Neugarten 1996). Older people share characteristics that differ from the young, including long experience, which gives them a better understanding of people and events, a commitment to the welfare of their children and thus to society at large. Therefore they tend to have a longer-term and a broader social outlook. Older persons in human societies also have a certain freedom to speak out and to take certain social risks that younger people are often unable to take (Neugarten 1996).

10.7. Minimal Life Spans

Understanding life span evolution requires an understanding of the conditions in which natural selection favors a particular life span, especially those that are extremely long or extremely short. Thus studies of short-lived species are nearly as important as studies of long-lived ones because these models can be used to test theories of aging and provide important comparative context. Studies of short-lived species also bring into sharp focus the evolutionary bottom line of reproduction—like any trait, a particular life span coevolved with other life history contexts to maximize fitness. One example of a "longevity minimalist" is the mayfly, the life history of which is shown in figure 10.4. The mayfly life history has several generic properties that favor the evolution of a short life span (Carey 2002), including: (1) synchronous emergence to concentrate adults so that the likelihood of finding a mate is maximized (this can be generalized as minimal mating requirement, including parthenogenesis); (2) inability to feed due to vestigial mouthparts, which preempts the need for individuals to forage for food (this trait

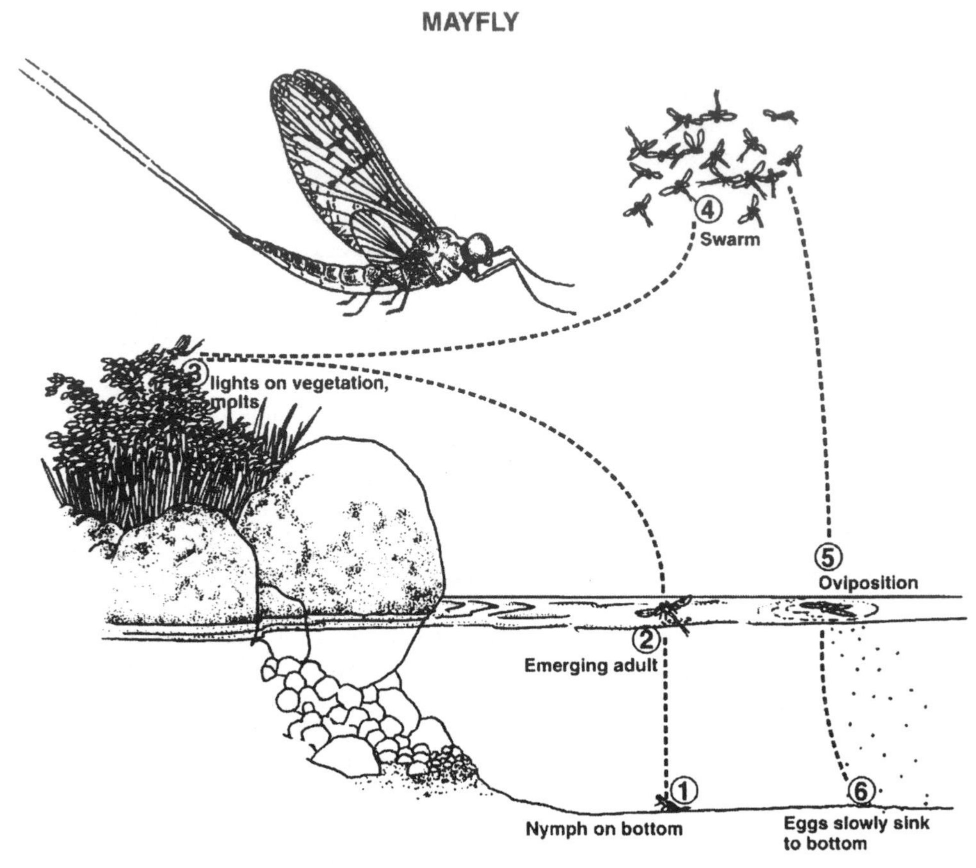

FIGURE 10.4. The life cycle of the mayfly. Members of this group are in the insect order Ephemeroptera, which is named after their ephemeral adulthood. After 1–3 years of development, depending on the species, fully grown nymphs come to the water surface and molt, and the winged forms fly a short distance to the shore where they alight on the vegetation, where they become part of a mating swarm the following evening. After a female successfully mates, she drops to the surface, where her abdomen ruptures and the eggs are dispersed (Carey 2002).

can be generalized as minimal food requirement); (3) general nymphal habitat that requires females to travel only a short distance (i.e., minimal energy requirements) to deposit their eggs; and (4) a single, large reproductive bout that eliminates the risk that some eggs are not deposited.

10.8. Absence of Life Span Limits

Although there has been considerable debate regarding whether the human life span (and life span in general) is fixed, the evidence now appears strongly to support the notion that there is not a single number that can be assigned to the life-span limit of species, particularly those with (so-called) iteroparous reproduction (as distinct from semelparous species like salmon, where individuals die soon after reproducing). The evidence can be outlined as follows.

First, the reliability theory of aging predicts the late-life mortality deceleration with subsequent leveling off as well as the late-life mortality plateaus, as an inevitable consequence of redundancy exhaustion at extreme old ages (Gavrilov and Gavrilova 2001). A deceleration of mortality at advanced ages is inconsistent with the concept of a specific life-span limit.

Second, the results of large-scale life table studies in virtually all species examined revealed, like the prediction from the reliability theory of aging, that mortality slows at older ages (Vaupel et al. 1998). As with the theory, this pattern of mortality is incompatible with a fixed life-span concept.

Third, if a single life-span limit existed for all species, a pattern of records of long-livers should begin to appear for at least some of the species for which longevity information is available. However, this is not the case—there is no clustering of longevity records from independent sources for thousands of species for which life-span data are available (Carey and Judge 2000a).

Fourth, the concept of a fixed species-specific life span is itself specious because of issues such as species strains, biotypes, sex, hybrids, and seasonal variants. For example, no single life span exists for honeybees (*Apis melifera*) even in theory since queens live 4–6 years, workers 6 weeks, and drones from a few days to several weeks. A similar problem exists for the domestic dog (*Canis familiaris*) with its scores of breed—the life span of smaller varieties (terriers) is substantially greater than that of the larger varieties (Great Danes) (Patronek et al. 1997).

Fifth, the human life span has been increasing for over a century, as is revealed by the records of extreme ages at death in Sweden for the last 130 years (Wilmoth et al. 2000). Thus the concept of a fixed life span in humans is not borne out in the data.

10.9. Humans

10.9.1. Life-Span Patterns: Humans as Primates

Estimates either based on regressions of anthropoid primate subfamilies or limited to extant apes indicate a major increase in longevity between *Homo habilis* (52–56 years) and *H. erectus* (60–63 years) occurring roughly 1.7 to 2 million years ago (Judge and Carey 2000). Predicted life spans for small-bodied *H. sapiens* is 66–72 years. From a Catarrhine (Old World monkeys and apes) comparison group, a life span of 91 years is predicted when contemporary human data are excluded from the predictive equation. For early hominids to live as long or longer than predicted was probably extremely rare; the important point is that the basic Old World primate design resulted in an organism with the potential to survive long beyond a contemporary mother's ability to give birth. This suggests that post-menopausal survival is not an artifact of modern life-style but may have originated between 1 and 2 million years ago coincident with the radiation of hominids out of Africa.

The general regression equation expresses the relationship of longevity to body and brain mass when 20 Old World anthropoid primate genera are the comparative group. Ninety-one years is the predicted longevity for a 50 kg primate with a brain mass of 1250 gm (conservative values for humans) when case deletion regressions methods are employed (each prediction is generated from the equation excluding the species in question), and 72 when humans are included within the predictive equation. When 6 genera of apes are used as the comparison group the regression equation is:

$$\log_{10} LS = 1.104 + 0.072(\log_{10} Mass) + 0.193(\log_{10} Brain), \qquad (10.1)$$

yielding a predicted human longevity of 82.3 years. Thus, a typical Old World primate with the body size and brain size of *Homo sapiens* can be expected to live between 72 and 91 years with good nutrition and protection from predation (figure 10.5).

10.9.2. Sex Life-Span Differentials

The general biodemographic principle to emerge from recent studies on male-female differences in life span is not that the female life span advantage is a universal "law" of nature. Rather the deeper principle is that the mortality response and, in turn, the life spans of the two sexes will always be different in similar environments. This is because the physiology, morphology, and behavior of males and females are different (Glucksmann 1974). Men have heavier bones, muscles, hearts, lungs, salivary glands, kidneys, and gonads in proportion to body weight, while females have proportionately heavier brains, livers, spleens, adrenals, thymus, stomach, and fat deposits. The basal metabolic rate of women is lower than men's. Men have more red corpuscles and hemoglobin per unit volume of blood than women. Sex differences are reflected in the rates of growth and maturation, metabolism, endocrine activity, blood formation, immune responses, disease patterns, behavior, and psychology. Males are generally more susceptible to reduction in food (Clutton-Brock 1994). Sexual dimorphism was highly pronounced in our *Australopithicus* ancestors suggesting that sexual selection via male-male competition had a strong influence on male-female differences in our progenitors. However, the height differences still persist as a residual of our

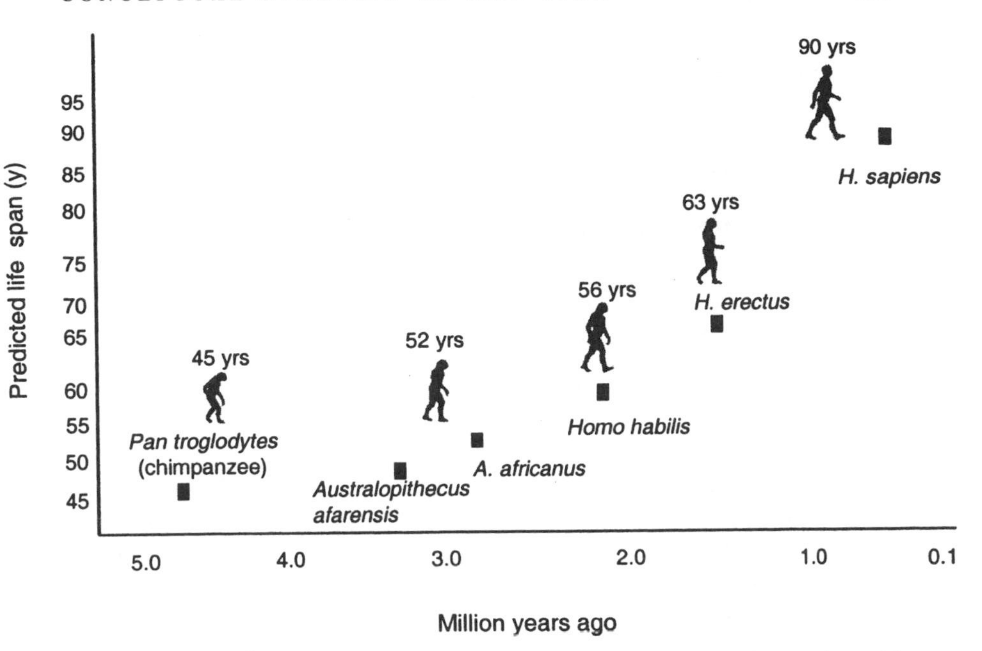

FIGURE 10.5. Predicted life spans of different hominoid species based on anthropoid subfamily values for body and brain mass regressions (redrawn from Judge and Carey 2000).

ancestry. In short, the mortality response of males and females in humans and virtually all animals differs because of differences in their physiology, reproductive biology, and behavior (Carey, Liedo, Orozco, Tatar, and Vaupel 1995a).

10.9.3. Age Classification

POSTNATAL AND TOTAL AGE

Age is as much a constituent of any living organism as is its physical self; its existence is identical to its duration. The time that an organism lives is measured, on one hand by hours, days, years, and decades according to chronological (solar) time, which flows evenly, and on the other hand, by rhythms and irreversible changes according to a physiological time scale that flows unevenly (Carrel 1931). Just as drying a rotifer can stop the stream of its duration or raising or lowering the temperature increase or decrease the rate of living of *Drosophila*, freezing a human embryo arrests its development. In all cases the changes in metabolism changes the rate of flow of physiological time and thus changes the chronological duration. In humans, the individual can be considered a single person who has two ages—*postnatal age*, which is measured from birth, and *total age*, which is measured from conception. The total age of a person who developed from an embryo not subject to metabolic arrest is that person's postnatal age plus the 9 months corresponding to the human gestation period. However, the total age of an individual who developed from an embryo that was metabolically arrested over a period of time is the postnatal age plus the duration of the period of metabolic arrest.

AGE CATEGORIES FOR HUMANS

Age and time can be considered in a number of different dimensions for humans (Neugarten and Datan 1973), including *historical time*, which shapes the social system, which, in turn, creates changing sets of age norms and a changing age-grade system that shapes the individual life cycle; *social time*, which refers to the dimension that underlies the age-grade system of a society (e.g., rites of passage); *psychological time*, which refers to the perceived increase in speed at which time progresses as a person grows older because each year that passes becomes a progressively smaller fraction of the accumulated years (Carrel 1931); and *life time*, which refers to the chronological age as a series of orderly events from infancy through childhood, adolescence, maturity, and old age. Demographers and sociologists have created new age categories and terms for the life times of seniors and elderly (Neugarten 1974). Thus persons classified as young-old, old, and oldest-old are considered in the age categories 55–65, 66–85, and over 85 years, respectively. The terms *septuagenarians*, *octogenarians*, *nonagenarian*, *centenarian*, and *supercentenarian* refer to persons in their 70s, 80s, 90s, 100s, and 110s, respectively, though *decacentenarian* and *dodecacentenarian* have recently been suggested for persons in their 110s and 120s, respectively (Allard et al. 1998). In general, distinctions among older people by age classifications are useful for understanding the human life cycle, as a framework for inter- and intragroup comparisons, and for health planners who are concerned with social and health characteristics of populations.

POST-DARWINIAN TAILS

The contemporary human life span of over 120 years is based on the maximal age recorded as consisting of two segments: (1) the Darwinian or "evolved" segment of 72–90 years; and (2) the post-Darwinian segment, which is the actifactual component that exceeds expectations from a primate of human body and brain size. Therefore the arguments that human life span has not changed in 100,000 years can be considered substantially correct when the "evolved" life span is considered. However, it is clear that this is not correct when the nonevolved segment of human life span is considered. There is evidence from Swedish death records that the record age in humans has been increasing for well over a century (Wilmoth et al. 2000).

EXCEPTIONAL AGES

Jeune (1995) has suggested that most of the literature on centenarians is based on the hypothesis of a secret of longevity, which is summarized in four assertions, each of which is debatable: maximum longevity is fixed, longevity is genetically determined, centenarians have always existed, and centenarians are qualitatively different. The first assertion is inconsistent with mortality patterns in both humans and nonhuman species. The second assertion is inconsistent with estimates of heritability estimates of longevity, which are between 0.2 and 0.3. The third assertion is improbable based on the estimated mortality levels that persisted before 1800. And the last assertion is tautological and thus unverifiable since the criteria for being different (i.e., being a centenarian) cannot be tested any other way than actually living to at least 100 years. Vaupel and Jeune (1995) claimed that supercentenarians did not exist prior to 1950 and centenarians not before 1800 in any population.

The oldest verified age to which a human being has ever lived, 122 years and 167 days, was that of the Frenchwoman Jeanne Calment, who was born on February 21, 1875 and died on August 4, 1997 (Allard et al. 1998). Other long-livers include Sarah Knauss, Mare Meiller, Chris Mortinson, and Charlotte Hughs, who lived 118.1, 117.6, 115.8, and 115.6 years respectively (Robine and Vaupel 2001). Madame Calment as well as these four long-livers all died between 1993 and 1999. Thus the five oldest persons whose ages of death have been verified died within the last decade.

10.10. Theory of Longevity Extension in Social Species: A Self-reinforcing Process

Improved health and increased longevity in societies may set in motion a self-perpetuating system of longevity extension (Carey and Judge 2001a). This positive feedback relationship may be one reason why human life span has been continuing to increase, and it is based on the demographic tenet that increased survival from birth to sexual maturity reduces the number of children desired by parents, *ceteris paribus*. Because of the reduced drain of childbearing and rearing, parents with fewer children remain healthier longer and raise healthier children with higher survival rates, which, in turn, fosters yet further reductions in fertility. Greater longevity of parents also increases the likelihood that they can contribute as grandparents to the fitness of both their children and grandchildren. And the self-reinforcing cycle continues.

The decline in mortality rates during the early stages of industrialization in the United States was probably one of the forces behind the expansion of educational efforts and growing mobility of people across space and between occupations. Whereas previous conditions of high mortality and crippling morbidity effectively reduced the prospective rewards to investment in education during the preindustrial period, prolonged expectancy for working life span must have made people more ready to accept the risks and costs of seeking their fortunes in distant places and in new occupations. The positive feedback of gains in longevity on future gains involves a complex interaction among the various stages of the life cycle with long-term societal implications in terms of the investment in human capital, intergenerational relations, and the synergism between technological and physiological improvements. In other words, long-term investment in science and education provides the tools for extending longevity, which make more attractive the opportunity cost of long-term investments in individual education, and thus help humans gain progressively greater control over their environment, their health, and overall quality of life.

Whereas the positive correlation between health and income per capita is very well known in international development, the health-income correlation is partly explained by a causal link running the other way—from health to income (Bloom and Canning 2000). In other words, productivity, education, investment in physical capital, and "demographic dividend" (positive changes in birth and death rates) are all self-reinforcing—these factors can contribute to health, and better health (and greater longevity) contributes to their improvements.

10.11. Future

Predictions of changes in human longevity in both the near- and medium-term are typically made using analytical methods such as extrapolation, relational modeling (i.e., reference population), mortality model (e.g., Gompertz), cause-of-death elimination, and stochastic methods (Ahlburg and Land 1992; Olshansky 1988; Tuljapurkar et al. 2000). The usual mortality component of population forecasts is provided by extrapolation of past trends in mortality, and the issue of which model to chose is to decide how to perform the extrapolation (Keyfitz 1982). Using both standard and novel forecasting techniques, demographers predict life spans in 2050 ranging from 80 to 83 years for the United States and from 83 to 91 years in Japan (Tuljapurkar et al. 2000). These forecasts are to be distinguished from forecasts based on expected medical discoveries or estimated danger of mass starvation. Thus as Keyfitz (1982) notes, the standard models cannot predict the "corners" (i.e., major changes); this is not what they are designed to do.

The future of human life span should be also be considered in the context of *possibility*—the constructive, scenario-building aspects of science that are akin to the conceptual arts (Gill 1986). Whereas predictions of changes in life expectancy are usually considered to point to *precise* results, the recognition of life-span possibilities is correlated with *vision* and includes the effects on longevity of scientific and medical breakthroughs. That is, prophecies that are not fixed and foreordained but rather based on the recognition of a present evolving toward a future of *multiple demographic alternatives*. These are not based on rates of change in life expectancy, per se, but rather on elements that will change the rate of change itself.

Integrating concepts related to different life-span *possibilities* into research and policy planning is important for several reasons. *First*, it will establish closer connections between the biological discoveries on the nature of aging and how these, at least in theory, might impact individuals and society. Currently there is very little exchange of ideas between biologists and demographers in this context. *Second*, considering possibilities may suggest different analytical models in which different types of scientific breakthroughs could be included in the predictions. For example, a breakthrough in therapeutic cloning would have sweeping implications for organ replacement (e.g., kidneys, livers) and hence for saving the lives of people with diseases that infect clonable organs. *Third*, a program to consider future life-span possibilities would provide a framework for a wide range of other disciplines (e.g., sociology, human development, economist, business) to engage in creative discussion about the nature of future society. *Fourth*, considering future possibilities for human life span will provide a foundation for policymakers to consider medical, economic, and political contingencies in which life-span increases are much greater than predicted by standard mortality modeling methods. Exploring a variety of possibilities based on potential scientific breakthroughs would ensure that policymakers have considered a wide variety of possible scenarios for the demographic future of both developed and the developing countries.

10.12. Scientific and Biomedical Determinants

Developments in four general areas of science and public health will likely determine the biological future of human life span:

1. Healthful living—the concept is to make vital choices regarding elimination of vices (smoking, drugs, alcohol abuse) and regarding proper diet, exercise, mental stimulation, and social support (Fries 1992; Verbrugge 1990; Willett 1994). Healthful living is framed around the concept of "successful aging," which consists of three components—avoiding disease and disability, maintaining mental and physical function, and continuing engagement with life (Rowe and Kahn 1998). The healthful living concept probably accounts for the substantial reductions in mortality over the past several decades and will serve as the foundation for further reductions in the future.

2. Disease prevention and cure—developing treatments and cures for debilitating diseases such as arthritis (Persidis 2000a), cancer (Greaves 2000; Pardoll 1998), cardiovascular diseases, stroke, and autoimmune diseases (Heilman and Baltimore 1998). New fields are being ushered in that will revolutionize medicine, including molecular and systems biology (Idelker et al. 2001); molecular cardiovascular medicine; gene therapy, where the genetic material is transferred to cure a disease (Pfeifer and Verma 2001); pharmacogenomics, where drug makers tailor a therapy to the individual's needs due to genetic nuances; and nanomedicine, which involves the monitoring, repair, construction, and control of human biological systems at the molecular level using nanodevices and nanostructures.

3. Organ replacement and repair—tissue engineering is emerging as a significant new technology (Colton 1999; Persidis 2000b), including skin equivalents and tissue-engineered bone, blood vessels, liver, muscle and nerve conduits, intact xenogeneic vessels such as a pig aorta, artificial thyroid tissues to produce T cells, and the manipulation of stem cells to produce a wide range of human tissues. Significant research and development is also underway on xenotransplantation (organs from other animals species into humans) (Cooper and Lanza 2000; Persidis 2000c; Steele and Auchincloss 1995), stem cell research (Blau et al. 2001), and therapeutic cloning (Colman 1999; Yang and Tian 2000).

4. Aging arrest and rejuvenation—identify aging processes in humans that can be studied in model organisms (Guarente and Kenyon 2000) and eventually develop therapeutic interventions (Hadley et al. 2001). Research on caloric restriction (CR) is designed to identify a CR-memetic (molecular mimic) with a known mechanism of action that produces effects on life span and aging similar to those of CR (Finch and Ruvkun 2001). Such research would implicate this mechanism as a likely mediator of CR's effects (Lee et al. 1999; Masoro and Austad 1996; Sohal and Weindruch 1996; Weindruch 1996). Recent results of studies on ovarian transplants in mice (Cargill et al. 2002) reveals yet a second method by which the rate of aging in mammals can be reduced. Reducing the rate of actuarial aging would increase longevity and life span far greater than would improvements from any other single mechanism.

The geneticist Aubrey de Grey (de Grey 2000) notes that we can be absolutely sure that human scientific knowledge and consequent technological and biomedical prowess will continue to advance for as long as civilization survives. He further notes that, inasmuch as we can also be sure that the complexity of the human body will remain constant, it is a mathematical certainty that human longevity will continue to increase.

10.13. Demographic Ontogeny

10.13.1. *Individuals*

The concept of *adaptive demography* that Wilson (1975) developed in the context of social insect evolution at both individual and societal levels applies equally well to humans. Future demographic changes that will likely occur at the individual level in humans include

1. Appearance of new extreme ages. Changes in record ages are important, not because the presence of these record-holders change society in any substantive way, but because they are the harbingers of the future and the extreme manifestations of improved health. Just as nonagenarians and centenarians probably first appeared in the 19th century and supercentenarians in the 20th century, it is virtually certain that persons living far beyond the current age record of 122 years will begin to appear in the 21st century; ages of 130 years will likely be reached before the end of the century.

2. Emergence of new functional and healthy age classes. Using equivalencies of age-specific mortality as proxies for similarities in health and functionality, the 70- and 60-year-olds of today have the same mortality risk as did the 61- and 45-year olds in 1900 (Bell et al. 1992). In other words, the frail elderly of yesterday are the more robust elderly of today. Although new age classes are added at the end of the life course with each new longevity record, a more important development is that large numbers of individuals are staying healthier longer and therefore are functional at more advanced ages.

3. Life-cycle and event-history modification. As life span and life expectancy increase, people plan their lives differently and thus change the timing, sequencing, duration, and spacing of key events including marriage and childbearing, education, working life, and retirement (Elder 1985; O'Rand and Krecker 1990; Settersten and Mayer 1997). Examples of how life events and, in turn, life trajectories can change are illustrated in figure 10.6.

10.13.2. *Societies*

Changes at the societal level will occur as the result of changes in event histories at the level of the individual, including

1. The family life cycle will probably continue to quicken as marriage, parenthood, empty nest, and grandparenthood all occur earlier, partly because of an ever-decreasing number of children.

2. A new rhythm of social maturity will impinge upon other aspects of family life in subtle ways, including the age of parents, which affects the degree of authoritarianism, the relative youth of parents and grandparents, which, in turn, affects

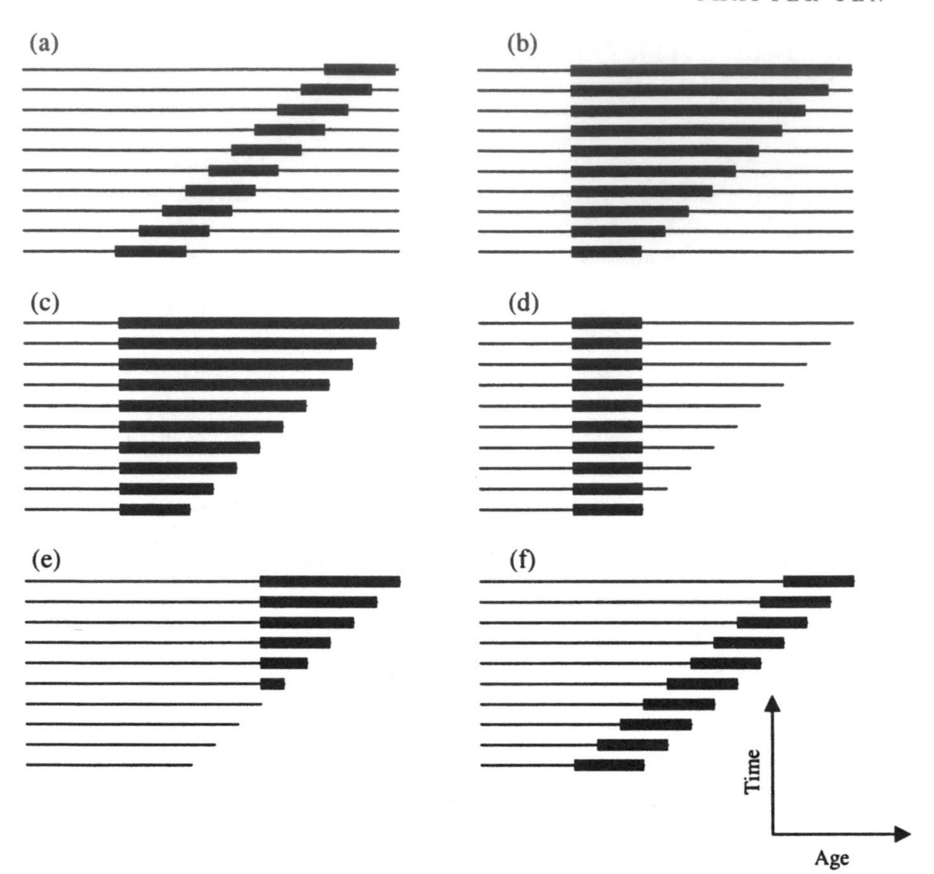

FIGURE 10.6. Examples of how the timing and duration of different life-course events may change with increasing longevity.

the patterns of health assistance, of flows of money in both directions, and of child care (Neugarten and Datan 1973).

3. The life cycle will continue to be fluid with an ever-increasing number of role transitions, the disappearance of traditional timetables, and the continued lack of synchrony among age-related roles.

4. New generational constellations will emerge including 4- and even 5-generation families with 2–3 middle generations and new contexts for sorting out needs and resources of the family.

5. The age at which full retirement benefits start will continue to rise in developed countries, to the mid-70s in the near term (Drucker 2001) and to the late-70s, early 80s, and beyond as life span increases.

6. Highly skilled and educated older people, instead of being retired, will be offered a choice of continuing relationships that preserve their skill and knowledge for the enterprise and yet gives them the flexibility and freedom that they expect and can afford. A model for this that is already in use comes from academia—the professor emeritus (Drucker 2001).

7. Education will become more of a seamless web in which primary and secondary education, undergraduate, graduate, and professional schooling, and apprenticeships, internships, on-job training, continuing education, and lifelong learning and enrichment will become a continuum (Duderstadt 1999; Duderstadt 2000).

Just as the presence of a stage and its duration in the life cycle are considered adaptations and thus of evolutionary value in ecological contexts (Bogin 2000; Bonner 1965), the evolution and emergence of new stages at the end of the human life course can also be considered adaptations in a biocultural context; the life history properties of every age class in the human life course shapes and, in turn, is shaped by every other age class.

In short, adding new age class extremes, expanding the duration of healthy living, and reconfiguring the event-history schedule creates new and unpredictable societal dynamics. Although the fate of the human species is impossible to predict far into the future, molecular biologist Lee Silver's (1998) perspectives on scientific and technological possibilities is probably the most insightful—look to the 4-billion-year creative history of life on Earth for ideas, most of which are locked within the genetic code of all living creatures. We have just begun the "century of biology" (Idelker et al. 2001), and it is a virtual certainty that, sooner or later, some of the deepest and most profound secrets of life will indeed be unlocked. It is also likely that some of these secrets will involve the keys to extending life span.

10.14. Postscript

Life span has a special character quite apart from mortality and longevity in that the concept extends beyond measures of biological efficiency and statistical expectations. It has multiple characteristics, including many etymological, semantic, and scientific nuances and is also a significant indicator of different ways of thinking about the evolution, ecology, and life course of humans. On one level the life-span concept is decoupled from death because it is explicitly a longevity rather than a mortality concept; it implies living not dying. However, on another level the concept is inextricably linked to death because it delimits an age that extends beyond what is unattainable by virtually the entire population; it implies life's ultimate age of closure.

Lovejoy (1981) noted that the social bonds, intelligence, learning ability, and intensive parenting that contributed to reducing mortality levels in primates both required and contributed to a long life-span in higher primates. Thus the most distinctive qualities of the human species including our language, our ability to teach and learn, our culture, and, most importantly, our humanity could not have emerged were it not for the tribal elders, the shamans, the seers, and the grandmothers. We impede evolution and the development of the human species if we neglect the aged because, as Hillman (1999) notes, it is the elderly who sheltered and promoted our civilization in the past and it is they who will continue to do so in the future.

Bibliography

Abbott, J. H., H. Abbey, D. R. Bolling, and E. A. Murphy. 1978. The familial component in longevity—as study of offspring of nonagenarians: III. Intrafamilial studies. *American Journal of Medical Genetics* 2:105–20.

Abernethy, J. D. 1979. The exponential increase in mortality rate with age attributed to wearing-out of biological components. *Journal of Theoretical Biology* 80:333–54.

Abramovitz, M. 1989. *Thinking About Growth. And Other Essays on Economic Growth and Welfare*. Cambridge: Cambridge University Press.

Abrams, P. A., and D. Ludwig. 1995. Optimality theory, Gompertz' law, and the disposable Soma theory of senescence. *Evolution* 49:1055–66.

Ahlburg, D. A., and K. C. Land. 1992. Population: forecasting: Guest Editors' introduction. *International Journal of Forecasting* 8:289–99.

Aigaki, T., and S. Ohba. 1984a. Effect of mating status on *Drosophila virilis* lifespan. *Experimental Gerontology* 19:267–78.

———. 1984b. Individual analysis of age-associated changes in reproductive activity and lifespan of *Drosophila virilis*. *Experimental Gerontology* 19:13–23.

Alexander, J., and W. H. Stimson. 1988. Sex hormones and the course of parasite infection. *Parasitology Today* 4:189–93.

Allard, M., V. Lébre, and J.-M. Robine. 1998. *Jeanne Calment: From Van Bogh's Time to Ours, 122 Extraordinary Years*. New York: Freeman.

Ames, B. N., M. K. Shigenaga, and T. M. Hagen. 1993. Oxidants, antioxidants, and the degenerative diseases of aging. *Proceedings of the National Academy of Science* 90:7915–22.

Andersen, P. K., O. Borgan, R. D. Gill, and N. Keiding. 1993. *Statistical Methods Based on Counting Processes*. New York: Springer.

Anderson, F. S. 1961. Effect of density on animal sex ratio. *Oikos* 12:1–16.

Angel, J. L. 1969. The bases of paleodemography. *American Journal Physical Anthropology* 30:427–38.

Anon. 1997. *Manual, S-plus version 4.03*. Seattle: Manual Statistical Sciences.

Anon. 2001. The regeneration gap. *Nature* 414:388–90.

Apfeld, J., and C. Kenyon. 1999. Regulation of lifespan by sensory perception in *Caenorhabditis elegans*. *Nature* 402:804–9.

Arita, L. H., and K. Y. Kaneshiro. 1983. Pseudomale courtship behavior of the female Mediterranean fruit fly, *Ceratitis capitata* (Wiedemann). *Proceedings of the Hawaiian Entomological Society* 24:205–10.

Ashburner, M. 1989. *Drosophila: A Laboratory Handbook*. Cold Spring Harbor, N.Y.: Cold Spring Harbor Laboratory.

Audesirk, T., and G. Audesirk. 1996. *Life on Earth*. 4th ed. Upper Saddle River, N.J.: Prentice Hall.

Austad, S. 1997. *Why We Age*. New York: Wiley.

Austad, S. J., and K. E. Fischer. 1992. Primate longevity: its place in the mammalian scheme. *American Journal of Primatology* 28:251–61.

Austad, S. N. 1989. Life extension by diet restriction in the bowl and doily spider, *Frontinella pyramitela. Experimental Gerontology* 24:83–92.

Austad, S. N., and K. E. Fischer. 1991. Mammalian aging, metabolisms, and ecology: evidence from the bats and marsupials. *Journal of Gerontology* 46:B47–B53.

Azbel, M. Y. 1996. Unitary mortality law and species-specific age. *Proceedings of the Royal Society of London* 263:1449–54.

Baker, G. T., and R. L. Sprott. 1988. Biomarkers of aging. *Experimental Gerontology* 23:223–39.

Balin, A. K., and M. M. Vilenchik. 1996. Oxydative damage. In *Encyclopedia of Gerontology*, ed. J. E. Birrin, 233–46. San Diego: Academic Press.

Ball, Z. B., R. H. Barnes, and M. B. Visscher. 1947. The effects of dietary caloric restriction on maturity and senescence, with particular reference to fertility and longevity. *American Journal of Physiology* 150:511–19.

Baltimore, D. 2001. Our genome unveiled. *Nature* 409:814–16.

Barber, G., and E. Starnes. 1949. The activities of house flies. *Journal of the New York Entomological Society* 57:203–14.

Barinaga, M. 1992. Mortality: Overturning received wisdom. *Science* 258: 398–99.

Barker, D.J.P. 1994. *Mothers, Babies, and Disease in Later Life*. London: BMJ Publishing.

Barrett, J. C. 1985. The mortality of centenarians in England and Wales. *Archives of Gerontology and Geriatrics* 4:211–18.

Bateman, M. A. 1972. The ecology of fruit flies. *Annual Review of Entomology* 17:493–518.

Becker, G. 1991. *A Treatise on the Family*. Cambridge: Harvard University Press.

Becker, G., and K. Murphy. 1992. The division of labor, coordination costs, and knowledge. *Quarterly Journal of Economics* 107:1137–60.

Begon, M. 1976. Temporal variations in the reproductive condition of *Drosophila obscura Fallen* and *D. subobscura Collin. Oecologia* 23:31–47.

Begon, M., J. L. Harper, and C. R. Townsend. 1996. *Ecology*. London: Blackwell Science.

Bell, F. C., A. H. Wade, and S. C. Goss. 1992. *Life Tables for the United States Social Security Area 1900–2080*. U.S. Department of Health and Human Services, Social Security Administration. Washington, D.C.: Office of the Actuary.

Bell, G. 1984. Measuring the cost of reproduction. II. The correlation structure of the life tables of five freshwater invertebrates. *Evolution* 38:314–26.

———. 1988. *Sex and Death in Protozoa*. Cambridge: Cambridge University Press.

Bell, G., and V. Koufopanou. 1986. The cost of reproduction. In R. Dawkins and M. Ridley *Oxford Surveys in Evolutionary Biology*, ed. R. Dawkins and M. Ridley, 83–131. Oxford: Oxford University Press.

Berkman, L. S., and S. L. Syme. 1979. Social networks, host resistance, and mortality: a nine year follow-up study of Alameda County residents. *American Journal of Epidemiology* 109:186.

Bertalanffy, L. v. 1960. Principles and theory of growth. In *Fundamental Aspects of Normal and Malignant Growth*, ed. W. N. Nowinske, 137–259. Amsterdam: Elsevier.

Beshers, S. N., and J. H. Fewell. 2001. Models of division of labor in social insects. *Annual Review of Entomology* 46:413–40.

Bilewicz, S. 1953. Experiments on the reproductive length of life of *Drosophila melanogaster*. *Folio Biologica* 1:177–94.

Birch, L. C. 1948. The intrinsic rate of natural increase of an insect population. *Journal of Animal Ecology* 17:15–26.

Blackburn, E. H. 2000. Telomere states and cell fates. *Nature* 408:53–56.

Blake, J. 1989a. *Family Size and Achievement*. Berkeley: University of California Press.

———. 1989b. Number of siblings and educational attainment. *Science* 245:32–36.

Blau, H. M., T. R. Brazelton, and J. M. Weimann. 2001. The evolving concept of stem cell: Entity or function? *Cell* 105:829–41.

Blay, S., and B. Yuval. 1997. Nutritional correlates of reproductive success of male Mediterranean fruit flies (*Diptera: Tephritidae*). *Animal Behavior* 54:59–66.

Bloom, D. E., and D. Canning. 2000. The health and wealth of nations. *Science* 287:1207–9.

Bloom, D. E., and J. G. Williamson. 1998. Demographic transitions and economic miracles in emerging Asia. *World Bank Economic Review* 12:419–55.

Bocquet-Appel, J. P. 1990. Familial transmission of longevity. *Annals of Human Biology* 17:81–95.

Boggs, C. L. 1986. Reproductive strategies of female butterflies: Variation in and constraints of fecundity. *Ecological Entomology* 11:7–15.

Boggs, C. L., and C. L. Ross. 1993. The effect of adult food limitation on life history traits in *Speyeria mormonia* (Lepidoptera: Nymphalidae). *Ecology* 74:433–41.

Bogin, B. 1990. The evolution of human childhood. *BioScience* 40:16–25.

———. 1997. Evolutionary hypotheses for human childhood. *Yearbook of Physical Anthropology* 40:63–89.

———. 2000. Evolution of the human life cycle. In *Human Biology: an Evolutionary and Biocultural Perspective*, ed. S. Stinson, B. Bogin, R. Huss-Ashmore and D. O'Rourke, 377–424, New York: Wiley-LISS.

Bogin, B., and B. H. Smith. 1996. Evolution of the human life cycle. *American Journal of Human Biology* 8:703–16.

Bonner, J. T. 1965. *Size and Cycle*. Princeton: Princeton University Press.

———. 1967. *The Cellular Slime Molds*. Princeton: Princeton University Press.

———. 1980. *The Evolution of Culture in Animals*. Princeton: Princeton University Press.

Borgerhoff-Mulder, M. 1998. The demographic transition: Are we any closer to an evolutionary explanation? *Trends in Ecology and Evolution* 13:266–70.

Bouletreau, J. 1978. Ovarian activity and reproductive potential in a natural population of *Drosophila melanogaster*. *Oecologia* 35:319–42.

Bourg, E. L., B. Thon, J. Legare, B. Desjardins, and H. Charbonneau. 1993. Reproductive life of French-Canadians in the 17–18th centuries: A search for a trade-off between early fecundity and longevity. *Experimental Gerontology* 28:217–32.

Bourgeois-Pichat, J. 1967. Social and biological determinants of human fertility in nonindustrial societies. *Proceedings of the American Philosophical Society* 111: 160–63.

Boyd, R., and P. Richerson. 1985. *Culture and the Evolutionary Process*. Chicago: University of Chicago Press.

Brody, J. A., and D. B. Brock. 1985. Epidemiologic and statistical characteristics of the United States elderly population. In *Handbook of the Biology of Aging*, ed. C. E. Finch and E. L. Schneider, 3–26. New York: Van Nostrand Reinhold.

Brooks, A., G. L. Lithgow, and T. E. Johnson. 1994. Mortality rates in a genetically heterogeneous population of *Caenorhabditis elegans*. *Science* 263:668–71.

Buchner, D. M., S. A. A. Beresford, E. B. Larson, A. Z. LaCroix, and E. H. Wagner. 1992. Effects of physical activity on health status in older adults. II. Intervention studies. *Annual Review of Public Health* 13:469–88.

Buikstra, J. E., and L. W. Konigsberg. 1985. Paleodemography: critiques and controversies. *American Anthropologist* 87:316–33.

Bundy, D.A.P. 1988. Sexual effects on parasite infection: Gender-dependent patterns of infection and disease. *Parasitology Today* 4:186–93.

Cain, M. 1977. The economic activities of children in a village in Bangladesh. *Population and Development Review* 3:201–8.

Cajal, S. R. 1999. *Advice for a Young Investigator*. Cambridge: MIT Press.

Calder, W. A. 1996. *Size, Function, and Life History*. New York: Dover.

Caldwell, J. C. 1976. Toward a restatement of demographic transition theory. *Population Development and Review* 2:321–66.

Calhoun, J. B. 1962. Population density and social pathology. *Scientific American* 206:139–48.

Calow, P. 1970. The cost of reproduction: a physiological approach. *Evolutionary Ecology* 54:23–40.

———. 1977. Ecology, evolution and energetics: A study in metabolic adaptation. *Advances in Ecological Research* 10:1–61.

Campbell, R. D. 1967. Tissue dynamics of steady state growth in Hydra littoralis. *Journal of Morphology* 121:19–28.

Cano, R. J., and M. Borucki. 1995. Revival and identification of bacterial spores in 25-to-40-million-year-old Dominican amber. *Science* 268:1060–64.

Carey, J. R. 1982. Demography and population dynamics of the Mediterranean fruit fly. *Ecological Modelling* 16:125–50.

———. 1984. Host-specific demographic studies of the Mediterranean fruit fly, *Ceratitis capitata*. *Ecological Entomology* 9:261–70.

———. 1989. Demographic analysis of fruit flies. In *Fruit Flies: Their Biology, Natural Enemies and Control*, ed. A. S. Robinson and G. Hooper, 253–65. Amsterdam: Elsevier.

———. 1993. *Applied Demography for Biologists*. New York: Oxford University Press.

———. 1997. What demographers can learn from fruit fly actuarial models and biology. *Demography* 34:17–30.

———. 1998. Population study of mortality and longevity with Gompertzian analysis. In *Methods in Aging Research*, ed. B. P. Yu, 3–24. Boco Ratan, Fl.: CRC Press.

———. 1999. Population study of mortality and longevity with Gompertzian analysis. In *Methods in Aging Research*, ed. B. P. Yu, 3–24. Boco Ratan, Fl.: CRC Press.

———. 2001a. Demographic mechanisms for the evolution of long life in social insects. *Experimental Gerontology* 36:713–22.

———. 2001b. Insect biodemography. *Annual Review of Entomology* 46:79–110.

Carey JR. 2002. Longevity minimalists: Life table studies of two northern Michigan adult mayflies. *Experimental Gerontology* 37: 567–70.

Carey JR. 2003 (forthcoming). Life span: A conceptual overview. *Population and Development Review.*

Carey, J. R., J. W. Curtsinger, and J. W. Vaupel. 1993a. Explaining fruit fly longevity. Response to technical comments. *Science* 260:1665–66.

———. 1993b. Fruit fly aging and mortality. Response to letters to the editor. *Science* 260:1567–69.

Carey, J. R., and C. A. Gruenfelder. 1997. Population biology of the elderly. In *Biodemography of Aging*, ed. K. Wachter and C. Finch, 127–60. Washington, D.C.: National Academy Press.

Carey, J. R., and D. S. Judge. 2000a. *Longevity Records: Life Spans of Mammals, Birds, Reptiles, Amphibians and Fish.* Odense: Odense University Press.

———. 2000b. Mortality dynamics of aging. *Generations* 24:19–24.

———. 2001a. Life span extension in humans is self-reinforcing: A general theory of longevity. Population and Development Review 27:411–36.

———. 2001b. Principles of biodemography with special reference to human longevity. *Population: An English Selection* 13:9–40.

Carey, J. R., D. Krainacker, and R. Vargas. 1986. Life history response of Mediterranean fruit fly females to periods of host deprivation. *Entomologia experimentalis et applicata* 42:159–67.

Carey, J. R., and P. Liedo. 1995a. Sex mortality differentials and selective survival in medfly cohorts: Implications for human sex mortality differentials. *The Gerontologist* 35:588–96.

———. 1995b. Sex-specific life table aging rates in large medfly cohorts. *Experimental Gerontology* 30:315–25.

———. 1999a. Measuring mortality and reproduction in large cohorts of the Mediterranean fruit fly. In *Studies of Aging—Springer Laboratory Manual*, ed. H. Sternberg and P. S. Timiras. 111–24. New York: Springer.

———. 1999b. Mortality dynamics of insects: Some general principles derived from aging research on the Mediterranean fruit fly. *American Entomologist* 45: 49–55.

Carey, J. R., P. Liedo, L. Harshman, X. Liu, H.-G. Müller, L. Partridge, and J.-L. Wang. 2002a. Food pulses increase longevity and induce cyclical egg production in Mediterranean fruit flies. *Functional Ecology* 16:313–25.

Carey, J. R., P. Liedo, L. Harshman, Y. Zhang, H.-G. Müller, L. Partridge, and J.-L. Wang. 2002b. A mortality cost of virginity at older ages in female Mediterranean fruit flies. *Experimental Gerontology* 37:507–12.

Carey, J. R., P. Liedo, H.-G. Müller, J.-L. Wang, and J.-M. Chiou. 1998a. Relationship of age patterns of fecundity to mortality, longevity, and lifetime reproduction in a large cohort of Mediterranean fruit fly females. *Journal of Gerontology: Biological Sciences* 53A:B245–51.

———. 1999. Mortality oscillations induced by periodic starvation alter sex-mortality differentials in Mediterranean fruit flies. *Journal of Gerontology: Biological Sciences* 54A:B424–31.

Carey, J. R., P. Liedo, H.-G. Müller, J.-L. Wang, B. Love, L. Harshman, and L. Partridge. 2001. Female sensitivity to diet and irradiation treatments underlies sex-mortality differentials in the Mediterranean fruit fly. *Journal of Gerontology: Biological Sciences* 56A:B1–B5.

Carey, J. R., P. Liedo, H.-G. Müller, J.-L. Wang, and J. W. Vaupel. 1998b. A simple graphical technique for displaying individual fertility data and cohort survival: Case study of 1000 Mediterranean fruit fly females. *Functional Ecology* 12: 359–63.

———. 1998c. Dual modes of aging in Mediterranean fruit fly females. Science 281:396–98.

Carey, J. R., P. Liedo, D. Orozco, M. Tatar, and J. W. Vaupel. 1995a. A male-female longevity paradox in medfly cohorts. *Journal of Animal Ecology* 64:107–16.

Carey, J. R., P. Liedo, D. Orozco, and J. W. Vaupel. 1992. Slowing of mortality rates at older ages in large medfly cohorts. *Science* 258:457–61.

Carey, J. R., P. Liedo, and J. W. Vaupel. 1995b. Mortality dynamics of density in the Mediterranean fruit fly. *Experimental Gerontology* 30:605–29.

Carey, J. R., P. Yang, and D. Foote. 1988. Demographic analysis of insect reproductive levels, patterns and heterogeneity: Case study of laboratory strains of three Hawaiian tephritids. *Entomologia Experimentalis et Applicata* 46:85–91.

Cargill, S., J. R. Carey, G. Anderson, and H.-G. Müller. 2002. Rejuvenation of reproductively-arrested mice. Submitted paper.

Carlson, K. A., and L. G. Harshman. 1999. Extended longevity lines of *Drosophila melanogaster*: Characterization of oocyte stages and ovariole numbers as a function of age. Manuscript.

Carlson, K. A., T. J. Nusbaum, M. R. Rose, and L. G. Harshman. 1998. Oocyte maturation and ovariole number in lines of *Drosophila melanogaster* selected for postponed senescence. *Functional Ecology* 12:514–20.

Carpenter, C. R. 1958. Territoriality: A review of concepts and problems. In *Behavior and Evolution*, ed. A. Roe and G. G. Simpson, 27–56. New Haven: Yale University Press.

Carrel, A. 1931. Physiological time. *Science* 74:618–21.

Carson, H. L., and H. D. Stalker. 1947. Reproductive diapause in *Drosophila robusta. Proceedings of the National Academy of Sciences* 34:124–29.

Carter, L. R., and R. D. Lee. 1992. Modeling and forecasting U.S. sex differentials in mortality. *International Journal of Forecasting* 8:393–412.

Casas, J., R. M. Nisbet, S. Swarbrick, and W. W. Murdoch. 2000. Eggload dynamics and oviposition rate in a wild population of a parasitic wasp. *Journal of Animal Ecology* 69:185–93.

Casper, R. C. 1995. Nutrition and its relationship to aging. *Experimental Gerontology* 30:299–314.

Caswell, H. 1989. *Matrix Population Models*. Sunderland, Mass.: Sinauer.

Caswell, H., and D. E. Weeks. 1986. Two-sex models: Chaos, extinction, and other dynamic consequences of sex. *American Naturalist* 128:707–35.

Chant, C. 1989. *Science, Technology, and Everyday Life, 1870–1950*. New York: Open University.

Chapman, R. F. 1969. *The Insects: Structure and Function*. New York: Elsevier.

Chapman, T., L. F. Liddle, J. M. Kalb, M. F. Wolfner, and L. Partridge. 1995. Cost of mating in *Drosophila melanogaster* females is mediated by accessory gland products. *Nature* 373:241–44.

Chapman, T., T. Miyatake, H. K. Smith, and L. Partridge. 1998. Interactions of mating, egg production and death rates in females of the Mediterranean fruit fly, *Ceratitis capitata. Proceedings of the Royal Society of London* B265:1879–94.

Chapman, T., and L. Partridge. 1996. Female fitness in *Drosophila melanogaster*: An interaction between the effect of nutrition and of encounter rate with males. *Proceedings of the Royal Society of London* B263:755–59.

Chapman, T., S. Trevitt, and L. Partridge. 1994. Remating and male-derived nutrients in *Drosophila melanogaster*. Journal of Evolutionary Biology 7:51–69.

Charlesworth, B. 1994. *Evolution in Age-structured Populations*. Cambridge: Cambridge University Press.

Charnov, E. 1982. *The Theory of Sex Allocation*. Princeton: Princeton University Press.

———. 1993. *Life History Invariants*. Oxford: Oxford University Press.

Chiang, C. L. 1984. *The Life Table and Its Applications*. Malabar, Fl.: Robert E. Krieger.

Chippindale, A. K., T.J.F. Chu, and M. R. Rose. 1996. Complex trade-offs and the evolution of starvation resistance in *Drosophila melanogaster. Evolution* 50:753–66.

Chippindale, A. K., A. M. Leroi, S. B. Kim, and M. R. Rose. 1993. Phenotypic plasticity and selection in *Drosophila* life-history evolution. I. Nutrition and the cost of reproduction. *Journal of Evolutionary Biology* 6:171–93.

Christensen, K., and J. W. Vaupel. 1996. Determinants of longevity: Genetic, environmental and medical factors. *Journal of Internal Medicine* 240:333–41.

Christenson, L. D., and R. H. Foote. 1960. Biology of fruit flies. *Annual Review of Entomology* 5:171–92.

Christian, J. J. 1970. Social subordination, population density, and mammalian evolution. *Science* 168:84–90.

Clark, A. G., and R. N. Guadalupe. 1995. Probing the evolution of senescence in *Drosophila melanogaster* with P-element tagging. *Genetica* 96:225–34.

Clarke, K. U., and J. B. Sardesai. 1959. An analysis of the effects of temperature upon growth and reproduction of *Dysdercus fasciatus* Sign (Hemiptera, Pyrrhocoridae). *Bulletin of Entomological Research* 50:387–405.

Clark, W. 1996. *Sex and the Origins of Death*. Oxford: Oxford University Press.

Clutton-Brock, T. H. 1991. *The Evolution of Parental Care*. Princeton: Princeton University Press.

———. 1994. The costs of sex. In *E. Balaban*, ed. R. V. Short, 347–62. Cambridge: Cambridge University Press.

Clutton-Brock, T. H., and G. R. Iason. 1986. Sex ratio variation in mammals. *The Quarterly Review of Biology* 61:339–74.

Coale, A. J., and P. Demeny. 1983. *Regional Model Life Table and Stable Populations*. 2nd ed. New York: Academic Press.

Coale, A. J., and E. E. Kisker. 1986. Mortality crossovers: Reality or bad data? *Population Studies* 40:389–401.

Coe, C. L. 1990. Psychobiology of maternal behavior in nonhuman primates. In

Mammalian Parenting, ed. N. A. Krasnegor and R. S. Bridges, 157–83. New York: Oxford University Press.

Collins, F. S. 1998. New goals for the U.S. human genome project: 1998–2003. *Science* 282:682–89.

———. 1999. Medical and societal consequences of the human genome project. *New England Journal of Medicine* 341:29–37.

Colman, A. 1999. Somatic cell nuclear transfer in mammals: Progress and application. *Cloning* 1:185–200.

Colton, C. K. 1999. Engineering a bioartificial kidney. *Nature Biotechnology* 17:421–22.

Comfort, A. 1961. The life span of animals. *Scientific American* 205:108–19.

———. 1979. *The Biology of Senescence*. 3rd ed. New York: Elsevier.

Connolly, K. 1968. The social facilitation of preening behaviour in *Drosophila melanogaster*. *Animal Behavior* 16:385–91.

Cooper, D.K.C., and R. P. Lanza. 2000. *Xeno: The Promise of Transplanting Animal Organs into Humans*. New York: Oxford University Press.

Courtice, A. C., and R.A.I. Drew. 1984. Bacterial regulation of abundance in tropical fruit flies (Diptera: Tephritidae). *Australian Journal of Zoology* 21:251–68.

Cox, D. R. 1972. Regression models and life tables (with discussion). *Journal of the Royal Statistical Society* B34:187–220.

Crimmins, E. M. 2001. Mortality and health in human life spans. *Experimental Gerontology* 36:885–97.

Crimmins, E. M., M. D. Hayward, and Y. Saito. 1994. Changing mortality and morbidity rates and the health status and life expectancy of the older population. *Demography* 31:159–75.

———. 1996. Differentials in active life expectancy in the older population of the United States. *Journal of Gerontology* 51B:S111–20.

Crimmins, E. M., Y. Saito, and D. Ingegneri. 1989. Changes in life expectancy and disability-free life expectancy in the United States. *Population and Development Review* 15:235–67.

Cruse, J. M. 1999. History of medicine: The metamorphosis of scientific medicine in the ever-present past. *American Journal of the Medical Sciences* 318:171–80.

Curtsinger, J. W. 1995. Density, mortality and the narrow view. *Genetica* 96:187–89.

Curtsinger, J. W., H. H. Fukui, D. R. Townsend, and J. W. Vaupel. 1992. Demography of genotypes: Failure of the limited life-span paradigm in *Drosophila melanogaster*. *Science* 258:461–63.

Daly, M. 1978. The cost of mating. *American Naturalist* 112:771–74.

Daly, M., and M. Wilson. 1983. *Sex, Evolution and Behavior*. Oxford: Oxford University Press.

Davey, K. G. 1999. *Insect reproduction, overview*. In *Encyclopedia of Reproduction*, ed. E. Knobil and J. D. Neill, 845–52. San Diego: Academic Press.

Davis, M. B. 1945. The effect of populaiton density on longevity in *Trogoderma creutz* (=*T. inclusa Lec.*). *Ecology* 26:353–62.

Dawkins R. 1989. *The Selfish Gene*. Oxford: Oxford University Press.

Deacon, T. 1997. *The Symbolic Species: The Co-evolution of Language and the Brain*. New York: Norton.

Dean, J. M. 1981. The relationship between lifespan and reproduction in the grasshopper *Melanoplus*. *Oecologia* 48:385–88.

Deevey, E.S.J. 1947. Life tables for natural populations of animals. *Quarterly Review of Biology* 22:283–14.

DeGregor, T. R. 1985. *A Theory of Technology*. Ames: Iowa State University Press.

de Grey, A.D.N.J. 2000. Gerontologists and the media: the dangers of over-pessimism. *Biogerontology* 1:369–70.

Demeny, P. 1968. Early fertility decline in Austria-Hungary: A lesson in demographic transition. *Daedalus* 97:502–22.

Demetrius, L. 2001. Mortality plateaus and directionality theory. *Proceedings of the Royal Society of London* B268:1–9.

DePaola, L. V. 1994. Dietary modulation of reproductive function. In *Modulation of Aging Processes by Dietary Restriction*, ed. B. P. Yu. Boco Raton, Fl.: CRC Press.

Diamond, J. 1992. *The Third Chimpanzee*. New York: HarperPerennial.

———. 1998. *Guns, Germs and Steel*. New York: Norton.

Dingle, H. 1968. The effect of population density on mortality and sex ratio in the milkweed bug, *Oncopeltus* and the cotton stainer, *Dysdercus* (Heteroptera). *American Naturalist* 100:465–70.

Doll, R. 1999. Risk from tobacco and potentials for health gain. *International Journal of Tuberculosis and Lung Disease* 3:90–99.

Drew, R. A., I.A.C. Courtice, and D. S. Teakle. 1983. Bacteria as a natural source of food for adult fruit flies (Diptera: Tephritidae). *Oecologia* 60:279–84.

Drucker, P. 2001. The next society. *Economist* 361:3–20.

Duderstadt, J. J. 1999. The twenty-first century university: A tale of two futures. In *Challenges Facing Higher Education at the Millennium*, ed. W. Z. Hirsch and L. E. Weber 37–55, Phoenix: Oryx.

———. 2000. Fire, ready, aim! University-decision making during an era of rapid change. Blion Colloquium II, La Jolla, California.

Dunbar, R.I.M. 1992. Neocortex size as a constraint on group size in primates. *Journal of Human Evolution* 20:469–93.

Durham, W. H. 1991. *Coevolution. Genes, Culture, and Human Diversity*. Stanford: Stanford University Press.

Economos, A. C., and J. Miquel. 1979. Analysis of population mortality kinetics with application to the longevity followup of the Navy's "1,000 Aviators." *Aviation Space and Environmental Medicine* 50:697–701.

Efron, B., and R. J. Tibshirani. 1993. *An Introduction to the Bootstrap*. London: Chapman and Hall.

Ehrlich, P., A. E. Launer, and D. D. Murphy. 1984. Can sex ratio be defined or determined? The case of a population of checkerspot butterflies. *The American Naturalist* 124:527–39.

Elder, G.H.J. 1985. Perspectives on the life course. In *Life Course Dynamics*, ed. G.H.F. Elder 23–49, Ithaca: Cornell University Press.

Ellison, C. G. 1991. Religious involvement and subjective well-being. *Journal of Health and Social Behavior* 32:80–99.

Elo, I. T., and S. H. Preston. 1992. Effects of early-life conditions on adult mortality: A review. *Population Index* 58:186–212.

Engelmann, F. 1968. Endocrine control of reproduction in insects. *Annual Review of Entomology* 13:1–26.

———. Engelmann, F. 1970. *The Physiology of Insect Reproduction*. Oxford: Pergamon Press.

Evans, F. C., and R. E. Smith. 1952. The intrinsic rate of natural increase for the human louse, *Pediculus humanus*. *American Naturalist* 86:299–310.

Evans, H. E. 1958. The evolution of social life in wasps. *Proceedings of the 10th International Congress of Entomology* 2:449–57.

Finch, C. 1990. *Longevity, Senescence, and the Genome*. Chicago: University of Chicago Press.

Finch, C. E., and G. Ruvkun. 2001. The genetics of aging. *Annual Review of Genomics and Human Genetics* 2:435–462.

Finch, C. E., and R. E. Tanzi. 1997. Genetics of aging. *Science* 278:407–11.

Fisher, R. A. 1958. *The Genetical Theory of Natural Selection*. New York: Dover.

Fogel, R. W. 1993. New sources and new techniques for the study of secular trends in nutritional status, health, mortality and the process of aging. *Historical Methods* 26:5–43.

———. 1994. Economic growth, population theory, and physiology: The bearing of long-term processes on the making of economic policy. *American Economic Review* 84:369–95.

Fogel, R. W., and D. L. Costa. 1997. A theory of technophysio evolution, with some implications for forecasting population, health care costs, and pension costs. *Demography* 34:49–66.

Foley, R. 1995. *Humans Before Humanity*. Cambridge, Mass.: Blackwell.

Foley, R. A. 1992. Evolutionary ecology of fossil hominids. In *Evolutionary Ecology and Human Behavior*, ed. E. Aldensmith and B. Winterholden. New York: Aldine de Gruyter.

Fowler, K., and L. Partridge. 1989. A cost of mating in female fruitflies. *Nature* 338:760–61.

Francis, R. C. 1995. Evolutionary neurobiology. *Trends in Ecology and Evolution* 10:276–81.

Frank, S. A. 1998. *Foundations of Social Evolution*. Princeton: Princeton University Press.

Freeman, R., and J. R. Carey. 1990. Interaction of host stimuli in the ovipositional response of the Mediterranean fruit fly (Diptera: Tephritidae). *Environmental Entomology* 19:1075–80.

Friedland, R. B. 1998. Life expectancy in the future: a summary of a discussion among experts. *North American Actuarial Journal* 2:48–63.

Fries, J. F. 1980. Aging, natural death, and the compression of morbidity. *New England Journal of Medicine* 303:130–35.

———. 1983. The compression of morbidity. *Milbank Memorial Fund Quarterly/Health and Society* 61:397–419.

———. 1987. Vis-à-vis: An introduction to the compression of morbidity. *Gerontologica Perspecta* 1:5–8.

———. 1992. Strategies for reduction of morbidity. *American Journal of Clinical Nutrition* 55:1257S–62S.

Fries, J. F., C. E. Koop, C. E. Beadle, P. P. Cooper, M. J. England, R. F. Greaves, J. J.

Sokolov, and D. Wright. 1993. Reduction in health care costs by reduction in need and demand for medical services. Manuscript.

Fukui, H. H., L. Xiu, and J. W. Curtsinger. 1993. Slowing of age-specific mortality rates in *Drosophila melanogaster*. *Experimental Gerontology* 28:585–99.

Futuma, D. J. 1998. *Evolutionary Biology*. Sunderland, Mass.: Sinauer.

Gaesser, G. A. 1999. Thinness and weight loss: Beneficial or detrimental to longevity? *Medicine and Science in Sports and Exercise* 31:1118–28.

Gaines, S. D., and M. W. Denny. 1993. The largest, smallest, highest, lowest, longest, and shortest: Extremes in ecology. *Ecology* 74:1677–92.

Galle, O. R., W. R. Gove, and J. M. McPherson. 1972. Population density and pathology: What are the relations for man? *Science* 176:23–30.

Gavrilov, L. A., and N. S. Gavrilova. 1991. *The Biology of Life Span: A Quantitative Approach*. Chur, Switzerland: Harwood Academic.

———. 2001. The reliability theory of aging and longevity. *Journal of Theoretical Biology* 213:527–45.

George, J. C., J. Bada, J. Zeh, L. Scott, S. E. Brown, T. O'Hara, and R. Suydam. 1999. Age and growth estimates of bowhead whales (*Balaena mysteicetus*) via aspartic acid racemization. *Canadian Journal of Zoology* 77:571–80.

Giesel, J. T. 1976. Reproductive strategies as adaptations to life in temporally heterogeneous environments. *Annual Review of Ecology and Systematics* 7:57–79.

Gill, S. P. 1986. The paradox of prediction. *Daedalus* 115:17–48.

Gillot, C. 1999. Male reproductive systems. In *Encyclopedia of Reproduction*, ed. E. Knobil and J. D. Neill, 41–48. San Diego: Academic Press.

Glucksmann, A. 1974. Sexual dimorphism in mammals. *Biological Reviews* 49:423–75.

Goldwasser, L. 2001. The biodemography of life span: Resources, allocation and metabolism. *Trends in Ecology and Evolution* 16:536–38.

Gompertz, B. 1825. On the nature of the function expressive of the law of mortality. *Philosophical Transactions* 27:513–85.

Goodman, L. A. 1953. Population growth of the sexes. *Biometrics* 9:212–25.

———. 1967. On the age-sex composition of the population that would result from given fertility and mortality conditions. *Demography* 4:423–41.

Graves, J. L. 1993. The costs of reproduction and dietary restriction: Parallels between insects and mammals. *Growth, Development and Aging* 57:233–49.

Greaves, M. 2000. *Cancer: The Evolutionary Legacy*. London: Oxford University Press.

Greenwood, P. J., and J. Adams. 1987. *The Ecology of Sex*. London: Edward Arnold.

Guarente, L., and Kenyon, C. 2000. Genetic pathways that regulate ageing in model organisms. *Nature* 408:255–62.

Gupta, P. D. 1973. Growth of U.S. population, 1940–1971, in the light of an interactive two-sex model. *Demography* 10:543–65.

Gurdon, J. B., and A. Colman. 1999. The future of cloning. *Nature* 402:743–46.

Hadley, E. C., C. Dutta, J. Finkelstein, T. B., Harris, M. A. Lane, G. S. Roth, S. S. Sherman, and P. E. Starke-Reed. 2001. Human implications of caloric restriction's effects on aging in laboratory animals: An overview of opportunities for research. *Journal of Gerontology*: 56A:5–6.

Hagen, K. 1952. Influence of adult nutrition upon fecundity, fertility and longevity of three tephritid species. Ph.D. diss., University of California, Berkeley.

Hagen, K. S., W. W. Allen, and R. L. Tassan. 1981. Mediterranean fruit fly: The worst may be yet to come. *California Agriculture* (March–April): 5–7.

Hagen, K. S., and G. L. Finney. 1950. A food supplement for effectively increasing the fecundity of certain Tephritid species. *Journal of Economic Entomology* 43:735.

Hairston, N.G.J., and C. E. Caceres. 1996. Distribution of crustacean diapause: Micro- and macroevolutionary pattern and process. *Hydrobiologia* 320:27–44.

Hakeem, A., R. Sandoval, M. Jones, and J. Allman. 1996. Brain and life span in primates. In *Handbook of the Psychology of Aging*, ed. J. Birren, 78–104. New York: Academic Press.

Hamilton, J. B. 1948. The role of testicular secretions as indicated by the effects of castration in man and by studies of pathological conditions and the short lifespan associated with maleness. In *Recent Progress in Hormone Research: The Proceedings of the Laurentian Hormone Conference*, ed. G. Pincus. New York: Academic Press.

Hamilton, J. B., and G. E. Mestler. 1969. Mortality and survival: Comparison of eunuchs with intact men and women in a mentally retarded population. *Journal of Gerontology* 24:395–471.

Hammer, M., and R. Foley. 1996. Longevity, life history and allometry: How long did hominids live? *Human Evolution* 11:61–66.

Harvey, P., A. Read, and D.E.L. Promislow. 1989. Life history variation in placental mammals: Unifying the data with theory. *Oxford Survey in Evolutionary Biology* 6:13–31.

Hauser, P. M., and O. D. Duncan. 1959. The nature of demography. In *The Study of Population*, ed. P. M. Hauser and O. D. Duncan, 29–44. Chicago: University of Chicago Press.

Hawkes, K., J. F. O'Connell, N.G.B. Jones, H. Alvarez, and E. L. Charnov. 1998. Grandmothering, menopause, and the evolution of life history traits. *Proceedings of the National Academy of Sciences USA* 953:1336–39.

Hawley, A. H. 1972. Population density and the city. *Demography* 9:521–29.

Hayflick, L. 1994. *How and Why We Age*. New York: Ballantine Books.

———. 2000. The future of ageing. *Nature* 408:267–69.

Hazzard, W. R. 1986. Biological basis of the sex differential in longevity. *Journal of American Geriatrics* 34:455–71.

———. 1990. The sex differential in longevity. In *Principles of Geriatric Medicine and Gerontology*, ed. W. R. Hazzard, R. Andres, W. L. Bierman, and J. P. Blass. New York: McGraw-Hill.

Hazzard, W. R., and D. Applebaum-Bowden. 1990. Why women live longer than men. In *American Clinical and Climatological Association* 101:168–89.

Heilman, C. A., and D. Baltimore. 1998. HIV vaccines—where are we going? *Nature Medicine* 4:532–34.

Hendrichs, J., G. Franz, and P. Rendo. 1995. Increased effectiveness and applicability of the sterile insect technique through male-only releases for control of Mediterranean fruit flies during fruiting seasons. *Journal of Applied Entomology* 119:371–77.

Hendrichs, J., and M. A. Hendrichs. 1990. Mediterranean fruit fly (Diptera: Tephritidae) in nature: Location and diel pattern of feeding and other activities on fruiting and nonfruiting hosts and nonhosts. *Annals of the Entomological Society of America* 83:632–41.

Hendrichs, J., C. R. Lauzon, S. S. Cooley, and R. J. Prokopy. 1993. Contribution of natural food sources to adult longevity and fecundity of *Rhagoletis pomonella* (Diptera: Tephritidae). *Annals of the Entomological Society of America* 86:250–64.

Hengartner, M. O. 1995. Life and death decisions: *ced*-9 and programmed cell death in *Caenorhabditis elegans. Science* 270:931.

Herrewege, J. V., and J. R. David. 1997. Starvation and desiccation tolerances in *Drosophila*: Comparison of species from different climatic origins. *Ecoscience* 4:151–57.

Herskind, A. M., M. McGue, N. V. Holm, T.I.A. Sorensen, B. Harvald, and J. W. Vaupel. 1996. The heritability of human longevity: A population-based study of 2872 Danish twin pairs born 1870–1900. *Human Genetics* 97:319–23.

Hill, K., and A. M. Hurtado. 1991. The evolution of premature reproductive senescence and menopause in human females: An evaluation of the "Grandmother hypothesis." *Human Nature* 2:313–50.

Hill, K., and H. Kaplan. 1999. Life history traits in humans: Theory and empirical studies. *Annual Review of Anthropology* 28:397–430.

Hillman, J. 1999. *The Force of Character*. New York: Ballantine Books.

Hirshfield, A. N., and J. A. Flaws. 1999. Reproductive senescence, nonhuman mammals. In *Encyclopedia of Reproduction*, ed. E. Knobil and J. D. Neill, 239–44. San Diego: Academic Press.

Hobcraft, J., J. Menken, and S. Preston. 1982. Age, period, and cohort effects in demography: A review. *Population Index* 48:4–43.

Hoffman, A. A., and P. A. Parson. 1989. An integrated approach to environmental stress tolerance and life history variation: Desiccation tolerance in *Drosophila. Biological Journal of the Linnean Society* 37:117–36.

Holden, C. 1987. Why do women live longer than men? *Science* 238:158–60.

Holehan, A. M., and B. J. Merry. 1985. Lifetime breeding studies in fully fed and dietary restricted female CFY Sprague-Dawley rates. 1. Effect of age, housing conditions and diet on fecundity. *Mechanisms of Ageing and Development* 33:19–28.

Holliday, R. 1989. Food, reproduction and longevity: Is the extended lifespan of calorie-restricted animals an evolutionary adaptation? *Bioessays* 10:125–27.

———. 1997. Understanding ageing. *Philosophical Transactions of the Royal Society of London* B352:1793–97.

———. 2001. Human ageing and the origins of religion. *Biogerontology* 2:73–77.

Horiuchi, S., and A. J. Coale. 1990. Age patterns of mortality for older women: An analysis using the age-specific rate of mortality change with age. *Mathematical Population Studies* 2:245–67.

Horiuchi, S., and J. R. Wilmoth. 1998. Decleration in the age pattern of mortality at older ages. *Demography* 35:391–412.

Hoyenga, K. B., and K. T. Hoyenga. 1981. Gender and energy balance: sex differences in adaptations for feast and famine. *Physiology and Behavior* 28:545–63.

Hoyert, D. L., D. K. Kochanek, and S. L. Murphy. 1999. National vital statistics reports. *Centers for Disease Control and Prevention* 47:1–47.

Hrdy, S. B. 1999. *Mother Nature : A History of Mothers, Infants, and Natural Selection*. New York: Pantheon Books.

Hsin, H., and C. Kenyon. 1999. Signals from the reproductive system regulate the lifespan of *C. elegans. Nature* 399:362–66.

Huebner, E. 1999. Female reproductive systems, insects. In *Encyclopedia of Reproduction*, ed. E. Knobil and J. D. Neill, 215–29. San Diego: Academic Press.

Hughes, K. A., and B. Charlesworth. 1994. A genetic analysis of senescence in *Drosophila. Nature* 367:64–66.

Hummer, R. A., R. G. Rogers, C. B. Nam, and C. G. Ellison. 1999. Religious involvement and U.S. adult mortality. *Demography* 36:273–85.

Idelker, T., T. Galitski, and L. Hood. 2001. A new approach to decoding life: Systems biology. *Annual Review of Genomics and Human Genetics* 2:343–72.

Jacome, I., M. Aluja, P. Liedo, and D. Nestel. 1995. The influence of adult diet and age on lipid reserves in the tropical fruit fly *Anastrepha serpentina* (Diptera: Tephritidae). *Journal of Insect Physiology* 41:1079–86.

Jazwinski, S. M. 1996. Longevity, genes, and aging. *Science* 273:54–59.

Jeune, B. 1995. In search of the first centenarians. In *Exceptional Longevity: From Prehistory to the Present*, ed. B. Jeune and J. W. Vaupel, 11–24, Odense: Odense University Press.

Johnson, R. A., and D. W. Wichem. 1998. *Applied Multivariate Statistical Analysis*. 4th ed. New York: Prentice-Hall.

Johnson, T. E. 1990. Increased life-span of age-1 mutants in *Caenorhabditis elegans* and lower Gompertz rate of aging. *Science* 249:908–12.

Jones, R., W. Perkins, and A. Sparks. 1975. *Heliothis zea*: Effects of population density and a marker dye in the laboratory. *Journal of Economic Entomology* 68:349–50.

Judge, D., and J. R. Carey. 2000. Post-reproductive life predicted by primate patterns. *Journal of Gerontology: Biological Sciences* 55A:B201–B209.

Kannisto, V. 1988. On the survival of centenarians and the span of life. *Population Studies* 42:389–406.

———. 1991. Frailty and survival. *Genus* 47:101–18.

Kannisto, V., J. Lauritsen, A. R. Thatcher, and J. W. Vaupel. 1994. Reductions in mortality at advanced ages: Several decades of evidence from 27 countries. *Population and Development Review* 20:793–810.

Kaplan, H. 1996. A theory of fertility and parental investment in traditional and modern Human societies. *Yearbook of Physical Anthropology* 39:91–135.

———. 1997. The evolution of the human life course. In *Between Zeus and the Salmon: The Biodemography of Longevity*, ed. K. Wachter and C. Finch, 175–211. Washington, D.C.: National Academy Press.

Kaplan, H., K. Hill, J. Lancaster, and A. M. Hurtado. 2000. A theory of human life history evolution: Diet, intelligence, and longevity. *Evolutionary Anthropology* 9:156–85.

Kass, L. R. 1983. The case for mortality. *American Scholar* 52:173–91.

Katz, S., L. G. Branch, M. H. Branson, J. A. Papsidero, J. C. Beck, and D. S. Greer. 1983. Active life expectancy. *New England Journal of Medicine* 309:1218–24.

Keyfitz, N. 1966. *Introduction to the Mathematics of Population*. Reading, Mass.: Addison-Wesley.

———. 1982. Choice of function for mortality analysis: Effective forecasting depends on a minimum parameter representation. *Theoretical Population Biology* 21:329–52.

———. 1985. *Applied Mathematical Demography*. New York: Springer-Verlag.

Keyfitz, N., and W. Flieger. 1990. *World Population Growth and Aging*. Chicago: University of Chicago Press.

Kinzey, W. G. 1997. *New World Primates: Ecology, Evolution, and Behavior*. New York: Aldine de Gruyter.

Kirk, K. L. 2001. Dietary restriction and aging: Comparative tests of evolutionary hypotheses. *Journal of Gerontology: Biological Sciences* 56A:B123–B129.

Kirkwood, T.B.L. 1977. Evolution of aging. *Nature* 270:301–4.

———. 1992. Comparative life spans of species: Why do species have the life spans they do? *American Journal of Clinical Nutrition* 55:1191S–95S.

———. 1997. The origins of human ageing. *Philosophical Transactions of the Royal Society of London* B353:1763–72.

Kirkwood, T.B.L., and M. R. Rose. 1991. Evolution of senescence: Late survival sacrificed for reproduction. *Philosophical Transactions of the Royal Society of London* 332:15–24.

Kitagawa, E., and P. M. Hauser. 1973. *Differential Mortality in the United States: A Study in Socioeconomic Epidemiology*. Cambridge: Harvard University Press.

Klein, J. P., and M. L. Moeschberger. 1997. *Survival Analysis*. New York: Springer.

Klemera, P., and S. Doubal. 1992. Human mortality at very advanced age might be constant. *Mechanisms in Ageing and Development* 98:167–76.

Köhler, G. and C. Milstein. 1975. Continuous cultures of fused cells secreting antibody of predefined specificity. *Nature* 256:495–97.

Kopec, S. 1928. On the influence of intermittent starvation on the longevity of the imaginal stage of *Drosophila melanogaster. Journal of Experimental Biology* 5:204–11.

Kowald, A., and T.B.L. Kirkwood. 1993. Explaining fruit fly longevity. *Science* 260:1664–65.

Krainacker, D. A., J. R. Carey, and R. Vargas. 1987. Effect of larval host on the life history parameters of the Mediterranean fruit fly, *Ceratitis capitata. Oecologia* 73:583–90.

Lamb, M. 1964. The effects of radiation on the longevity of female *Drosophila subobscura. Journal of Insect Physiology* 10:487–97.

Lamb, M. J. 1977. *Biology of Ageing*. New York: Wiley.

Landes, D. S. 1998. *The Wealth and Poverty of Nations*. New York: Norton.

Lee, C.-K., R. G. Klopp, R. Weindruch, and T. A. Prolla. 1999. Gene expression profile of aging and its retardation by caloric restriction. *Science* 285:1390–93.

Lee, E. T. 1992. *Statistical Methods for Survival Data Analysis*. New York: Wiley.

Lee, R. D. 1997. Intergenerational relations and the elderly. In *Between Zeus and the Salmon. The Biodemography of Longevity*, ed. K. W. Wachter and C. E. Finch, 213–33. Washington, D.C.: National Academy Press.

Leslie, P. H., and T. Park. 1949. The intrinsic rate of natural increase of *Tribolium castaneum Herbst. Ecology* 30:469–77.

Leslie, P. H., and R. M. Ransom. 1940. The mortality, fertility and rate of natural increase of the vole (*Microtus agrestis*) as observed in the laboratory. *Journal of Animal Ecology* 9:27–52.

Liedo, P., and J. R. Carey. 2000. Unpublished life table data on medflies collected at the Moscamed medfly rearing facility in Tapachula, Mexico.

Lints, F. A., M. Bourgois, A. Delalieux, J. Stoll, and C. V. Lints. 1983. Does the female life span exceed that of the male? *Gerontology* 29:336–52.

Ljungquist, B., S. Berg, J. Lanke, G. E. McClearn, and N. L. Pedersen. 1998. The effect of genetic factors for longevity: A comparison of identical and fraternal twins in the Swedish twin resistry. *Journal of Geronology* 53A:M441–46.

Lopez, A. D., and L. T. Ruzicka. 1983. *Sex Differentials in Mortality*. Canberra: Australian National University Press.

Lotka, A. J. 1928. The progeny of a population element. *American Journal of Hygiene* 8:875–901.

Lovejoy, O. 1981. The origin of man. *Science* 211:341–50.

Luckinbill, L. S., and M. J. Clare. 1986. A density threshold for the expression of longevity in *Drosophila melanogaster*. *Heredity* 56:329–35.

Luckinbill, L. S., J. L. Graves, A. Tomkiw, and O. Sowirka. 1988. A qualitative analysis of some life-history correlates of longevity in *Drosophila melanogaster*. *Evolutionary Ecology* 2:85–94.

Makeham, W. H. 1867. On the law of mortality. *Journal of the Institute of Actuaries* 13:325–58.

Malick, L. E., and J. F. Kidwell. 1966. The effect of mating status, sex, and genotype on longevity in *Drosophila melanogaster*. *Genetics* 54:203–209.

Malthus, T. R. [1799] 1988. *Population: The First Essay*. Reprint. Ann Arbor: University of Michigan Press.

Manly, B.G.J. 1997. *Randomization, Bootstrap and Monte Carlo Methods in Biology*. London: Chapman and Hall.

Mansingh, A. 1971. Physiological classification of dormancies in insects. *Canadian Entomologist* 103:983–1009.

Manton, K. G., and K. C. Land. 2000. Active life expectancy estimates for the U.S. elderly population: A multidimensional continuous-mixture model of functional change applied to completed cohorts, 1982–1996. *Demography* 37:253–65.

Manton, K. G., S. S. Poss, and S. Wing. 1979. The Black/White mortality crossover: Investigation from the perspective of the components of aging. *The Gerontologist* 19:291–300.

Manton, K. G., and E. Stallard. 1984. *Recent Trends in Mortality Analysis*. Orlando: Academic Press.

———. 1996. Longevity in the United States: Age and sex-specific evidence on life span limits from mortality patterns 1960–1990. *Journal of Gerontology* 51A: B362–75.

Manton, K. G., E. Stallard, and H. D. Tolley. 1991. Limits to human life expectancy: Evidence, prospects, and implications. *Population and Development Review* 17: 603–37.

Markowska, A. L., and S. J. Breckler. 1999. Behavioral biomarkers of aging: Illustration of a multivariate approach for detecting age-related behavioral changes. *Journal of Gerontology: Biological Sciences* 12:B549–66.

Masaki, S. 1980. Summer diapause. *Annual Review of Entomology* 25:1–25.

Masoro, E. J. 1988. Minireview: Food restriction in rodents: An evaluation of its role in the study of aging. *Journal of Gerontology* 43:B59–64.

Masoro, E. J., and S. N. Austad. 1996. The evolution of the antiaging action of dietary restriction: A hypothesis. *Journal of Gerontology*: Biological Sciences 51A:B387–91.

Matthews, R. W., and J. R. Matthews. 1978. *Insect Behavior*. New York: Wiley.

McAdam, A. G., and J. S. Millar. 1999. Dietary protein constraint on age at maturity: An experimental test with wild deer mice. *Journal of Animal Ecology* 68:733–40.

McCullagh, P., and J. A. Nelder. 1986. *Generalized Linear Models*. London: Chapman and Hall.

McCurdy, S. A. 1994. Epidemiology of disaster: The Donner Party (1846–1847). *Western Journal of Medicine* 160:338–42.

McGue, M., J. W. Vaupel, N. Holm, and B. Harvald. 1993. Longevity is moderately heritable in a sample of Danish twins born 1870–1880. *Journal of Gerontology* 48:B237–44.

McHenry, H. M. 1994. Behavioral ecological implications of early hominid body size. *Journal of Human Evolution* 27:77–87.

Medawar, P. B. 1952. *An Unsolved Problem of Biology*. London: H. K. Lewis.

———. 1955. The definition and measurement of senescence. In *CIBA Foundation Colloquia on Ageing*, ed. G.E.W. Wolstenholme, M. P. Cameron, and J. Etherington. Boston: Little, Brown.

———. 1957. *The Uniqueness of the Individual*. New York: Dover.

Merry, B. J., and A. M. Holehan. 1979. Onset of puberty and duration of fertility in rats fed a restricted diet. *Journal of Reproduction and Fertility* 57:253–59.

Mertz, D. B. 1975. Senescent decline in flour beetle strains selected for early adult fitness. *Physiological Zoology* 48:1–23.

Michod, R. E. 1999. Individuality, immortality, and sex. In *Levels of Selection in Evolution*, ed. L. Keller, 53–74. Princeton: Princeton University Press.

Milgram, S. 1970. The experience of living in cities. *Science* 167:1461–68.

Miller, J. D. 1997. Reproduction in sea turtles. In *The Biology of Sea Turtles*, ed. P. L. Lutz, and J. A. Musick, 51–81. Boca Raton: CRC Press.

Minchella, D. J., and P. T. Loverde. 1981. A cost of increased early reproduction in the snail *Biomphalaria glabrata*. *American Naturalist* 118:876–81.

Montagu, A. 1974. *The Natural Superiority of Women*. New York: Collier.

Montgomery, M. R., and B. Cohen. 1998. *From Death to Birth: Mortality Decline and Reproductive Change*. Washington, D.C.: National Academy Press.

Moriyama, I. M. 1956. Development of the present concept of cause of death. *American Journal of Public Health* 46:436–41.

Morris, D. 1955. The causation of pseudomale and pseudofemale behavior—a further comment. *Behaviour* 8:46–56.

Moyle, P. B., B. Herbold. 1987. Life-history patterns and community structure in stream fishes of Western North America: Comparisons with Eastern North America and Europe. In *Community and Evolutionary Ecology of North American Stream Fishes.*, ed. W. J. Matthews and D. C. Heins, 25–34. Norman: University of Oklahoma Press.

Mueller, L. D., J. L. Graves, and M. R. Rose. 1993. Interactions between density-dependent and age-specific selection in *Drosophila melanogaster. Functional Ecology* 7:469–79.

Mueller, L. D., P. Guo, and F. J. Ayala. 1991. Density-dependent natural selection and trade-offs in life history traits. *Science* 253:433–35.

Müller, H.-G., J. R. Carey, D. Wu, and J. W. Vaupel. 2001. Reproductive potential determines longevity of female Mediterranean fruitflies. *Proceedings of the Royal Society of London* B268:445–50.

Müller, H.-G., J. L. Wang, and W. B. Capra. 1997a. From life tables to hazard rates: The transformation approach. *Biometrika* 84:881–92.

Müller, H.-G., J.-L. Wang, W. B. Capra, P. Liedo, and J. R. Carey. 1997b. Early mortality surge in protein-deprived females causes reversal of sex differential of life expectancy in Mediterranean fruit flies. *Proceedings of the National Academy of Sciences USA* 94:2762–65.

Murchie, R. 1967. *The Seven Mysteries of Life* New York: Houghton Mifflin Company.

Neugarten, B. 1974. Age groups in American society and the rise of the young-old. *Annals of the American Academy of Political and Social Science* 415:187–98.

———. 1996. Family and community support systems. Paper presented to Committee #7, White House Conference on Aging, Washington, D.C., November 1981. Published in *The Meanings of Age*, ed. B. Neugarten, 355–76. Chicago: University of Chicago Press.

Neugarten, B., and N. Datan. 1973. Sociological perspectives on the life cycle. In *Life-span Developmental Psychology*, ed. P. B. Baltes, and K. W. Schaie, 53–79. New York: Academic Press.

Neuweiler, G. 1999. Neuroethology, vertebrates. In *Encyclopedia of Neuroscience*, ed. G. Adelman and B. H. Smith, 1345–52. Amsterdam: Elsevier.

Norris, M. J. 1933. Contributions toward the study of insect fertility. II. Experiments on the factors influencing fertility in *Ephestia kuhniella. Proceedings of the Zoological Society of London* 903:342–54.

Notestein, F. 1945. Population—the long view. In *Food for the World*, ed. T. W. Schultz. Chicago: University of Chicago Press.

Nusbaum, T. J., J. L. Graves, L. D. Mueller, and M. R. Rose. 1993. Fruit fly aging and mortality. *Science* 260:1567.

Nutton, V. 1996. The rise of medicine. In *The Cambridge Illustrated History of Medicine*, ed. R. Porter 52–81. Cambridge: Cambridge University Press.

Olshansky, S. J. 1988. On forecasting mortality. *The Milbank Quarterly* 66:482–530.

———. 1995. Mortality crossovers and selective survival in human and nonhuman populations. *The Gerontologist* 35:583–87.

Olshansky, S. J., and B. A. Carnes. 1997. Ever since Gompertz. *Demography* 34:1–15.

Olshansky, S. J., B. A. Carnes, and C. K. Cassel. 1993. Fruit fly aging and longevity. *Science* 260:1565–66.

Olshansky, S. J., B. A. Carnes, and A. Desesquelles. 2001. Prospects for human longevity. *Science* 291:1491–92.

Omran, A. R. 1971. The epidemiologic transition: A theory of the epidemiology of population change. *Milbank Memorial Fund Quarterly* 49:509–538.

O'Rand, A. M., and M. L. Krecker. 1990. Concepts of the life cycle: Their history, meanings, and uses in the social sciences. *Annual Review of Sociology* 16:241–62.

Ovchinnkov, I. V., A. Gotherstrom, G. P. Romanova, V. M. Kharitonov, K. Liden, and W. Goodwin. 2000. Molecular analysis of Neanderthal DNA from the northern Caucasus. *Nature* 404:490–93.

Papadopoulos, N. T., J. R. Carey, B. I. Katsoyannos, N. A. Kouloussis, H.-G. Müller, and X. Liu. 2002. Supine behaviour predicts time-to-death in male Mediterranean fruit flies. *Proceedings of the Royal Society of London: Biological Sciences* 269: 1633–37.

Pardoll, D. M. 1998. Cancer vaccines. *Nature Medicine* 4:525–31.

Partridge, L. 1986. Sexual activity and life span. In *Insect Aging*, ed. K.-G. Collatz and R. S. Sohal. Berlin: Springer-Verlag.

———. 1987. Is accelerated senescence a cost of reproduction? *Functional Ecology* 1:317–320.

Partridge, L., and R. Andrews. 1985. The effect of reproductive activity on the longevity of male *Drosophila melanogaster* is not caused by an acceleration of ageing. *Journal of Insect Physiology* 31:393–95.

Partridge, L., and M. Farquhar. 1981. Sexual activity reduced lifespan of male fruitflies. *Nature* 294:580–82.

Partridge, L., and K. Fowler. 1992. Direct and correlated responses to selection on age at reproduction in *Drosophila melanogaster*. *Evolution* 46:76–91.

Partridge, L., A. Green, and K. Fowler. 1987. Effects of egg production and of exposure to males on female survival in *Drosophila melanogaster*. *Journal of Insect Physiology* 33:745–49.

Partridge, L., and P. H. Harvey. 1985. Costs of reproduction. *Evolutionary Biology* 316:20.

———. 1988. The ecological context of life history evolution. *Science* 241:1449–55.

Partridge, L., and L. D. Hurst. 1998. Sex and conflict. *Science* 281:2003–8.

Patronek, G. J., D. J. Waters, and L. T. Glickman. 1997. Comparative longevity of pet dogs and humans: Implications for gerontology research. *Journal of Gerontology* 52A:B171–78.

Patterson, R. 1957. On the causes of broken wings of the house fly. *Journal of Economic Entomology* 50:104–5.

Pearl, R. 1931. Studies on human longevity. IV. The inheritance of longevity. Preliminary report. *Human Biology* 3:245–69.

Pearl, R., and J. R. Miner. 1935. Experimental studies on the duration of life. XIV. The comparative mortality of certain lower organisms. *Quarterly Review of Biology* 10:60–79.

Pearl, R., J. R. Miner, and S. L. Parker. 1927. Experimental studies on the duration of life. XI. Density of population and life duration in *Drosophila*. *The American Naturalist* 61:289–318.

Pearl, R., and S. L. Parker. 1922. Experimental studies on the duration of life. IV. Data on the influence of density of population on duration of life in *Drosophila*. *American Naturalist* 56:312–21.

———. 1924a. Experimental studies on the duration of life IX. New life tables for Drosophila. *American Naturalist* 58:71–82.

———. 1924b. Experimental studies on the duration of life. X. The duration of life of *Drosophila melanogaster* in the complete absence of food. *American Naturalist* 58:193–218.

Pearl, R., S. L. Parker, and B. M. Gonzalez. 1923. Experimental studies on the duration of life. VII. The Mendelian inheritance of duration of life in crosses of wild type and quintuple stocks of *Drosophila melanogaster*. *American Naturalist* 57: 153–92.

Perls, T. T., L. Alpert, and R. C. Fretts. 1997. Middle-aged mothers live longer. *Nature* 389:133.

Perls, T. T., E. Burbrick, C. G. Wagner, J. Vijg, and L. Kruglyak. 1998. Siblings of centenarians live longer. *Nature* 351:1560.

Persidis, A. 2000a. Cancer multidrug resistance. *Nature Biotechnology* 18:IT18–20.

———. 2000b. Tissue engineering. *Nature Biotechnology* 18:IT56–58.

———. 2000c. Xenotransplantation. *Nature Biotechnology* 18:IT53–55.

Peters, R. H. 1991. *A Critique for Ecology*. Cambridge: Cambridge University Press.

Petersen, W., and R. Petersen. 1986. *Dictionary of Demography*. New York: Greenwood Press.

Peterson, W. 1975. *Population*. New York: Macmillan.

Pfeifer, A., and I. M. Verma. 2001. Gene therapy: Promises and problems. *Annual Review of Genomics and Human Genetics* 2:177–211.

Pollak, R. A. 1986. A reformulation of the two-sex problem. *Demography* 23:247–59.

———. 1987. The two-sex problem with persistent unions: A generalization of the birth matrix-mating rule model. *Theoretical Population Biology* 32:176–87.

Pressat, R. 1985. *The Dictionary of Demography*. Oxford: Blackwell.

Preston, S. H. 1976. *Mortality Patterns in National Populations*. New York: Academic Press.

———. 1978. *The Effects of Infant and Child Mortality on Fertility*. New York: Academic Press.

Preston, S. H., and M. R. Haines. 1991. *Fatal Years: Child Mortality in Late Nineteenth-Century America*. Princeton: Princeton University Press.

Preston, S. H., N. Keyfitz, and R. Schoen. 1972. *Causes of Death: Life Tables for National Populations*. New York: Seminar Press.

———. 1978. *Causes of Death: Life Tables for National Populations*. New York: Seminar Press.

Price, D. L., S. S. Sisodia, and C. H. Kawas. 1999. Aging of the brain and Alz heimer's disease. In *Encyclopedia of Neuroscience*, ed. G. Adelman and B. H. Smith, 41–43. Amsterdam: Elsevier.

Price, E. O. 1984. Behavioral aspects of animal domestication. *Quarterly Review of Biology* 59:1–32.

Prokopy, R. J., B. D. Roitberg, and R. I. Vargas. 1994. Effects of egg load on finding and acceptance of host fruit in *Ceratitis capitata* flies. *Physiological Entomology* 19:124–32.

Promislow, D.E.L. 1991. Senescence in natural populations of mammals: A comparative study. *Evolution* 45:1869–87.

Promislow, D.E.L., and P. H. Harvey. 1990. Living fast and dying young: A comparative analysis of life-history variation among mammals. *Journal of the Zoological Society of London* 220:417–37.

Promislow, D.E.L., M. Tatar, S. Pletcher, and J. R. Carey. 1999. Below-threshold mortality: implications for studies in evolution, ecology and demography. *Journal of Evolutionary Biology* 12:314–28.

Prout, T., and F. McChesney. 1985. Competition among immatures affects their adult fertility: Population dynamics. *American Naturalist* 126:521–58.

Quiring, D. T., and J. N. McNeil. 1984. Influence of intraspecific larval competition and mating on the longevity and reproductive performance of females of the leaf miner *Agromyza frontella* (Rondani) (Diptera: Agromyzidae). *Canadian Journal of Zoology* 62:2197–203.

Ragland, S. S., and R. S. Sohal. 1973. Mating behavior, physical activity and aging in the housefly, *Musca domestica*. *Experimental Gerontology* 8:135–45.

———. 1975. Ambient temperature, physical activity, and aging in the housefly, *Musca domestica*. *Experimental Gerontology* 10:279–89.

Ranter, S. C., and R. Boice. 1975. Effects of domestication on behaviour. In *The Behaviour of Domestic Animals*, ed. E.S.E. Hafez, 3–19. New York: Macmillan.

Read, A., and P. Harvey. 1989. Life history differences among the Eutherian radiations. *Journal of Zoology* 219:329–53.

Reznick, D. 1985. Costs of reproduction: An evaluation of the empirical evidence. *Oikos* 44:257–67.

Richards, M., R. Hardy, D. Kuh, and M.E.J. Wadsworth. 2001. Birth weight and cognitive function in the British 1946 birth cohort: Longitudinal population based study. *British Medical Journal* 322:199–203.

Ricklefs, R. 1979. *Ecology*. 2nd ed. New York: Chiron Press.

Riedman, M. 1990. *The Pinnipeds*. Berkeley: University of California Press.

Riggs, J. E., and R. J. Millecchia. 1992. Mortality among the elderly in the U.S.: 1856–1987. Demonstration of the upper boundary to Gompertzian mortality. *Mechanisms in Ageing and Development* 62:191–99.

Rivero-Lynch, A. P., and H. C. J. Godfray. 1997. The dynamics of egg production, oviposition and resorption in a parasitoid wasp. *Functional Ecology* 11:184–88.

Robine, J.-M., and M. Allard. 1998. The oldest human. *Science* 279:1834–35.

Robine, J.-M., and K. Ritchie. 1993. Explaining fruit fly longevity. *Science* 260: 1664.

Robine, J.-M., and J. W. Vaupel. 2001. Supercentenarians: Slower ageing individuals or senile elderly? *Experimental Gerontology* 36:915–30.

Rockstein, M. 1957. Longevity of male and female house flies. *Journal of Gerontology* 12:253–56.

Rockstein, M., J. A. Chesky, M. H. Levy, and L. Yore. 1981. Effect of population density upon life expectancy and wing retention in the common house fly, *Musca domestica*. *Gerontology* 27:13–19.

Rockstein, M., and H. M. Lieberman. 1959. A life table for the common house fly, *Musca domestica*. *Gerontologia* 3:23–36.

Roff, D. A. 1992. *The Evolution of Life Histories*. New York: Chapman & Hall.

Rogers, A. 1992. Heterogeneity and selection in multistate population analysis. *Demography* 29:31–38.

Rogers, R. G. 1995. Sociodemographic characteristics of long-lived and healthy individuals. *Population Development Review* 21:33–58.

Roitberg, B. D. 1989. The cost of reproduction in rosehip flies, *Rhagoletis basiola*: Eggs are time. *Evolutionary Ecology* 3:183–88.

Rose, M. 1991. *Evolutionary Biology of Aging*. New York: Oxford University Press.

Rose, M., and B. Charlesworth. 1980. A test of evolutionary theories of senescence. *Nature* 287:141–42.

Rosenheim, J. A. 1996. An evolutionary argument for egg limitation. *Evolution* 50:2089–94.

———. 1999. The relative contributions of time and eggs to the cost of reproduction. *Evolution* 53:376–85.

Ross, J. A. 1982. *International Encyclopedia of Population*. Vol. 2. New York: Free Press.

Rowe, J. W., and R. L. Kahn. 1998. *Successful Aging*. New York: Pantheon Books.

Ryder, N. B. 1959. Fertility. In *The Study of Population*, ed. P. M. Hauser and O. D. Duncan, 400–426. Chicago: University of Chicago Press.

Sacher, G. A. 1978. Longevity and aging in vertebrate evolution. *Bioscience* 28:497–501.

———. 1980a. The constitutional basis for longevity in the Cetacea: Do the whales and the terrestrial mammals obey the same laws? *Report of the International Whaling Commission* (Special Issue) 3:209–13.

———. 1980b. Theory in Gerontology, Part I. *Annual Review of Gerontology and Geriatrics* 1:3–25.

———. 1982. Evolutionary theory in gerontology. *Perspectives in Biology and Medicine* 25:339–53.

Scheibel, A. B. 1999. Aging of the brain. In *Encyclopedia of Neuroscience*, ed. G. Adelman and B. H. Smith, 38–40, Amsterdam: Elsevier.

Schoen, R. 1978. A standardized two-sex stable population. *Theoretical Population Biology* 14:357–70.

Schwarz, A. J., A. Zambada, D.H.S. Orozco, J. L. Zavala, and C. O. Calkins. 1989. Mass production of the Mediterranean fruit fly at Metapa, Mexico. *Florida Entomologist* 68:467–77.

Service, P. M., E. W. Hutchinson, M. D. MacKinley, and M. R. Rose. 1985. Resistance to environmental stress in *Drosophila melanogaster* selected for postponed senescence. *Physiology and Zoology* 58:380–389.

Settersten, R.A.J., and K. U. Mayer. 1997. The measurement of age, age structuring, and the life course. *Annual Review of Sociology* 23:233–61.

Sgro, C. M., and L. Partridge. 1999. A delayed wave of death from reproduction in *Drosophila. Science* 286:2521–24.

Shelly, T., and T. S. Whittier. 1995. Effect of sexual experience on the mating success of males of the Mediterranean fruit fly, *Ceratitis capitata* Wiedemann (Diptera: Tephritidae). *Proceedings of the Hawaiian Entomological Society* 32:91–94.

Silver, L. 1998. *Remaking Eden*. New York: Avon Books.

Skloot, R. 2001. Cells that save lives are a mother's legacy. *New York Times*, November 17, pp. A21, A23.

Smaragdova, N. P. 1930. Data on the influence of the density of population on the duration of life of the workers of the honey-bee. *Bee World* 11:103–5.

Smith, A. [1776] 1961. *The Wealth of Nations*. Reprint. Stand, England: Methuen.

Smith, D.W.E. 1989. Is greater female longevity a general finding among animals? *Biology Review* 64:1–12.

———. 1993. *Human Longevity*. New York: Oxford University Press.

Smith, D.W.E., and H. R. Warner. 1989. Does genotypic sex have a direct effect on longevity? *Experimental Gerontology* 24:277–88.

Smith, J. M. 1958. The effects of temperature and of egg-laying on the longevity of *Drosophila subobscura. Journal of Experimental Biology* 35:832–42.

Smith, J. M., and E. Szathmary. 1999. *The Major Transitions in Evolution*. Oxford: Oxford University Press.

Snowdon, D. A., R. L. Kane, W. L. Beeson, G. L. Burke, J. M. Sprafka, J. Potter, H. Iso, D. R. Jacobs, and R. L. Phillips. 1989. Is early natural menopause a biologic marker of health and aging? *American Journal of Public Health* 79:709–14.

Sohal, R. S., and P. B. Buchan. 1981. Relationship between physical activity and life span in the adult housefly, *Musca domestica. Experimental Gerontology* 16: 157–62.

Sohal, R. S., and R. Weindruch. 1996. Oxidative stress, caloric restriction, and aging. *Science* 273:59–63.

Sonleitner, F. J. 1961. Factors affecting egg cannibalism and fecundity in populations of adult *Tribolium castaneum* Herbst. *Physiological Zoology* 34:233–35.

Southern, R. W. 1953. *The Making of the Middle Ages*. New Haven: Yale University Press.

Stearns, S. C. 1992. *The Evolution of Life Histories*. Oxford: Oxford University Press.

Steele, D.J.R., and H. Auchincloss. 1995. Xenotransplantation. *Annual Review of Medicine* 46:345–60.

Stolnitz, G. J. 1957. A century of international mortality trends. *Population Studies* 10:17–42.

Strehler, B. L., and A. S. Mildvan. 1960. General theory of mortality and aging. *Science* 132:14–21.

Strickberger, M. W. 1996. *Evolution*. 4th ed. Sudbury, Mass.: Jones and Partlet.

Swedlund, A. C., R. S. Meindl, J. Nydon, and M. Gradie. 1983. Family patterns in longevity and longevity patterns of the family. *Human Biology* 55:115–29.

Szirmai, A. 1997. *Economic and Social Development*. London: Prentice-Hall.

Taeuber, C., and I. Rosenwaike. 1992. A demographic portrait of America's oldest-old. In *The Oldest Old*, ed. R. M. Suzman, D. P. Wilis, and K. G. Manton 17–42. New York: Oxford University Press.

Tanner, J. T. 1966. Effects of population density on growth rates of animal populations. *Ecology* 47:733–45.

Tatar, M., and J. R. Carey. 1994a. Genetics of mortality in the bean beetle, *Callosobruchus maculatus. Evolution* 48:1371–76.

———. 1994b. Sex mortality differentials in the bean beetle: Reframing the question. *American Naturalist* 144:165–75.

———. 1995. Nutrition mediates reproductive trade-offs with age-specific mortality in the beetle, *Calosobruchus maculatus. Ecology* 76:2066–73.

Tatar, M., J. R. Carey, and J. W. Vaupel. 1993. Long-term cost of reproduction with and without accelerated senescence in *Callosobruchus maculatus*: Analysis of age-specific mortality. *Evolution* 47:1302–12.

———. 1994. Long-term cost of reproduction with and without accelerated senescence in *Callosobruchus maculatus*: Analysis of age-specific mortality. *Evolution* 47:1302–12.

Thatcher, A. R. 1992. Trends in numbers and mortality at high ages in England and Wales. *Population Studies* 46:411–26.

Thomas, J. H. 1994. The mind of a worm. *Science* 264:1698–99.

Thomas, L. 1988. On the science and technology of medicine. *Daedalus* 117:299–316.

Thun, M. J., R. Peto, A. D. Lopez, J. H. Monaco, S. L. Henley, C.W.J. Heath, and R. Doll. 1997. Alcohol consumption and mortality among middle-aged and elderly U.S. adults. *New England Journal of Medicine* 337:1705–14.

Timiras, P. 1994. Introduction: Aging as a stage in the life cycle. In *Physiological Basis of Aging and Geriatrics*, ed. P. Timiras, 1–5. Boca Raton, Fl.: CRC Press.

Tinsley, R. C. 1989. The effects of host sex on transmission success. *Parasitology Today* 5:190–95.

Trivers, R. L. 1972. Parental investment and sexual selection. In *Sexual Selection and the Descent of Man 1871–1971*, ed. B. Campbell, 136–79. Chicago: Aldine.

Troiano, R. P., J.E.A. Frongillo, J. Sobal, and D. A. Levitsky. 1996. The relationship between body weight and mortality: A quantative analysis of combined information from existing studies. *International Journal of Obesity* 20:63–75.

Trounson, A. 1986. Preservation of human eggs and embryos. *Fertility and Sterility* 46:1–12.

Tufte, E. R. 1983. *The Visual Display of Quantitative Information*. Cheshire, Conn.: Graphics Press.

Tuljapurkar, S. 1989. An uncertain life: Demography in random environments. *Theoretical Population Biology* 35:227–94.

Tuljapurkar, S., and C. Boe. 1998. Mortality change and forecasting: How much and how little do we know. *North American Actuarial Journal* 2:13–47.

Tuljapurkar, S., N. Li, and C. Boe. 2000. A universal pattern of mortality decline in the G7 countries. *Nature* 405:789–92.

Tuomi, J., T. Hakala, and E. Haukioja. 1983. Alternative concepts of reproductive effort, costs of reproduction, and selection in life-history evolution. *American Zoologist* 34:25–34.

Turner, J. S. 2000. *The Extended Organism: The Physiology of Animal-built Structures*. Cambridge: Harvard University Press.

Vargas, R. 1989. Mass production of tephritid fruit flies. In *World Crop Pests. Fruit Flies: Their Biology, Natural Enemies and Control*, ed. A. S. Robinson and G. Hooper, 141–52. Amsterdam: Elsevier.

Vargas, R. I., and J. R. Carey. 1989. Comparison of demographic parameters for wild and laboratory-adapted Mediterranean fruit fly (Diptera: Tephritidae). *Annals of the Entomological Society of America* 82:55–59.

Vargas, R. I., W. A. Walsh, D. Kanehisa, E. B. Jang, and J. W. Armstrong. 1997. Demography of four Hawaiian fruit flies (Diptera: Tephritidae) reared at five constant temperatures. *Annals of the Entomological Society of America* 90:162–68.

Vaupel, J. W. 1997. The average French baby may live 95 to 100 years. In *Longevity: To the Limits and Beyond*, ed. J.-M. Robine, J. W. Vaupel, B. Jeune, and M. Allard. Berlin: Springer-Verlag.

Vaupel, J. W., and J. R. Carey. 1993. Compositional interpretations of medfly mortality. *Science* 260:1666–67.

Vaupel, J. W., J. R. Carey, K. Christensen, T. E. Johnson, A. I. Yashin, et al. 1998. Biodemographic trajectories of longevity. *Science* 280:855–60.

Vaupel, J. W., and B. Jeune. 1995. The emergence and proliferation of centenarians. In *Exceptional Longevity: From Prehistory to the Present*, ed. B. Jeune and J. W. Vaupel, 109–16. Odense: Odense University Press.

Vaupel, J. W., K. G. Manton, and E. Stallard. 1979. The impact of heterogeneity in individual frailty on the dynamics of mortality. *Demography* 16:439–54.

Vaupel, J. W., and A. I. Yashin. 1985. Heterogeneity's ruses: Some surprising effects of selection on population dynamics. *American Statistician* 39:176–85.

Vaupel, J. W., A. I. Yashin, and K. G. Manton. 1988. Debilitation's aftermath: Stochastic process models of mortality. *Mathematical Population Studies* 1:21–48.

Verbrugge, L. M. 1990. Pathways of health and death In *Women, Health, and Medicine in America*, ed. R. D. Apple, New York: Garland.

Visscher, M. B., J. T. King, and Y.C.P. Lee. 1952. Further studies on influence of age and diet upon reproductive senescence in Strain A female mice. *American Journal of Physiology* 170:72–76.

Vreeland, R. H., W. D. Rosenzweig, and D. W. Powers. 2000. Isolation of a 250 million-year-old halotolerant bacterium from a primary salt crystal. *Nature* 407: 897–900.

Wachter, K. 1997. Between Zeus and the salmon. Introducing the biodemography of longevity. In *Biodemography of Aging*, ed. K. Wachter and C. Finch. Washington, D.C.: National Academy Press.

———. 1999. Evolutionary demographic models for mortality plateaus. *Proceedings of the National Academy of Sciences* 96:10544–47.

Wagner, G. P., and L. Altenberg. 1996. Complex adaptations and the evolution of evolvability. *Evolution* 50:967–76.

Waite, L. J. 1995. Does marriage matter? *Demography* 32:483–507.

Wake, M. H. 1998. Amphibian reproduction, overview. In *Encyclopedia of Reproduction*, edited by E. Knobil and J. D. Neill, 161–66. San Diego: Academic Press.

Waldron, I. 1992. Recent trends in sex mortality ratios for adults in developed countries. *Social Science Medicine* 36:451–62.

Walle, F. v. d. 1986. Infant mortality and the European demographic transition. In *The Decline of Fertility in Europe*, ed. A. J. Coale and S. C. Watkins, 201–33. Princeton: Princeton University Press.

Wang, J. L., H.-G. Müller, and W. B. Capra. 1998. Analysis of oldest-old mortality: Life tables revisited. *Annals of Statistics* 26:126–63.

Wang, J. L., H.-G. Müller, W. B. Capra, and J. R. Carey. 1994. Rates of mortality in populations of *Caenorhabditis elegans*. *Science* 266:827–28.

Warburg, M. S., and B. Yuval. 1996. Effects of diet and activity on lipid levels of adult Mediterranean fruit flies. *Physiological Entomology* 21:151–58.

Washburn, S. L. 1981. Longevity in primates. In *Aging: Biology and Behavior*, ed. J. L. McGaugh and S. B. Kiesler, 11–29. New York: Academic Press.

Webster, R. P., and J. G. Stoffolano. 1978. The influence of diet on the maturation of the reproductive system of the apple maggot, *Rhagoletis pomonella*. *Annals of the Entomological Society of America* 71:844–49.

Webster, R. P., J. G. Stoffolano, and R. J. Prokopy. 1979. Long-term intake of protein and sucrose in relation to reproductive behavior of wild and laboratory cultured *Rhagoletis pomonella*. *Annals of the Entomological Society of America* 72:41–46.

Weibull, W. 1951. A statistical theory of wide applicability. *Journal of Applied Mechanics* 18:293–97.

Weindruch, R. 1996. Caloric restriction and aging. *Scientific American* 40:46–52.

Werren, J. H., and E. L. Charnov. 1978. Facultative sex ratios and population dynamics. *Nature* 272:349–50.

West-Eberhard, M. J. 1992. Adaptation: Current usages. In *Keywords in Evolutionary Biology*, ed. E. F. Keller and E. A. Lloyd 13–18. Cambridge: Harvard University Press.

Westendorp, R.G.J., and T.B.L. Kirkwood. 1998. Human longevity at the cost of reproductive success. *Nature* 396:743–46.

Wheeler, W. M. 1928. *The Social Insects*. New York: Harcourt Brace.

White, T.C.R. 1978. The importance of a relative shortage of food in animal ecology. *Oecologia* 33:71–86.

Whittier, T. S., F. Y. Nam, T. E. Shelly, and K. Y. Kaneshiro. 1994. Male courtship success and female discrimination in the Mediterranean fruit fly (Diptera: Tephritidae). *Journal of Insect Behavior* 7:159–70.

Widdowson, E. M. 1976. The response of the sexes to nutritional stress. *Nutritional Society Proceedings* 35:175–80.

Willett, W. C. 1994. Diet and health: What should we eat? *Science* 264:532–37.

Williams, G. C. 1957. Pleiotropy, natural selection, and the evolution of senescence. Evolution 11:398–411.

Williamson, D. L., S. Mitchell, and S. T. Seo. 1985. Gamma irradiation of the Mediterranean fruit fly (Diptera: Tephritidae): Effects of puparial age under induced hypoxia on female sterility. *Annals of the Entomological Society of America* 78: 101–6.

Wilmoth, J. R. 1998. The future of human longevity: A demographer's perspective. *Science* 280:395–97.

Wilmoth, J. R., L. J. Deegan, H. Lundstrom, and S. Horiuchi. 2000. Increase of maximum life-span in Sweden, 1861–1999. *Science* 289:2366–68.

Wilmoth, J. R., and H. Lundstrom. 1996. Extreme longevity in five countries. *European Journal of Population* 12:63–93.

Wilson, E. O. 1971. *The Insect Societies*. Cambridge: Harvard University Press.

———. 1975. *Sociobiology: The New Synthesis*. Cambridge, Mass.: Belknap.

———. 1998. *Consilience*. New York: Knopf.

Wingard, D. L. 1984. The sex differential in morbidity, mortality, and lifestyle. *Annual Review of Public Health* 5:433–58.

Wood, J. W. 1994. *Dynamics of Human Reproducton: Biology, Biometry, Demography*. Hawthorne, N.Y.: Aldine de Gruyter.

Yang, X., and X. C. Tian. 2000. Cloning adult animals—What is the genetic age of the clones? *Cloning* 2:123–28.

Yu, B. P. 1994. *Modulation of Aging Processes by Dietary Restriction*. Boca Raton, Fl.: CRC Press.

Zuk, M. 1990. Reproductive strategies and disease susceptibility: An evolutionary viewpoint. *Parasitology Today* 6:231–33.

Zvelebil, M. 1994. Plant use in the Mesolithic and the transition to farming. *Proceedings of the Prehistoric Society* 60:35–74.

Index

NORTHERN MICHIGAN UNIVERSITY
3 1854 007 532 370